T0276826

Quantum Chemistry

Quantum Chemistry

Edited by **Axel Hou**

New York

Published by NY Research Press,
23 West, 55th Street, Suite 816,
New York, NY 10019, USA
www.nyresearchpress.com

Quantum Chemistry
Edited by Axel Hou

International Standard Book Number: 978-1-63238-379-2 (Hardback)

Printed in the United States of America.

Contents

	Preface	**VII**
Part 1	**Theories in Quantum Chemistry**	**1**
Chapter 1	**Numerical Solution of Linear Ordinary Differential Equations in Quantum Chemistry by Spectral Method** Masoud Saravi and Seyedeh-Razieh Mirrajei	**3**
Chapter 2	**Composite Method Employing Pseudopotential at CCSD(T) Level** Nelson Henrique Morgon	**11**
Part 2	**Electronic Structures and Molecular Properties**	**23**
Chapter 3	**Quantum Chemistry and Chemometrics Applied to Conformational Analysis** Aline Thaís Bruni and Vitor Barbanti Pereira Leite	**25**
Chapter 4	**Quantum Chemical Calculations for some Isatin Thiosemicarbazones** Fatma Kandemirli, M. Iqbal Choudhary, Sadia Siddiq, Murat Saracoglu, Hakan Sayiner, Taner Arslan, Ayşe Erbay and Baybars Köksoy	**53**
Chapter 5	**Elementary Molecular Mechanisms of the Spontaneous Point Mutations in DNA: A Novel Quantum-Chemical Insight into the Classical Understanding** Ol'ha O. Brovarets', Iryna M. Kolomiets' and Dmytro M. Hovorun	**87**
Part 3	**Molecules to Nanodevices**	**131**
Chapter 6	**Quantum Transport and Quantum Information Processing in Single Molecular Junctions** Tomofumi Tada	**133**

Chapter 7 **Theoretical Study for High Energy Density Compounds from Cyclophosphazene** 159
Kun Wang, Jian-Guo Zhang, Hui-Hui Zheng, Hui-Sheng Huang and Tong-Lai Zhang

Chapter 8 **Charge Carrier Mobility in Phthalocyanines: Experiment and Quantum Chemical Calculations** 185
Irena Kratochvilova

Permissions

List of Contributors

Preface

Every book is a source of knowledge and this one is no exception. The idea that led to the conceptualization of this book was the fact that the world is advancing rapidly; which makes it crucial to document the progress in every field. I am aware that a lot of data is already available, yet, there is a lot more to learn. Hence, I accepted the responsibility of editing this book and contributing my knowledge to the community.

This book deals with the field of quantum chemistry providing comprehensive information regarding theoretical basis and novel applications. Until early 20th century, many questions regarding molecules like their existence in stable and rigid networks between atoms and changing into different types of molecules had been unanswered. With the discovery of quantum mechanics, this scenario changed. Such queries were especially important, given the fact that they form the basis of life. Quantum mechanics is the theory for small particles such as electrons and nuclei. Heitler and London applied these principles to the hydrogen molecule in 1927. This breakthrough resulted in the explanation of the chemical bonding between hydrogen atoms. Since then, quantum chemistry has been critically important in the understanding of molecular properties such as stability, reactivity, and applicability for devices in various spheres.

While editing this book, I had multiple visions for it. Then I finally narrowed down to make every chapter a sole standing text explaining a particular topic, so that they can be used independently. However, the umbrella subject sinews them into a common theme. This makes the book a unique platform of knowledge.

I would like to give the major credit of this book to the experts from every corner of the world, who took the time to share their expertise with us. Also, I owe the completion of this book to the never-ending support of my family, who supported me throughout the project.

Editor

Part 1

Theories in Quantum Chemistry

1

Numerical Solution of Linear Ordinary Differential Equations in Quantum Chemistry by Spectral Method

Masoud Saravi[1] and Seyedeh-Razieh Mirrajei[2]
[1]*Islamic Azad University, Nour Branch, Nour,*
[2]*Education Office of Amol, Amol,*
Iran

1. Introduction

The problem of the structure of hydrogen atom is the most important problem in the field of atomic and molecular structure. Bahr's treatment of the hydrogen atom marked the beginning of the old quantum theory of atomic structure, and wave mechanics had its inception in Schrodinger 's first paper, in which he gave the solution of the wave equation for the hydrogen atom. Since the most differential equations concerning physical phenomenon could not be solved by analytical method hence, the solutions of the wave equation are based on polynomial (series) methods. Even if we use series method, some times we need an appropriate change of variable, and even when we can, their closed form solution may be so complicated that using it to obtain an image or to examine the structure of the system is impossible. For example, if we consider Schrodinger equation, i.e.,

$$\varphi'' + (2mEh^{-2} - \alpha^2 x^2)\varphi = 0,$$

we come to a three-term recursion relation, which work with it takes, at least, a little bit time to get a series solution. For this reason we use a change of variable such as

$$\varphi = e^{-\alpha x^2/2} f(x),$$

or when we consider the orbital angular momentum, it will be necessary to solve

$$\frac{d^2 s}{d\theta^2} + cot\theta \frac{ds}{d\theta} + \left(\frac{c}{h^2} - \frac{m^2}{sin^2\theta}\right) s = 0.$$

As we can observe, working with this equation is tedious. Another two equations which occur in the hydrogen atom wave equations, are Legendre and Laguerre equations, which can be solved only by power series methods.

In next section, after a historical review of spectral methods we introduce Clenshaw method, which is a kind of spectral method, and then solve such equations in last section. But, first of all, we put in mind that this method can not be applied to atoms with more electrons. With the increasing complexity of the atom, the labour of making calculations increases tremendously. In these cases, one can use variation or perturbation methods for overcoming such problems.

2. Historical review

Spectral methods arise from the fundamental problem of approximation of a function by interpolation on an interval, and are very much successful for the numerical solution of ordinary or partial differential equations. Since the time of Fourier (1882), spectral representations in the analytic study of differential equations have been used and their applications for numerical solution of ordinary differential equations refer, at least, to the time of Lanczos.

Spectral methods have become increasingly popular, especially, since the development of Fast transform methods, with applications in problems where high accuracy is desired.

Spectral methods may be viewed as an extreme development of the class of discretization schemes for differential equations known generally as the *method of weighted residuals* (MWR) (Finlayson and Scriven (1966)). The key elements of the MWR are the trial functions (also called expansion approximating functions) which are used as basis functions for a truncated series expansion of the solution, and the test functions (also known as weight functions) which are used to ensure that the differential equation is satisfied as closely as possible by the truncated series expansion. The choice of such functions distinguishes between the three most commonly used spectral schemes, namely, Galerkin, Collocation(also called Pseudo-spectral) and Tau version. The Tau approach is a modification of Galerkin method that is applicable to problems with non-periodic boundary conditions. In broad terms, Galerkin and Tau methods are implemented in terms of the expansion coefficients, where as Collocation methods are implemented in terms of physical space values of the unknown function.

The basis of spectral methods to solve differential equations is to expand the solution function as a finite series of very smooth basis functions, as follows

$$y_N(x)=\sum_{n=0}^{N}a_n\phi_n(x)\,, \tag{1}$$

in which, one of our choice of ϕ_n, is the eigenfunctions of a singular Sturm-Liouville problem. If the solution is infinitely smooth, the convergence of spectral method is more rapid than any finite power of *1/N*. That is the produced error of approximation (1), when $N\to\infty$, approaches zero with exponential rate. This phenomenon is usually referred to as "spectral accuracy". The accuracy of derivatives obtained by direct, term by term differentiation of such truncated expansion naturally deteriorates. Although there will be problem but for high order derivatives truncation and round off errors may deteriorate, but for low order derivatives and sufficiently high-order truncations this deterioration is negligible. So, if the solution function and coefficient functions of the differential equation are analytic on $[a,b]$, spectral methods will be very efficient and suitable. We call function y

is analytic on $[a,b]$ if is infinitely differentiable and with all its derivatives on this interval are bounded variation.

3. Clenshaw method

In this section, we are going to introduce Clenshaw method. For this reason, first we consider the following differential equation:

$$Ly=\sum_{0}^{M} f_{M-i}(x)D^{i}y=f(x),\ x\in[-1,1], \tag{2}$$

$$\mathrm{B}y=C\ , \tag{3}$$

where $L=\sum_{0}^{M} f_{M-i}(x)D^{i}$, and f_i , $i=0,1,...,M,f$, are known real functions of x,D^{i} denotes i^{th} order of differentiation with respect to x,B is a linear functional of rank M and $C\in\Re^{M}$. Here (3) can be initial, boundary or mixed conditions. The basis of spectral methods to solve this class of equations is to expand the solution function, y , in (2) and (3) as a finite series of very smooth basis functions, as given below

$$y_N(x)=\sum_{n=0}^{N} a_n T_n(x)\ , \tag{4}$$

where, $\{T_n(x)\}_0^N$ is sequence of Chebyshev polynomials of the first kind. By replacing y_N in (2), we define the residual term by $r_N(x)$ as follows

$$r_N(x)=Ly_N-f\ . \tag{5}$$

In spectral methods, the main target is to minimize $r_N(x)$, throughout the domain as much as possible with regard to (3), and in the sense of pointwise convergence. Implementation of these methods leads to a system of linear equations with $N+1$ equations and $N+1$ unknowns $a_0, a_1,...,a_N$.

The Tau method was invented by Lanczos in 1938. The expansion functions $\phi_n(n=1,2,3,...)$ are assumed to be elements of a complete set of orthonormal functions. The approximate solution is assumed to be expanded in terms of those functions as $u_N=\sum_{n=1}^{N+m} a_n\phi_n$, where m is the number of independent boundary constraints $Bu_N=0$ that must be applied. Here we are going to use a Tau method developed by Clenshaw for the solution of linear ODE in terms of a Chebyshev series expansion.

Consider the following differential equation:

$$P(x)y''+Q(x)y'+R(x)y=S(x)\,,x\in(-1,1)\,,$$
$$y(-1)=\alpha\;,\;y(1)=\beta\,. \tag{6}$$

First, for an arbitrary natural number N, we suppose that the approximate solution of equations (6) is given by (4). Our target is to find $\underline{a}=(a_0,a_1,...,a_N)^t$. For this reason, we put

$$P(x)\cong\sum_{i=0}^{N}\xi_iT_i(x),$$
$$Q(x)\cong\sum_{i=0}^{N}\gamma_iT_i(x), \tag{7}$$
$$R(x)\cong\sum_{i=0}^{N}\lambda_iT_i(x).$$

Using this fact that the Chebyshev expansion of a function $u\in L_w^2(-1,1)$ is

$u(x)=\sum_{k=0}^{\infty}\hat{u}_kT_k(x)\,;\;\hat{u}_k=\frac{2}{\pi c_k}\int_{-1}^{1}u(x)T_k(x)w(x)dx$, we can find coefficients ξ_i,γ_i and λ_i as follows:

$$\xi_i=\frac{2}{\pi c_i}\int_{-1}^{1}\frac{P(x)T_i(x)}{\sqrt{1-x^2}}dx$$
$$\gamma_i=\frac{2}{\pi c_i}\int_{-1}^{1}\frac{Q(x)T_i(x)}{\sqrt{1-x^2}}dx \tag{8}$$
$$\lambda_i=\frac{2}{\pi c_i}\int_{-1}^{1}\frac{R(x)T_i(x)}{\sqrt{1-x^2}}dx,$$

where, $c_0=2$ and $c_i=1$ for $i\geq 1$.

To compute the right-hand side of (8) it is sufficient to use an appropriate numerical integration method. Here, we use $(N+1)$- point Gauss-Chebyshev-Lobatto quadrature $x_j=\cos\frac{\pi j}{N},w_j=\frac{\pi}{\tilde{c}_jN},0\leq j\leq N$,

where $\tilde{c}_0=\tilde{c}_N=2$ and $\tilde{c}_j=1$ for $j=1,2,...,N-1$.

Note that, for simplicity of the notation, these points are arranged in descending order, namely, $x_N<x_{N-1}<...<x_1<x_0$, with weights

$$w_k=\frac{\pi}{N}\;,\quad 1\leq k\leq N-1\,,$$
$$=\frac{\pi}{2N}\,,\quad k=0\,,k=N\,,$$

and nodes $x_k = \cos\frac{\pi k}{N}$, $k = 0, 1, \ldots, N$. That is, we put:

$$\xi_i \cong \frac{\pi}{N} \sum_{k=0}^{N}{}'' P(\cos(\frac{k\pi}{N})) T_i(\cos(\frac{k\pi}{N})),$$

and using $T_i(x) = \cos(i \cos^{-1} x)$, we get

$$\xi_i \cong \frac{\pi}{N} \sum_{k=0}^{N}{}'' P(\cos(\frac{k\pi}{N})) \cos(\frac{\pi i k}{N}),$$

where, notation $\sum''$ means first and last terms become half .Therefore, we will have :

$$\xi_i \cong \frac{\pi}{N} \sum_{k=0}^{N}{}'' P(\cos(\frac{k\pi}{N})) \cos(\frac{\pi i k}{N}),$$

$$\gamma_i \cong \frac{\pi}{N} \sum_{k=0}^{N}{}'' Q(\cos(\frac{k\pi}{N})) \cos(\frac{\pi i k}{N}), \tag{9}$$

$$\lambda_i \cong \frac{\pi}{N} \sum_{k=0}^{N}{}'' R(\cos(\frac{k\pi}{N})) \cos(\frac{\pi i k}{N}).$$

Now, substituting (4) and (9) in equations (6), and using the fact that

$$y'(x) \cong \sum_{m=0}^{N} a_m^{(1)} T_m(x), \; a_m^{(1)} = \frac{2}{c_m} \sum_{\substack{p=m+1 \\ m+p=\text{odd}}}^{N} p a_p, \; m = 0, 1, \ldots, N-1, \; a_N{}^{(1)} = 0,$$

$$y''(x) \approx \sum a_m^{(2)} T_m(x), \; a_m^{(2)} = \frac{1}{c_m} \sum_{\substack{p=m+2 \\ m+p=\text{even}}}^{N} p(p^2 - m^2) a_p, \; m = 0, 1, \ldots, N-2, \; a_{N-1}^{(2)} = a_N^{(2)} = 0,$$

we get

$$\sum_{i=0}^{N}\sum_{m=0}^{N} \xi_i a_m^{(2)} T_i(x) T_m(x) + \sum_{i=0}^{N}\sum_{m=0}^{N} \gamma_i a_m^{(1)} T_i(x) T_m(x) + \sum_{i=0}^{N}\sum_{m=0}^{N} \lambda_i a_m T_i(x) T_m(x) = S(x), \tag{10}$$

$$\begin{aligned} &\sum_{i=0}^{N} a_i T_i(-1) = \alpha, \\ &\sum_{i=0}^{N} a_i T_i(1) = \beta. \end{aligned} \tag{11}$$

Now, we multiply both sides of (10) by $\frac{2}{\pi c_j}\frac{T_j(x)}{\sqrt{1-x^2}}$, and integrate from -1 to 1, we obtain

$$\frac{2}{\pi c_j}\sum_{i=0}^{N}\sum_{m=0}^{N}[\xi_i a_m^{(2)}+\gamma_i a_m^{(1)}+\lambda_i a_m]\int_{-1}^{1}\frac{T_i(x)T_m(x)T_j(x)}{\sqrt{1-x^2}}dx$$

$$=\frac{2}{\pi c_j}\int_{-1}^{1}\frac{S(x)T_j(x)}{\sqrt{1-x^2}}dx, j=0,1,...,N-2, \tag{12}$$

where,

$$\int_{-1}^{1}\frac{T_i(x)T_m(x)T_j(x)}{\sqrt{1-x^2}}dx = \begin{cases} \pi & , i=m=j=0 \ , \\ \frac{\pi}{2}\delta_{i,m} & , i+m>0 \ , j=0 \ , \\ \frac{\pi}{4}(\delta_{j,i+m}+\delta_{j,|i-m|}) \ , & j>0 \ , \end{cases} \tag{13}$$

with, $\delta_{i,j}=1$,when $i=j$, and zero when $i\neq j$.

We can also compute the integrals in the right-hand side of (12) by the method of numerical integration using $N+1$ -point Gauss-Chebyshev-Lobatto quadrature. Therefore, substituting (13) in (12) and using the fact that $T_i(\pm1)=(\pm1)^i$, equations (12) and (11) make a system of $N+1$ equations for $N+1$ unknowns, $a_0, a_1, ..., a_N$, hence we can find $(a_0, a_1, ..., a_N)^t$ from this system.

4. Numerical examples

As we mentioned the important problem in the field of atomic and molecular structure, is solution of wave equation for hydrogen atom. In this section we will solve Schrodinger, Legendre and Laguerre equations, which occur in the hydrogen atom wave equations, by Clenshaw method and observe the power of this method comparing with usual numerical methods such as Euler's or Runge-Kutta's methods. We start with Schrodinger's equation.

Example 1. Let us consider

$$\varphi'' + (2mEh^{-2} - \alpha^2x^2)\varphi = 0.$$

Assume $\alpha = 2, mEh^{-2} = -1$, with $\varphi(0) = 1, \varphi(1) = e$. The exact solution is $\varphi(x) = e^{-x^2}$.

Here interval is chosen as [0,1], but using change of variable such as $t = \frac{x+1}{2}$ we can transfer interval [0,1] to [-1,1].

We solve this equation by Clenshaw method and compare the results for different values of *N*. The results for *N*=4, 7, 10, 13, respectively, were:

$$1.660 \times 10^{-2}, 4.469 \times 10^{-5}, 5.901 \times 10^{-8}, 7.730 \times 10^{-11}.$$

As we expected when N increases, errors decrease.

Example 2. Consider Legendre's equation given by

$$(1-x^2)y''-2xy'+\lambda(\lambda+1)y=0.$$

As we know, this equation for $\lambda=2$, and boundary conditions $y(\pm 1)=-2$ has solution $y(x)=1-3x^2$. The results for N=4, 6, 10 were:

$$5.5511\times 10^{-17}, 2.2204\times 10^{-16}, 2.7756\times 10^{-17}.$$

Since our solution is a polynomial then for $N>3$, we come to a solution with error very closed to zero. If such cases you find the error is not zero but closed to it, is because of rounding error. We must put in our mind that the results by this method will be good if the exact solution is a polynomial.

We end this section by solving Laguerre's equation.

Example 3. Consider

$$xy''+(1-x)y'+\lambda y=0.$$

Suppose $\lambda=2$ and boundary conditions are given by $y(-1)=\frac{7}{2}, y(1)=-\frac{1}{2}$.

The exact solution is $(x)=1-2x+{x^2}/{2}$.

Here we have again a polynomial solution, so we expect a solution with very small error. We examined for different values of N such as N=2, 3 and get the results 0 and 3×10^{-17}, respectively.

Results in these examples show the efficiency of Clenshaw method for obtaining a good numerical result.

In case of singularity, one can use pseudo-spectral method. Some papers also modified pseudo-spectral method and overcome the problem of singularity even if the solution function was singular.

5. References

Babolian. E, Bromilow. T. M, England. R, Saravi. M, *'A modification of pseudo-spectral method for olving linear ODEs with singularity'*, AMC 188 (2007) 1260-1266.

Babolian. E, Delves. L .M, *A fast Galerkin scheme for linear integro-differential equations*, IMAJ. Numer. Anal, Vol.1, pp. 193-213, 1981.

Canuto. C, Hussaini. M. Y, Quarteroni. A, Zang. T. A, *Spectral Methods in Fluid Dynamics*, Springer- Verlag,NewYork,1988.

Delves. L. M, Mohamed. J. L, *Computational methods for integral equations*, Cambridge University Press, 1985.

Gottlieb. D, Orszag. S. A, *Numerical Analysis of Spectral Methods, Theory and Applications*, SIAM,Philadelphia,1982.

Lanczos. C, *Trigonometric interpolation of empirical and analytical functions,* J. Math. Phys. 17 (1938) 123-129.

Levine. Ira N, *Quantum Chemistry,* 5th ed, City University of NewYork, Prentice-Hall Publication, 2000.

Pauling. L, Wilson. E.B, *Quantum Mechanic,* McGraw-Hill Book Company, 1981.

2

Composite Method Employing Pseudopotential at CCSD(T) Level

Nelson Henrique Morgon
Universidade Estadual de Campinas
Brazil

1. Introduction

Thermochemical data are among the most fundamental and useful information of chemical species which can be used to predict chemical reactivity and relative stability. Thus, it is not surprising that an important goal of computational chemistry is to predict thermochemical parameters with reasonable accuracy (Morgon, 1995a). Reliability is a critical feature of any theoretical model, and for practical purposes the model should be efficient in order to be widely applicable in estimating the structure, energy and other properties of systems, as isolated ions, atoms, molecules(Ochterski et al., 1995), or gas phase reactions(Morgon, 2008a).

What is the importance of these studies?

For instance gas phase reactions between molecules and ions, and molecules and electrons are known to be important in many scientifically and technologically environments. On the cosmic scale, the chemistry that produces molecules in interstellar clouds is dominated by ion-molecule reactions. Shrinking down to our own planet, the upper atmosphere is a plasma, and contains electrons and various positive ions. Certain anthropogenic chemical compounds (including SF_6 and perfluorocarbons) can probably not be destroyed within the troposhere or stratosphere, but may be removed by reactions with ions or electrons in the ionosphere. Recent years have seen a massive growth in the industrial use of plasmas, particularly in the fabrication of microelectronic devices and components. The chemistry within the plasma, much of which involves ion-molecule and electron-molecule reactions, determines the species that etch the surface, and hence the outcome and rate of an etching process. Much of the chemistry that is often labelled as 'organic' or 'inorganic' involves ion-molecule reactions, usually carried out in the presence of a solvent. For instance, S_N2 reactions, such as OH^- + CH_3Cl, fall into this category. To gain a clearer picture of how these (gas phase) reactions occur, it is advantageous to study them removed from the (very great) perturbations due to the solvent.

So, the need for thorough studies of ion-molecule and electron-molecule reactions are thus well established, ranging from the astrophysical origins of molecules, through the survival of the earth's atmosphere, to modelling the plasmas that underpin many advanced processing technologies. There is intrinsic interest too in the studies, as they help to explore the nature and progress of binary encounters between molecules and ions, and molecules and electrons. At the most basic level answers are needed to the following questions - how fast does a reaction proceed? and what are the products of the reaction? What determines which reactions occur? and what products are formed? Beyond these may come questions about the detailed dynamics of the reaction, such as how changing the energy of the reactants may influence the progress and outcome of the reaction.

Many powerful experimental techniques have been developed to give the basic data of reaction rate coefficients and products (usually just the identification of the ion product). These results are part of the raw data needed to understand and model the complex chemistry occurring in the diverse environments identified above. There is much information that is not directly available from the experimental data. This includes identification of the neutral products of a reaction, knowledge of the thermochemistry of the reaction, and characterization of the pathway that connects reactants to products. By invoking some general rules, the experimental observations can be used to provide partial answers. Thus the fast flow techniques that are used to provide much of the experimental data on ion-molecule reactions can only detect the occurrence of very rapid reactions ($k >= 10^{-12}\ cm^3\ s^{-1}$), which places an upper bound on the exothermicity of the reaction of +20 kJ mol^{-1}. In cases where there are existing reliable enthalpies of formation of each of the species in a proposed pathway to an observed ion product, this rule can test whether the suggested neutral products may be correct. In other cases, where the enthalpy of formation of just one of the species involved in a reaction (usually the product ion) is unknown, the observation of a specific reaction pathway can be used to place a bound on the previously unkown enthalpy of formation. Finally for reactions which are known to be exothermic, if the experimenal rate coefficient is observed to be less than the capture theory rate coefficient, then it is usual to conclude that there must be some bottleneck or barrier to the reaction.

What can theoretical calculations add to the experimental data?

Three important and fundamental gas-phase thermochemical properties from a theoretical and experimental point of view are the standard heat of formation ($\Delta_f H^o_{gas}$), the electron (EA) and proton (PA) affinities. Thus, it is not surprising that an important goal of computational chemistry is to predict such thermochemical parameters with reasonable accuracy, which can be useful in the gas phase reaction studies. Proton transfer reactions are also of great importance in chemistry and in biomolecular processes of living organisms(Ervin, 2001). Absolute values of proton affinities are not always easy to obtain and are often derived from relative measurements with respect to reference molecules. Relative proton affinities are usually measured by means of high pressure mass spectrometry, with triple quadrupole and ion trap mass spectrometers (Mezzache et al., 2005) or using ion mobility spectrometry (Tabrizchi & Shooshtari, 2003). The importance and utility of the EA extend well beyond the regime of gas-phase ion chemistry. A survey of examples illustrates the diversity of areas in which electron affinities play a role: silicon, germanium clusters, interstellar chemistry, microelectronics, and so on.

The standard heat of formation, which measures the thermodynamic stability, is useful in the interpretation of the mechanisms of chemical reactions (Badenes et al., 2000).

On the other hand, theoretical calculations represent one attempt to study absolute values of electron or proton affinity and other thermochemical properties (Smith & Radom, 1991). However, accurate calculations of these properties require sophisticated and high level methods, and great amount of computational resources. This is particularly true for atoms of the 2nd, 3rd, ..., periods and for calculating properties like the proton and electron affinity of anions. Gaussian-n theories (G1, G2, G3, and G4) (Curtiss et al., 1997; 1998; 2000; 2007) have given good results for properties like proton and electron affinities, enthalpies of formation, atomization energies, and ionization potentials. These theories are a composite technique in which a sequence of well-defined ab initio molecular orbital calculations is performed to arrive at a total energy of a given molecular species. There are other techniques that have been demonstrated to predict accurate thermochemical properties of chemical species, and are alternative to the Gaussian-n methods: the Correlation Consistent Composite Approach (ccCA)(DeYonker et al., 2006), the Multireference Correlation Consistent Composite Approach (MR-ccCA)(Oyedepo & Wilson, 2010), the Complete Basis Set Methods (CBS) and its versions: CBS-4M, CBS-Lq, CBS-Q, CBS-QB3, CBS-APNO(Montgomery Jr. et al., 2000; Nyden & Petersson, 1981; Ochterski et al., 1996; Peterson et al., 1991), and Weizmann Theories (W1 to W4)(Boese et al., 2004; Karton et al., 2006; Martin & De Oliveira, 1999; Parthiban & Martin, 2001).

Recently, we have implemented and tested a pseudopotential to be used with the G3 theory for molecules containing first-, second-, and non-transition third-row atoms (G3CEP) (Pereira et al., 2011). The final average total absolute deviation using this methodology and the all-electron G3 were 5.39 kJ mol^{-1} and 4.85 kJ mol^{-1}, respectively. Depending on the size of the molecules and the type of atoms considered, the CPU time was drastically decreased.

2. Computational methods

In this chapter we have developed a computational model similar to version of the G2(MP2, SVP) theory (Curtiss et al., 1996). Both theories are based on the additivity approximations to estimate the high level energy for the extended function basis set. While G2(MP2,SVP) is based on the additivity approximation to estimate the QCISD(T) energy for the extended 6-311+G (3df,2p) basis set: E[QCISD(T)/ 6-311+G(3df,2p)] E[QCISD(T)/6-31G(d)] + E[MP2/6-311+G(3df,2p)] - E[MP2/6-31G(d)]. Our methodology employs CCSD(T) energies in addition to the the valence basis sets adapted for pseudopotential (ECP) (Stevens et al., 1984) using the Generator Coordinate Method (GCM) procedure (Mohallem & Dreizler, 1986; Mohallem & Trsic, 1985).

The present methodology which relies on small basis sets (representation of the core electrons by ECP) and an easier and simpler way for correcting the valence region (mainly of anionic systems) appears as an interesting alternative for the calculation of thermochemical data such as electron and proton affinities or heat of formation for larger systems.

2.1 Development of basis sets

The GCM has been very useful in the study of basis sets(Morgon, 1995a;b; 2006; 2008b; 2011; Morgon et al., 1997). It considers the monoelectronic functions $\psi(1)$ as an integral transform,

$$\psi(1) = \int_0^{\infty} f(\alpha)\,\phi(\alpha,1)\,d\alpha \qquad (1)$$

where $f(\alpha)$ and $\phi(\alpha,1)$ are the weight and generator functions respectively (gaussian functions are used in this work), and α is the generator. The existence of the weight functions (graphical display of the linear combination of basis functions) is an essential condition for the use of GCM. Analysis of the behavior of the weight functions by the GCM permits the atomic basis set to be adapted in such a way as to yield a better description of the core electrons (represented by ECP) and the valence orbitals (corrected by addition of the extra diffuse functions), in the molecular environment. With the exception of some simple systems the analytical expression of the weight functions is unknown. Thus, an analytical solution of the integral transform (Eq. 1) is not viable in most cases, and suggests the need of numerical techniques to solve Eq. 1 (Custodio, Giordan, Morgon & Goddard, 1992; Custodio, Goddard, Giordan & Morgon, 1992). The solution can be carried out by an appropriate choice of discrete points on the generate coordinate, represented by:

$$\alpha_{i,(k)} = exp[\Omega_{o,(k)} + (i-1) \cdot \Delta\Omega_{(k)}], \quad i = 1, 2, 3, \ldots N_{(k)} \tag{2}$$

The discretization of the set is defined by the following parameters: an initial value (Ω_o), an increment ($\Delta\Omega$), and by the number of primitives used (N) for a given orbital k (s, p, d, ...). The search for the best representation is obtained using the total energy of the electronic ground state as the minimization criterion.

The SIMPLEX search method (Nelder & Mead, 1965) can be adapted to the any electronic structure program to provide the minimum energy of the ground state of the atom corresponding to the optimized discretization parameters.

The basic procedure consists of the following steps:

(a) search of the optimum discretized parameter set for the atoms using the GCM for variation on the generator coordinate space. The core electrons are represented by a pseudopotential. The discretization parameters (Ω_o and $\Delta\Omega$) are defined with conjunction with this ECP;

(b) the minimum energy criterion is observed and the characteristics of the atomic orbital weight functions are analyzed;

(c) extra functions (polarizaton or diffuse funcions - s, p, d and f) are obtained by observing the convergent behavior of the weight functions of the outer atomic orbitals (s and p). These extra functions are needed for the correct description of the electronic distribution in an anion (diffuse character of electronic cloud).

To the heavy atoms f type polarization functions are not available in these valence basis sets for this kind of ECP(Stevens et al., 1984). So, it was need to define the value of these f functions for Br and I atoms. The determination of the best value was carried out considering the smaller difference between the PAs (experimental and theoretical) values considering the Br^- and I^- anions. The f exponent values found are 0.7 and 0.3 for Br and I atoms, respectively.

In fact two sets of basis functions are used, a small basis and a larger basis, with extra diffuse and polarization functions, B0 and B1, respectivally. Calculations with basis B1 are naturally much more expensive than those employing basis B0, so it is important to have computational schemes that perform the minimum number of calculations using basis B1.

For instance, the B0 basis set is defined as: (31) for H; (311/311) for C, O, F, S, and Cl; and (411/411) Br and I. For more refined energy calculations (B1 basis set are used), this set was augmented with additional diffuse and polarization functions (p for H and s, p, d, and f for heavy atoms) to yield a (311/11) set for H; (311/311/11/1) for C, O, F, S, and Cl; and (411/411/11/1) for Br and I atoms.

2.2 Molecular calculations

In many problems to be addressed by electronic structure methodology, high accuracy is of crucial importance. In order to obtain energies that may approach chemical accuracy ($\approx$10 kJ.mol^{-1}) calculations must take account of electron correlation. The 'ideal' methodology would have been a multi-reference configuration interaction all electron calculation with several large, flexible basis sets to enable extrapolation to the complete basis set limit. Depend on the size of the systems, performing accurate calculations (methodology and basis sets) represents a significant challenge.

Morgon *et al.* (Morgon, 1998; Morgon et al., 1997; Morgon & Riveros, 1998) have been developing techniques to tackle such problems. These are centered around the use of effective core potentials, in which the inner electrons are represented by an effective potential derived from calculations on atoms. The electronic wavefunction itself then only contains the outer electrons.

The procedure to the molecular calculations employing this methodology is:

(a) optimization of the molecular geometries and vibrational analysis are carried out at HF/B0 level. The harmonic frequencies confirm that the stationary points correspond to minima and are used to compute the zero-point energies;

(b) further optimization is carried out at MP2/B0 level;

(c) at the MP2 equilibrium geometry corrections to the total energies are performed at higher level of theory. First, this is carried out at CR-CCSD[T]/B0 level (Completely Renormalized Coupled-Cluster with Single and Double and Perturbative Triple excitation) (Kowalski & Piecuch, 2000) (or at CCSD(T)/B0 level for EA calculations), and later by addition of extra functions (*s*, *p*, *d*, and *f*) at MP2/B1 level.

Thus, these results coupled to additive approximations for the energy yield an effective calculation at a high level of theory,

$$\approx E[CR-CCSD[T]/(B1)] = E[CR-CCSD[T]/(B0)] + \\ + E[MP2/(B1)] - E[MP2/(B0)] + ZPE[HF/(B0)] * scal \quad (3)$$

where *scal* (0.89) is the scaling factor on the vibrational frequencies.

The CR-CCSD[T] (Completely Renormalized Coupled-Cluster with Single and Double and Perturbative Triple excitation) methodology refers to size-extensive left eigenstate completely renormalized (CR) coupled-cluster (CC) singles (S), doubles (D), and noniterative triples (T). This approach is abbreviated as CR-CCL and is appropriately described by Piecuch(Piecuch et al., 2002) and Ge(Ge et al., 2007).

An alternative model was developed for the study of heat of formation. This model employs valence basis sets aug-CCpVnZ (n = 2, 3, and 4) (Dunning, 1989). These basis sets were adapted to ECP using the GCM and are identified by ECP+ACCpVnZm (m = modified). The energies are obtained through the extrapolations to the complete basis set limit (CBS) using Peterson mixed exponential/Gaussian function extrapolation scheme (Feller & Peterson, 1999).

$$E(MP2) = E_{CBS} + B\,exp[-(x-1)] + C\,exp[-(x-1)^2] \quad (4)$$

where x = 2, 3, and 4 come from ECP+ACCpV2Zm, ECP+ACCpV3Zm and ECP+ACCpV4Zm energies, respectively.

For this electronic property, molecular calculations consist of:

(a) optimization of the molecular geometries and vibrational analysis are carried out at HF/ECP+ACC2Zm level. The harmonic frequencies are employed to characterize the local minima and to compute the zero-point energies;

(b) further optimization is carried out at MP2/ECP+ACC2Zm level;

(c) at the MP2 equilibrium geometry corrections to the total energies are performed at higher level of theory. First, at CR-CCL/ECP+ACC2Zm level, and calculations by addition of extra functions ($s, p, d,$ and f) at MP2/ECP+ACC3Zm and MP2/ECP+ACC4Zm levels. The E[MP2/ECP+ACC5Zm] is estimated throught the Eq. 4.

(d) Finally, the results are coupled through additive approximations, and the energy corresponds to an effective calculation at a high level of theory,

$$\begin{aligned} \approx E[CR-CCL/ECP+ACC5Zm] &= E[CR-CCL/ECP+ACC2Zm] + \\ &+ E[MP2/ECP+ACC5Zm] - E[MP2/ECP+ACC2Zm] - \\ &+ ZPE[HF/ECP+ACC2Zm] * scal + E(HLC) \end{aligned} \quad (5)$$

where *scal* is the scaling factor on ZPE.

The method also includes an empirical higher-level correction (HLC) term. This term is given by either Eq. 6 or Eq. 7 depending on whether the species is a molecule or an atom:

$$HLC_{molec.} = -C \cdot n_\beta - D \cdot (n_\alpha - n_\beta) \quad (6)$$

$$HLC_{atom} = -A \cdot n_\beta - B \cdot (n_\alpha - n_\beta) \quad (7)$$

In these equations, n_α and n_β are the numbers of α and β electrons, respectively. The parameters A (4.567mH), B (2.363mH), C (4.544mH), and D (2.337mH) were obtained by fitting to the experimental data of heats of formation.

It is important to note that since it only depends on the number of α and β valence electrons, the HLC cancels entirely from most reaction energies, except when the reactions involve a mixture of atoms and molecules (as in heats of formation and bond dissociation energies) and/or when spin is not conserved (Lin et al., 2009).

It should be also noted that the absolute values of the calculated energies have no real significance, as no common energy zero for different atoms has been used. This arises from how the effective core potentials are constructed. When differences are formed, the differences between the zeros cancel, to leave for instance the difference in energy between the products and reactants of a reaction.

Additionally, an alternative approach for molecular geometry optimization and harmonic frequencies calculation can be considered through the use of the DFT (B3LYP, M06, ...).

3. Results and discussion

3.1 Basis sets

The existence of the weight functions (graphic representations of the linear combination of the atomic orbitals) is the fundamental condition to use the GCM. The analysis of the behavior of the weight functions by the GCM allows the fitting of the atomic basis sets in order to get a better description of the electrons in the molecular environment. The analysis of the weight functions of the outermost atomic orbitals suggested the need for improvements of the basis sets for the heavy atoms. Observing the plots of the weight functions of the outermost orbitals it is possible to establish the best fit of the basis sets. Using the atom of Cl as an example, the representation of the weight function of the atomic orbital 3s is shown in Fig. 1. The continuous line represents the plot of the weight function of the original primitive basis set for the all electron system. The dashed line represents the same weight function obtained using the pseudopotential. This figure shows that the 10 inner functions (large α) have a contribution close to zero towards the description of the weight function. The vertical solide line cutoff of the basis set indicates precisely where the pseudopotential starts to represent the core atomic region.

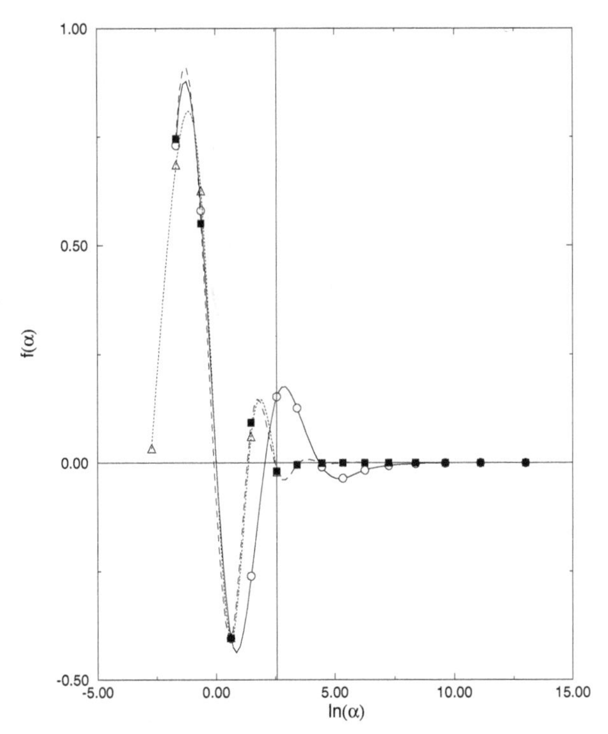

Fig. 1. Weight functions for the 3s atomic orbital of Cl in systems with all electrons (ae, continuous line), with the pseudopotential (pp, dashed line), with the addition of one diffuse function plus pseudopotential (dif+pp, dot-dashed line) and the cutoff point line represented by the vertical solid line.

3.2 Proton affinity

Proton affinity is a very sensitive property of the electronic structure and it is appropriate to test our methodology. Table 1 shows the results of the acidity (in kJ.mol^{-1}) for a set of anions systems. These results were obtained using the B0 and B1 basis sets and Eq. 3. A comparison is also presented with experimental results. One can observe that our theoretical results are very close to the experimental errors (well within 5 kJ.mol^{-1}) with root mean square deviation of 4.14 kJ mol^{-1}.

System	$PA_{Calc.}$	$PA^{a}_{Exp.}$
F^-	1556.17 (-2.17)	1554.0
CH_2F^-	1710.21 (0.79)	1711 ± 17
Cl^-	1395.56 (-0.56)	1395.0
CH_2Cl^-	1658.18 (-1.18)	1657 ± 13
Br^-	1354.80 (-1.80)^b	1353.0 ± 8.80
CHClBr	1563.4 (0.82)	1560.0 ± 13.00
CH_2Br^-	1640.54 (2.46)	1643 ± 13
I^-	1315.10	1315.0
CH_2I^-	1617.47 (-1.47)	1616 ± 21
Acetamide,2,2,2-trichloride	1493,6	1436 ± 8.8
Formamide-N,N-dimethyl	1671.34 (-1.34)	1670.0 ± 17.00
Formamide	1510.13 (-5.13)	1505.0 ± 8.80
methyl cyclopentanol	(2.88)	1559.00 ± 8.40
2,methyl cyclopentanol	(2.99)	1558.00 ± 8.40
CH_2CHCHI^-	1555.24 (-4.24)	1551.0 ± 8.8
δ_{rmsd}	4.14	

[a]Experimental values from NIST Webbase - `http://webbook.nist.gov/chemistry/`.
[b]($PA_{Exp.}$ - $PA_{Calc.}$).

Table 1. Proton Affinity (kJ.mol^{-1}) calculated with the method given by Eq. 3 and comparison with experimental values.

3.3 Electron affinity

In the Table 2 are the electron affinities calculated with the CCSD(T)/B1 energy from Eq. 3. It also shows a comparison between our results and the experimental values, where experimental data are available. The root mean square deviation (δ_{rmsd}) calculated is 0.15 eV.

The use of pseudopotential is competitive, mainly in systems containing S, Cl, and Br atoms. The computational time is almost constant for analogous systems with Cl and Br atoms, because in these cases we have an equal number of outer electrons. In calculations involving all electrons the computational performance is totally different and increases with the number of electrons. The CR-CCSD[T]/ECP computational demand is decreased by 10% when compared with all-electron calculations. For molecules containing Cl, Br or I atom the time is drastically decreased (Morgon, 2006).

3.4 Heat of formation

In the Table 3 are the heats of formation calculated with the CCR-CL/ ECP+ACC5Zm from Eq. 6. It also shows a comparison between our results and the experimental values, where experimental data are available. The average error using this methodology with respect to experimental results is closer to 10 kJ.mol^{-1}.

System	$EA^{a}_{Calc.}$	$EA^{b}_{Exp.}$
Br	3.36 (-)[c]	-
Br_2	2.53 (-0.11)	2.42
CBr_3	2.49 (0.08)	2.57 ± 0.12
CCl_3	2.22 (-0.05)	2.17 ± 0.10
CF_3	1.74 (0.08)	1.82 ± 0.05
CF_3COO	4.56 (-0.10)	4.46 ± 0.18
CH_2=CH	0.71 (-0.04)	0.67 ± 0.02
CH_2=CHCHBr	0.96 (-)	-
CH_2=CHCHF	0.50 (-)	-
$CH_2=CCH_3$	0.69 (-0.13)	0.56 ± 0.19
$CH_2=NO_2$	2.40 (0.07)	2.48 ± 0.01
CH_2Br	0.91 (-0.12)	0.79 ± 0.14
CH_2BrCOO	3.99 (-0.01)	3.98 ± 0.16
CH_2Cl	0.70 (0.04)	0.74 ± 0.16
CH_2ClCOO	4.02 (-0.11)	3.91 ± 0.16
CH_2F	0.17 (0.08)	0.25 ± 0.18
CH_2FCOO	4.55 (-0.76)	3.79 ± 0.16
CH_3	0.08 (0.00)	0.08 ± 0.03
CH_3CH_2O	1.75 (-0.03)	1.71
CH_3CH_2S	2.02 (-0.07)	1.95
CH_3O	1.53 (0.05)	1.57
CH_3S	1.75 (0.12)	1.87
$CHBr_2$	1.74 (-0.03)	1.71 ± 0.08
$CHBr_2COO$	4.37 (-0.11)	4.26 ± 0.16
CHClBr	1.64 (-0.17)	1.47 ± 0.04
CHF_2COO	4.25 (-0.10)	4.15 ± 0.16
Cl	3.60 (0.01)	3.61
Cl_2	2.55 (-0.15)	2.40 ± 0.20
F	3.34 (0.06)	3.40
F_2	2.92 (0.08)	3.00 ± 0.07
Formamide-N,N-dimethyl	0.65 (-)	-
Formamide	3.34 (-)	-
HS	2.27 (0.05)	2.32
NH_2	0.70 (0.07)	0.77 ± 0.01
OH	1.77 (0.06)	1.83
iPrO	1.87 (0.01)	1.87
nPrO	1.74 (0.05)	1.79 ± 0.03
δ_{rmsd}	0.15	

[a]The higher level correction (HLC) from G3 theory was added to the final energy.
[b]Experimental values from NIST Webbase - `http://webbook.nist.gov/chemistry/`.
[c]($EA_{Exp.}$ - $EA_{Calc.}$).

Table 2. Electron Affinity (eV) calculated with the method given by Eq. 3, and comparison with experimental values.

Molecule	Point Group	Ground State	$\Delta_f H^o_{gas}$ calc	$\Delta_f H^o_{gas}$ exp[a]
OF	$C_{\infty v}$	$^2\Pi$	110.999 (2.21)[c]	108.78
OF_2	C_{2v}	$^1A_1$	28.582 (4.06)	24.52
OCl	$C_{\infty v}$	$^2\Pi$	103.482 (2.26)	101.22
OCl_2	C_{2v}	$^1A_1$	92.341 (4.48)	87.86
SF	$C_{\infty v}$	$^2\Pi$	8.661 (-4.31)	12.97
SF_2	C_{2v}	$^1A_1$	-289.573 (7.08)	-296.65
SF_3	C_s	$^2A'$	-512.605 (-9.57)	-503.03
SF_4	C_{2v}	$^1A_1$	-775.362 (-12.20)	-763.16
SF_5	C_{4v}	$^2A_1$	-887.770 (20.68)	-908.45
SF_6	O_h	$^1A_{1g}$	-1218.986 (1.48)	-1220.47
SCl	$C_{\infty v}$	$^2\Pi$	153.843 (-2.62)	156.47
SCl_2	C_{2v}	$^1A_1$	-19.262 (-1.69)	-17.57
SCl_3	C_s	$^2A'$	22.346 (-14.47)	36.82[b]
SCl_4	C_{2v}	$^1A_1$	-15.301 (-12.38)	-2.92[b]
SCl_5	C_{4v}	$^2A_1$	-38.76 (-2.78)	-35.98[b]
SCl_6	O_h	$^1A_{1g}$	-87.54 (-4.7)	-82.84[b]

[a] Experimental values from NIST Webbase - `http://webbook.nist.gov/chemistry/`.
[b] Ref. (Ditter & Niemann, 1982).
[c] ($\Delta_f H_{gas}$ *Calc.* - $\Delta_f H_{gas}$ *Exp.*).

Table 3. Heats of formation (in kJ mol^{-1}) with the method given by Eq. 6, and comparison with experimental values.

4. Conclusions

The proton and electron affinities and the heats of formation of some simple systems obtained by the procedure outlined in this paper are in very good agreement with experimental values. These results can be compared with those obtained by sophisticated and computationally more expensive calculations. ECP-based methods have been shown to be powerful, and of affordable computational cost for the systems addressed in this work. This is due to three features:

1) the number of steps employed during the calculations,
2) the smaller basis sets used in our methodology, and
3) the use of ECP.

The use of adapted basis functions for atoms by the Generator Coordinate Method along with the use of the pseudopotential allows a high quality calculation at a lower computational cost.

The present methodology - Eqs. 3 and 6, which relies on small basis sets (representation of the core electrons by ECP) and an easier and simpler way for correcting the valence region (mainly of anionic systems) appears as an interesting alternative for the calculation of thermochemical data such as electron and proton affinities and enthalpies of formation for larger systems.

The CCSD(T)/B1 method have been shown to be powerful, and of affordable computational cost for the systems containing atoms of the 2nd and 3rd periods.

5. Acknowledgments

I would like to thank the computational facilities of Chemistry Institute at UNICAMP and the financial support from Conselho Nacional de Desenvolvimento Científico e Tecnológico (CNPq) and Fundação de Amparo à Pesquisa de São Paulo (FAPESP).

6. References

Badenes, M. P., Tucceri, M. E. & Cobos, C. J. (2000). *Zeitschrift für Physikalische Chemie* 214: 1193.

Boese, A. D., Oren, M., Atasoylu, O., Martin, J. M. L., Kállay, M. & Gauss, J. (2004). *J. Chem. Phys.* 120: 4129–4141.

Curtiss, L. A., Raghavachari, K., Redfern, P. C. & Pople, J. A. (1997). *J. Chem. Phys.* 106: 1063.

Curtiss, L. A., Raghavachari, K., Redfern, P. C. & Pople, J. A. (1998). *J. Chem. Phys.* 109: 7764–7776.

Curtiss, L. A., Raghavachari, K., Redfern, P. C. & Pople, J. A. (2000). *J. Chem. Phys.* 112: 7374.

Curtiss, L. A., Redfern, P. C. & Raghavachari, K. (2007). *J. Chem. Phys.* 126: 084108.

Curtiss, L. A., Redfern, P. C., Smith, B. J. & Radom, L. (1996). *J. Chem. Phys.* 104: 5148.

Custodio, R., Giordan, M., Morgon, N. H. & Goddard, J. D. (1992). *Int. J. Quantum Chem.* 42: 411.

Custodio, R., Goddard, J. D., Giordan, M. & Morgon, N. H. (1992). *Can. J. Chem.* 70: 580.

DeYonker, N. J., Cundari, T. R. & Wilson, A. K. (2006). *J. Chem. Phys.* 124: 114104.

Ditter, G. & Niemann, U. (1982). *Phillips J. Res.* 37: 1.

Dunning, T. H. (1989). *J. Chem. Phys.* 90: 1007.

Ervin, M. K. (2001). *Chem. Rev.* 101: 391.

Feller, D. & Peterson, K. A. (1999). *J. Chem. Phys.* 110: 8384.

Ge, Y., Gordon, M. S. & Piecuch, P. (2007). *J. Chem. Phys.* 174: 174106.

Karton, A., Rabinovich, E., Martin, J. M. L. & Ruscic, B. (2006). *J. Chem. Phys.* 125(14): 144108.

Kowalski, K. & Piecuch, P. (2000). *J. Chem. Phys.* 18: 113.

Lin, C. Y., Hodgson, J. L., Namazian, M. & Coote, M. L. (2009). *J. Phys. Chem. A* 113: 3690.

Martin, J. M. L. & De Oliveira, G. (1999). *J. Chem. Phys.* 111: 1843–1856.

Mezzache, S., Bruneleau, N., Vekey, K., Afonso, C., Karoyan, P. Fournier, F. & Tabet, J.-C. (2005). *J. Mass Spectrom.* 40: 1300.

Mohallem, J. R. & Dreizler, R. M. Trsic, M. (1986). *Int. J. Quantum Chem. - Symp.* 20: 45.

Mohallem, J. R. & Trsic, M. (1985). *Z. Phys. A - Atoms and Nuclei* 322: 538.

Montgomery Jr., J. A., Frisch, M. J., Ochterski, J. W. & Petersson, G. A. (2000). *J. Chem. Phys.* 112: 6532.

Morgon, N. H. (1995a). *J. Phys. Chem. A* 99: 17832.

Morgon, N. H. (1995b). *J. Phys. Chem.* 99: 17832.

Morgon, N. H. (1998). *J. Phys. Chem. A* 102: 2050.

Morgon, N. H. (2006). *Int. J. Quantum Chem.* 106: 2658.

Morgon, N. H. (2008a). *J. Braz. Chem. Soc.* 19: 74.

Morgon, N. H. (2008b). *Int. J. Quantum Chem.* 108: 2454.

Morgon, N. H. (2011). *Int. J. Quantum Chem.* 111: 1555–1561.

Morgon, N. H., Argenton, A. B., Silva, M. L. P. & Riveros, J. M. (1997). *J. Am. Chem. Soc.* 119: 1708.

Morgon, N. H. & Riveros, J. M. (1998). *J. Phys. Chem. A* 102: 10399.

Nelder, J. A. & Mead, R. (1965). *Computer J.* 7: 308.

Nyden, M. R. & Petersson, G. A. (1981). *J. Chem. Phys.* 75: 1843.

Ochterski, J. W., Petersson, G. A. & Montgomery Jr., J. A. (1996). *J. Chem. Phys.* 104: 2598.
Ochterski, J. W., Petersson, G. A. & Wiberg, K. B. (1995). *J. Amer. Chem. Soc.* 117: 11299.
Oyedepo, G. A. & Wilson, A. K. (2010). *J. Chem. Phys.* 114: 8806.
Parthiban, S. & Martin, J. M. L. (2001). *J. Chem. Phys.* 114: 6014–6029.
Pereira, D. H., Ramos, A. F., Morgon, N. H. & Custodio, R. (2011). *J. Chem. Phys.* 135: 034106.
Peterson, K. A., Tensfeldt, T. G. & Montgomery Jr., J. A. (1991). *J. Chem. Phys.* 94: 6091.
Piecuch, P., Kucharski, S. A., Kowalski, K. & Musial, M. (2002). *Comp. Phys. Commun.* 149: 71.
Smith, B. J. & Radom, L. (1991). *J. Phys. Chem.* 95: 10549.
Stevens, W. J., Basch, H. & Krauss, M. (1984). *J. Chem. Phys.* 81: 6026.
Tabrizchi, M. & Shooshtari, S. (2003). *J. Chem. Thermodyn.* 35: 863.

Part 2

Electronic Structures and Molecular Properties

3

Quantum Chemistry and Chemometrics Applied to Conformational Analysis

Aline Thaís Bruni[1] and Vitor Barbanti Pereira Leite[2]
[1]*Departamento de Química, Faculdade de Filosofia, Ciências e Letras de Ribeirão Preto, Universidade de São Paulo*
[2]*Departamento de Física, Instituto de Biociências, Letras e Ciências Exatas, Universidade Estadual Paulista, São José do Rio Preto*
Brazil

1. Introduction

1.1 Conformational analysis: Early history and Importance

Molecular structure plays a special role in science. Knowledge of the atomic arrangement is essential in order to be able to elucidate chemical properties and processes. The first advances in determining molecular structure occurred in nineteenth century. Around 1812, Jean-Baptiste Biot, a French physicist, discovered optical activity by observing polarized light shifting when crossing a quartz crystal. He observed that the light was displaced to the right in some cases and to the left in others. The conclusion was that rotation of polarized light by quartz is an inherent property of the crystal. Interested in the phenomenon, Biot noticed in further studies that similar effects were found when polarized light passed through certain liquids such as natural oils (lemon extract and laurel), alcoholic solutions of camphor, some sugars and tartaric acid. (Drayer, 1993; Cintas, 2007; Gal, 2011) Biot's observations were very important in laying foundation for the concept of optical activity.

In 1948, Louis Pasteur discovered molecular chirality when studying a mixture of tartaric acid crystals.(Gal, 2007) He patiently performed the manual separation of tartarate enantiomer crystals (Cintas, 2007) and observed that each solution made with them was able to displace polarized light in one direction. He concluded that compounds with non-superimposable molecular asymmetry have identical chemical properties despite the inverse behavior related to polarized light. Pasteur argued that the optical activity of organic solutions is related to molecular geometry. This insight was far ahead of the organic structural theory of the time.(Drayer, 1993) Although Pasteur was the first to show a relationship between optical activity and molecular symmetry, he was not able to say exactly how a molecule could be right- or left-handed. The main advances in this idea occurred in 1874 when a theory of organic structure in three dimensions was independently and simultaneously developed by Jacobus Henricus van't Hoff in Holland, and Joseph Achille Le Bel in France. (Drayer, 1993; Cintas, 2007)

In 1865, August Kekulé proposed his theory of the benzene molecular structure and proposed that the carbon atom has valence 4.(Brush, 1999) His principal idea was that the carbon atom is tetravalent and can form valence bonds with other carbon atoms yielding to chains. These carbon chains can sometimes have closed arrangements, forming rings. (Drayer, 1993)

Van't Hoff and Le Bel proposed that the four valences of the carbon atom were not planar, but directed into three-dimensional space. Van't Hoff specifically proposed that the spatial arrangement was tetrahedral. Later, he used the tetrahedron as a graphic representation of the valence arrangement around the carbon atom and also used this model to explain the physical property of optical activity.(Ramberg & Somsen, 2001) A compound containing a four different substituted carbon - described by Van't Hoff as asymmetric carbon - would be capable of existing in two distinctly different nonsuperimposable forms. Finally, he stated that the asymmetric carbon atom was the cause of molecular asymmetry and optical activity.(Drayer, 1993)

Le Bel, in turn, also published his stereochemical ideas in 1874, but with a different approach to the problem from that presented by Van't Hoff. His hypothesis was not based on the tetrahedral model for the carbon atom and the fixed valences between the atoms. His investigation was into the asymmetry as a whole, without evaluating the individual atoms. The full system was considered in his evaluation, and his interpretation could be inserted into the field that is currently understood as molecular asymmetry. He mentions the tetrahedral carbon atom only in special cases, and not as a general principle. Many molecules confirm Le Bel's concepts of molecular asymmetry. Allenes, spiranes, and biphenyls are some examples of asymmetric molecules that do not contain any asymmetric carbons.

Van't Hoff's and Le Bel's different approaches can be explained by the origin of their formation. Van't Hoff, based on Kekulé tetrahedron models, suggested the concept of the asymmetric carbon atom. On the other hand, Le Bel based his investigations on Pasteur's considerations of the connections between optical rotation and molecular structure.(Drayer, 1993)

The historical development of conformational search does not end here and has many other important aspects and particularities. Our goal was just to give a basic outline of the initial concepts and how they influence current conformational understanding. Despite the historical progress in conformational studies, the advances in structure determination has been relatively recent and have been made possible by the development of analytical instruments and computational tools. Early structural studies were applied only to small molecules or substructures that could be expressed in terms of a few settings.(Allen et al., 2010)

Currently, a great evolution is occurring in mechanisms for determining and understanding molecular structures. The relationship between geometry and energy is experimentally measurable and gives an idea of the balance between energy factors involved in each structure. (Pietropaolo et al., 2011) Reactivity and other properties are directly linked to the conformational arrangement of molecules.(Hunger & Huttner, 1999) Every chemical property must be understood according to its molecular structure and atomic connections. (Pietropaolo et al., 2011) Indeed, knowledge of structural arrangement is important since it underlies studies in chemical reactions and other molecular behaviors. There are experimental techniques for the structural determination, such as X-ray, magnetic resonance, infrared, mass spectroscopy and others.

In this chapter we will discuss theoretical methods for molecular conformational determination. The field that concerns ways to mimic the behavior of molecules and molecular systems is molecular modeling. It seeks a simplified or idealized description of molecular systems, making it possible to produce three-dimensional representations that provide insights into their behavior. As computer tools have enjoyed a spectacular increase in last decades, theoretical methods are invariably associated with computer modeling. This has become a powerful tool for evaluating molecular structure, from which special chemical information about molecular behavior can be inferred. (Pietropaolo et al., 2011)

2. Statement of the problem

For the theoretical and computational determination of molecular properties it is necessary to previously determine the minimum energy structure of the system being studied. A central issue is to probe the equilibrium configuration of the molecular system. The way that energy varies with the coordinates is usually referred to as the potential energy surface. At the atomic level, the interaction energy between atoms is essentially ruled by quantum mechanics, which provides the basic elements and methods used in molecular modeling. However, the potential energy surface can be addressed with different degrees of approximation, *i.e., ab initio*, effective potentials or even more coarse-grained potentials. Irrespective of the details with which the system is considered, one usually faces the problem of a highly dimensional system with the occurrence of multiple minima. Low energy minima play an important role in determining molecular properties, and the determination of these minima conformational states is a non-trivial task, usually referred to as energy minimization method for exploring the energy surface.

If four or more atoms are connected in chain by single bonds we can suppose that there is considerable flexibility in the molecule. The existence of hindered rotation about a single bond is one of the fundamental concepts in conformational analysis.(Mo & Gao, 2007) The understanding of the connections between the atoms is related to the internal coordinate parameters, *i.e.*, bond length, bond angle and dihedral angle, and is essential in designing molecular models.

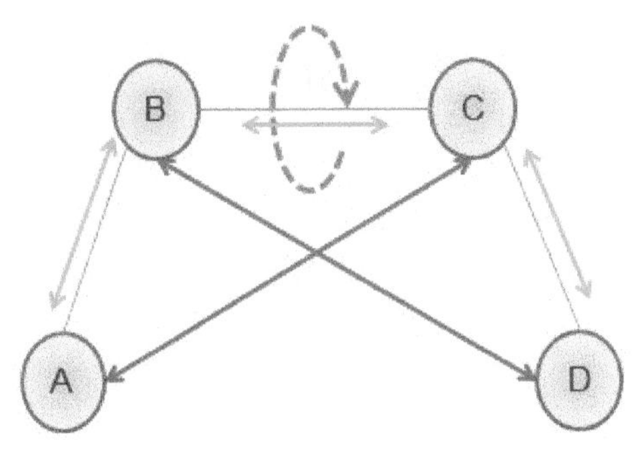

Fig. 1. Model for a generic molecule with four different atoms.

For instance, let us consider a molecule composed of four different atoms which are single-bond linked (Figure 1). The green arrows represent the bond stretch and the average value is the bond length; the red arrows correspond to the angle formed by three sequential atoms, *i.e.*, the angle bond; the curved blue arrow indicates the free rotation around the only single bond able to perform changes in the molecule conformation, as shown in Figure 2:

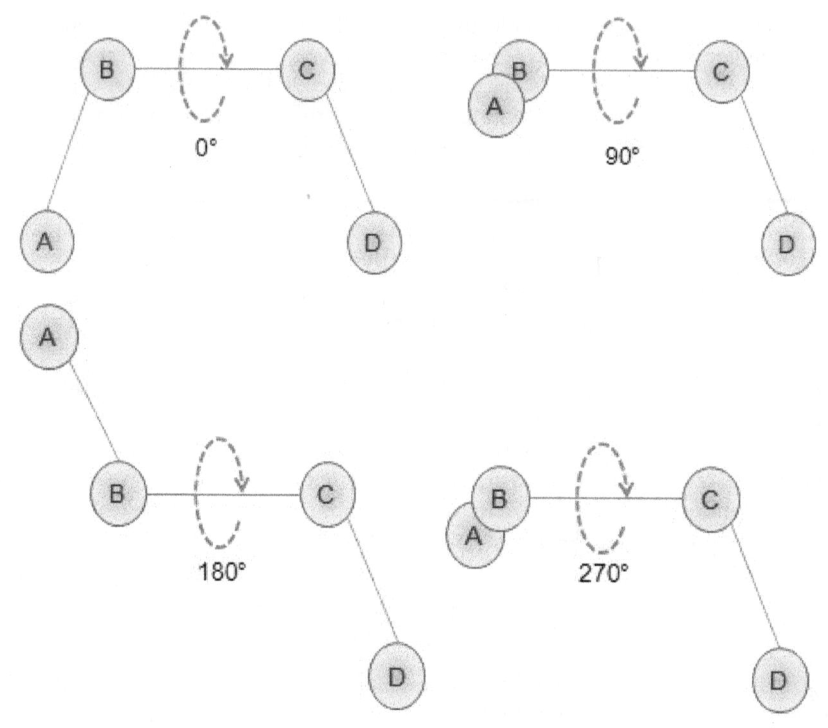

Fig. 2. 90° rotation around the single bond.

In other words, different conformations are obtained when a dihedral angle is rotated. A dihedral angle is that composed by the planes formed by the sequence of three atoms (Figure 3):

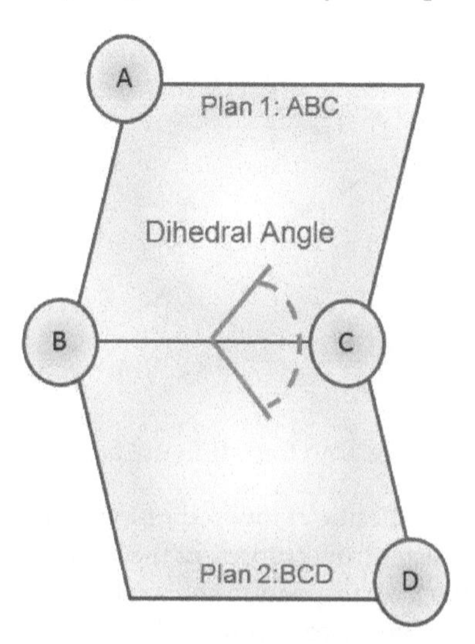

Fig. 3. Dihedral angle representation for the molecule ABCD.

The main questions on molecular modeling are concerned with a good way of finding the global minimum energy structure. Important information is also concerned with the behavior of the conformational space. It is not only necessary to know which is the global minimum, but also the whole shape of the potential energy surface (PES). The main characteristic of this conformational phase space is that it is exponentially large, and a computationally hard problem. (Fraenkel, 1993) One problem that illustrates this difficulty is the protein folding, in which one searches for the global energy minimum structure associated with its functional conformation. If a method can be used to describe the relevant potential energy surface of a given molecule, it can also accurately elucidate its behavior against many interest situations.

Many techniques have been presented, and it is not the goal of this work to make a deep study on them. A good discussion of practical methods is given by Leach.(Leach, 2001) We intend to give a brief idea of the most popular techniques used for investigating the molecular structure computationally. For conformational sampling, one can imagine a hierarchy of methods with different computational costs.(Seabra et al., 2009) However, there is no sovereign truth about what is the best method for performing a conformational analysis. Each situation must be evaluated. The best method is one that has the best fit to the problem studied; in practical terms it will provide the answer as quickly as possible, using the least amount of computer resources.

3. Stochastic methods of conformational analysis

The literature reports many methods for trying to solve multiconformational problems, and most of them are based on stochastic approaches. Put simply, stochastic methods work with random variables, such as initial conformations for the search or the steps probing the configuration phase space. The simple criterion for establishing a minimum energy conformation is that the first derivatives of the energy E with respect to each variable (x_i) is zero and the second derivatives are all positive:

$$\frac{\partial E}{\partial x_i} = 0 \text{ and } \frac{\partial^2 E}{\partial x_i^2} > 0.$$

The algorithms that search for minimum energy states can be classified into two groups: those which use derivatives of the energy with respect to the coordinates, and those which do not. The most used derivative minimization methods are the steepest descent, line search in one dimension and conjugate gradient methods.(Leach, 2001) These algorithms are very useful for conducting local (restricted) searches of minima, or downhill searches to the nearest minimum, since they are not able to overcome energy barriers. They are often used in combination with other stochastic methods.

In the remainder of this section we discuss examples of stochastic methods. Havel, Kuntz and Crippen described distance geometry algorithms in conformational analysis.(Havel et al., 1983b) Given the impossibility of examining all possible conformations, they introduced a method which is capable of finding global optima without considering all possible solutions by means of combinatorial optimization. The method is known as *brunch and bound* and involves logical tests that allow whole classes of solutions to be eliminated without examining them one by one. The method converts a set of distance ranges (or "bounds") into a set of Cartesian coordinates that are consistent with those bounds. (Spellmeyer et al., 1997) The efficiency of a *branch and bound* algorithm depends on how effective these tests are compared to the time required to perform them. (Havel et al., 1983a; Havel et al., 1983b) In another study, Havel *et al* presented the basic theorems of distance geometry in Euclidean space. They proposed new algorithms and described refinements to the existing ones. All these algorithms were similar because they utilize geometric principles in order to interpret structural relationships. (Havel et al., 1983b)

According to Leach and Smellie,(Leach & Smellie, 1992) distance geometry is a method for searching conformational space in which a structure is initially formulated in terms of interatomic distances. Any molecular system can be described as the set of minimum and maximum interatomic distances between all pairs of atoms in the molecule. The complete conformational space of the molecule is contained within this space. In distance geometry, a matrix is defined as the set of minimum and maximum distances, and then used to create a series of conformers that are consistent with those distances.(Spellmeyer et al., 1997)

Another tool for performing conformational searches is the genetic algorithm, a stochastic method first introduced by Holland in 1975. Genetic algorithm (GA) is a method applied to solve problems using a natural evolution process simulation. It is a stochastic method developed in analogy to Darwin's theory of evolution in order to perform the optimization.(Brodmeier & Pretsch, 1994; Lucasius, 1993; Nair & Goodman, 1998)

Genetic algorithm is commonly used for studying a large-scale space of possible solutions. The goal is to identify the best solutions within that space without the need to evaluate all possibilities.(Yanmaz et al., 2011) The GA is the optimization of a large number of possible solutions using a randomly generated population. When applied to conformational analysis, the population of interest consists of different conformations. The biological evolution of this population is simulated. A population of trial solutions is iteratively manipulated by a series of genetic operators to satisfy an objective function. The adjustment is calculated, and a new population is generated according to operators, such as selective reproduction, recombination and mutation. The process is repeated until the minimum energy structures are obtained.(Lucasius & Kateman, 1994; Beckers et al., 1996; Beckers et al., 1997)

Artificial Neural Networks (ANN) are another example of stochastic methods used in conformational analysis. This method is based on concepts of the behavior of the human brain. Although artificial neural networks are primitive compared to their biological counterparts, they exhibit some interesting properties which make them useful as multivariate tools in various fields of research. During the last decade, ANN have been successfully applied in non-linear modeling, classification, signal processing and process control.(Derks & Buydens, 1996) The properties of a molecule are intimately linked to the conformations that it adopts and so an understanding of the conformational space is important in rationalizing and predicting its behavior.(Jordan et al., 1995)

Among the most popular stochastic methods for covering the conformational space are Monte Carlo (MC) and Molecular Dynamics (MD). They are similar in the sense that both procedures include the same representation of molecules and use classical force fields for the potential energy terms, under periodic boundary conditions. The main purpose of these methods is to sample the phase space and to use the force fields ability to represent the conformational space near minima and connecting transition structures.(Jorgensen & TiradoRives, 1996; Grouleff & Jensen, 2011) However, large differences are found in sampling and configuring space available to the system. For MC, a new configuration is generated by selecting a random molecule or part of it, rotating it, translating it, and performing an internal structural variation. These changes do not necessarily need to follow a realistic physical trajectory. The acceptance of the new configuration is, however, determined by the Metropolis sampling algorithm. The sampling criterion is set in a way that enhances the likelihood of probing low energy conformations. Application over enough configurations yields properly Boltzmann-weighted averages for structure and thermodynamic properties. For MD, given a set of initial conditions (position and velocities of all atoms), new configurations are generated by application of Newton's equations of motion, so that the new atomic positions and velocities of all atoms are determined simultaneously over a small time step. In both cases, the force field controls the total energy (MC) and forces (MD), which determines the evolution of the systems. (Jorgensen & TiradoRives, 1996)

Examples of problems related to large systems are the interaction between drug and the receptor, and protein behavior and folding. Molecular docking procedures are capable of predicting the three-dimensional structure of macromolecular complexes and their binding affinity. The information required is simple and corresponds to the structures of the receptor and ligand and the presumable interfacing region between them. Besides the simplicity of these docking procedures, they have low computational costs. However, molecular plasticity and solvation effects are not, or are only approximately, taken into account in these approaches. Free energy simulations may be then used to investigate the molecular association process and to predict binding affinity. (Biarnés et al., 2011)

It is important to realize that sometimes the probing of PES addresses singular questions, which involve association of several methods, also called hybrid methods. A particular well-known tailored one is the quantum mechanics/molecular dynamics approach, also known as QM/MM approach. This is a molecular simulation method that combines the strength of both QM (high accuracy in specific regions) and MD fast calculations (in not so crucial regions), in such a way that it efficiently allows the study of chemical processes in solution and in proteins.

When stochastic methods are used to find minimum energy conformations, asymptotic states in restricted regions of the phase space are probed. This means that there is no end point in the search, and the convergence cannot be assured.

4. Systematic search in conformational analysis

As seen before, stochastic techniques use different heuristics to randomly cover the conformational space. These algorithms apply a perturbation to the initial conformer and minimum energy conformation is associated with the lowest energy state that is found through out this procedure. They provide a sampling of energy minima structures and the shape of the PES is obtained in an indirect way.

Beyond the stochastic methods there are procedures that do not work with random choice to cover conformational space. These classes of methods are described as deterministic and are capable of searching the conformational map in a systematic way, providing a direct knowledge of PES shape. These searches divide conformational space into quantized units and apply algorithms to search this discrete space or define a set of heuristic rules that are used to drive the search.(Smellie et al., 2003)

Systematic methods are those that explore all conformational space at some fixed degree of resolution. To perform the systematic search, a molecule must be numerically described by its atoms' internal coordinates. The internal coordinates are bond length, angle bond and dihedral (torsion) angle. For a given initial structure the systematic conformational search is conducted by regular variation in dihedral angles (Figure 2).

Although a systematic search can obtain the morphology of a molecule's energetic behavior directly, this method is not feasible for evaluating complex systems. (Beusen et al., 1996) Systematic search is most usefully applied for molecules with few degrees of freedom.(Li Manni et al., 2009)

According to literature (Beusen et al., 1996), to cover the PES corresponding to the conformational space, different molecular structures must be systematically generated by rotating the torsion angles around the single bonds between 0° and 360°. The number of conformations is given by:

$$\text{Number of conformations} = s^N \tag{1}$$

where N is the number of free rotation angles, and s is the number of defining steps according to the angle increment:

$$s = \frac{360°}{\theta_i} \tag{2}$$

with θi being the dihedral increment of angle i.

An examination of equation (1) reveals that the number of conformations generated will exponentially increase in proportion to the number of bonds with free rotation in the molecule under study. A problem arises if the number of steps is large, *i.e.*, when a very refined surface is required by small angle increments. This problematic behavior of the systematic study of PES, described as combinatorial explosion, is the major restriction involved in this kind of search. Figures 4 and 5 illustrate how combinatorial explosion works. In Figure 4, we have a representation of the system growth where many single bonds can be rotated.

The combinatorial explosion problem is represented by Figure 5. The number of branches to be considered is shown by the ramification achieved according the number of angles (A, B, C, D...) and will depend on the dihedral increment chosen. Due to the problem involved in combinatorial explosion, systematic search becomes nonviable for studying large molecules, since the number of degrees of freedom increases. A useful strategy for reducing the dimensionality of the conformational space is to perform systematic conformational searches on small portions of the molecule (either as isolated fragments or *in situ*). Using these optimal parts, one builds the conformation of the whole molecule with only limited additional searching of the relative conformations of the fragments. Approaches that incorporate this principle are known as "build-up" methods. (Beusen et al., 1996; Izgorodina et al., 2007) There are some strategies for overcoming the combinatorial explosion. We will focus our discussion on procedures that involve chemometrical approaches.

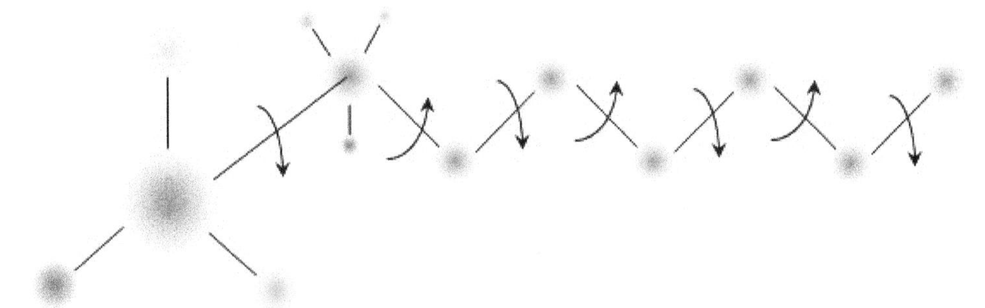

Fig. 4. A general structure with many single bonds.

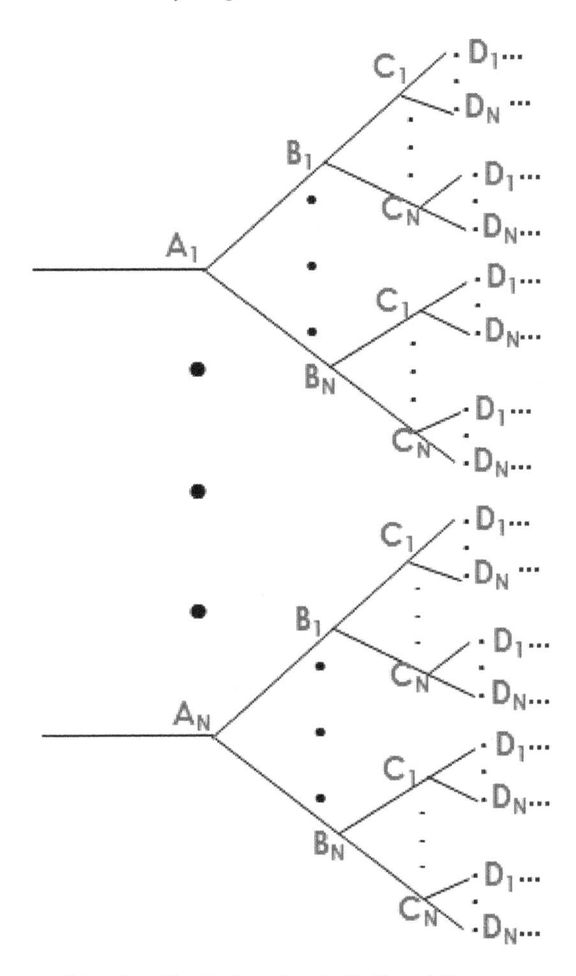

Fig. 5. Branches generated by the dihedral angles A, B, C and D.

5. Chemometrics and structure determination

A conformational search, independently of the method chosen, usually involves large amounts of data. Sometimes, data achieved from a given methodology must be explored by an additional technique. According to Geladi (2003) *"data exploration means taking a look at the data to find interesting phenomena, often without prior expectations. As a result, outliers, clustering of objects and gradients between clusters may be detected."*(Geladi, 2003)

Chemometrics has been used extensively in recent years for exploring chemical problems by means of computer tools and statistical observations. The literature presents many definitions for chemometrics. For our purposes, this field of knowledge is better defined as a combination of two definitions found in the literature:

a. According to Wold (1995), chemometrics can be understood as a way *"to get chemically relevant information out of measured chemical data, how to represent and display this information, and how to get such information into data"*;(Wold, 1995)
b. For Beeb (1998), chemometrics corresponds to "*the entire process whereby data (e.g., numbers in a table) are transformed into information used for decision making.*" (Beebe, 1998)

The above definitions indicate that chemometrics offers a broad approach to chemical measurement sciences. It is not restricted to the actual experimental analysis but also considers what happens before and after it. (Massart et al., 2004) It is the goal of chemometrics to extract the information from the data. (Ramos et al., 1986) Chemometrical approaches have been applied to conformational analysis for handling special difficulties of large amounts of data generated both by stochastic and by systematic searches.

Among the various chemometrical techniques, Principal Component Analysis (PCA) is the most commonly used for conformational problems. In many ways, it forms the basis for multivariate data analysis. PCA is a multivariate method of analysis whose main concern is to reduce the dimensions needed to portray accurately the characteristics of a large dimensional data matrix.(Beebe, 1998; Wold et al., 1987) This mathematical procedure consists of eliminating a large number of correlated variables without changing the characteristics of the original data-set that contribute most to its variance.

For an easy graphical representation, consider a two-dimensional set of variables as shown in Figure 6 (a).

PCA can be performed on the original variables as shown in Figure 6 (b) and new axes, called Principal Components, arise to account for the maximum variation. A subsequent rotation (Figure 6(c)) is made on these new PC axes in order to rewrite the original variables in terms of this new axes-system. Each PC is constructed as a linear combination of variables:

$$P_i = \sum_{j=1}^{v} c_{i,j} x_j \qquad (3)$$

where Pi is the ith principal component and ci,j is the coefficient of the variable xi,j. (Leach, 2001) There are v such variables. The first principal component PC1 is chosen in order to maximize the data variance of the axis. The second and subsequent ones are chosen to be orthogonal to each other and account for the maximum variance in the data not yet described by previous principal components. A variety of algorithms can be used to calculate the principal components. The most commonly employed approach is singular value decomposition SVD. (Golub & Loan, 1996)

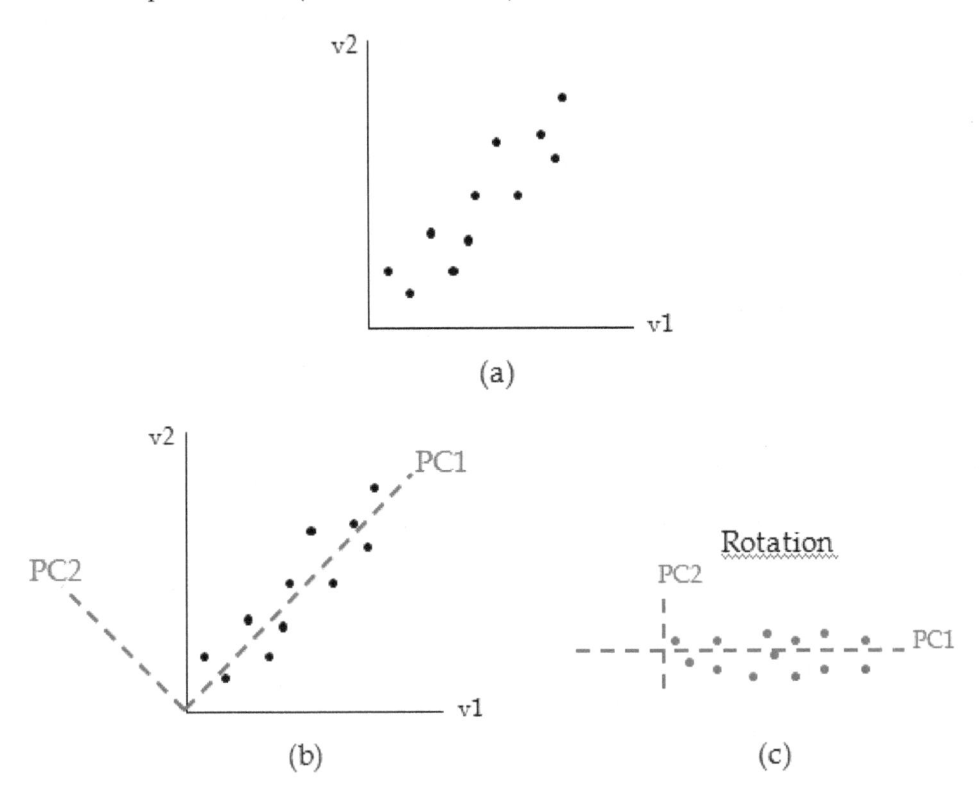

Fig. 6. PCA procedure: (a) original data set; (b) PCA on original data set and (c) Variables according to the new PC coordinates.(Beebe, 1998)

A matrix of arbitrary size can be decomposed into the product of three matrices in such a way that:

$$\mathbf{X} = \mathbf{USV}_t \quad (4)$$

where **U** and **V** are square orthogonal matrices. The matrix **U** (whose columns are the eigenvectors of $\mathbf{XX}_t$) contains the coordinates of samples along the PC axes. The **V** matrix (which contains the eigenvectors of the correlation matrix $\mathbf{X}_t\mathbf{X}$) contains the information about how the original variables were used to make the new axes[ci,j coefficients in eq. (3)]. The **S** matrix is a diagonal matrix that contains the eigenvalues of the correlation matrix (standard deviations) or singular values of each of the new PCs. The diagonalization of symmetric matrices (such as $\mathbf{XX}_t$ and $\mathbf{X}_t\mathbf{X}$) and SVD are fundamental problems in linear algebra (Golub & Loan, 1996), for which computationally efficient software has been developed and can be used on a routine basis (Hanselma et al., 1997) for very large-size matrices.

In chemistry, PCA was introduced by Malinowski around 1960 under the name Principal Factor Analysis, and further developed after 1970.(Malinowski, 2003) Principal Component Analysis can be used for crystallographic structure data; in its general form, conformational analysis is applied to multivariate numerical problems. (Allen et al., 2010) Many studies report on the use of PCA for handling Molecular Dynamics data. Among them, we highlight the application that uses PCA for mapping potential energy surfaces, by the quantitative visualization of a macromolecular energy funnel. (Becker, 1998) Other examples where PCA can be applied in molecular structure determination can also be found in recent studies. (Das et al., 2011; Araujo-Andrade et al., 2010; Kiralj et al. 2007; Oblinsky et al., 2009; Silva et al., 2011)

6. Pairs of dihedral angles-systematic analysis

There is a variety of theoretical methods that are capable of locating minimum energy structures in the potential energy surface. The problem of stochastic methods is that there is no natural end point for the conformational search. In some cases, only a small subset of conformational space is explored and the convergence of the system is not guaranteed. Only Systematic Conformational Analysis maps the conformational space completely. We stress the principal difficulty inherent in this method is the combinatorial explosion. In a previous study (Bruni et al., 2002), a new methodology was introduced for controlling the combinatorial explosion through a systematic reduction in the size of the system by means of chemometrics.

This method consists of a small systematic conformational analysis, in which the conformational space is studied by rotation of the important free rotation in pairs, described as Pairs of Dihedral Angles-Systematic Analysis – PDA-SA. The main objective is to reduce the dimension of the investigated system. The idea is to address the conformational space in small portions, evaluating PES in combinations of angles in pairs. If the problem of combinatorial explosion is controlled, the conformational space can be sufficiently refined in the regions of minimum energy, taking care to minimize the information lost. The energy surfaces are obtained for each pair of angles and the number of conformations is given by Equation 5:

$$\text{Number of conformations} = s^2 \frac{N(N-1)}{2} \tag{5}$$

where s and N have the same meaning as in Equation 1. The number of conformations, in this case, is given by the combinatory arrangement of the N dihedrals in pairs.

The main observation of the comparison between equations (1) and (5) is that the number of conformations as given by Eq. (1) increases exponentially with the number of bonds with free rotation, while from Eq. (5), the number of studied conformations increases quadratically with *N*. As the number of free rotation angles increases, the difference in the number of conformers generated by these two equations becomes more evident.

The computational procedure for PDA-SA can be organized in five basic steps:

1. **Molecular Building:** The interest molecule must be defined in terms of its internal coordinates: bond length, angle bond and dihedral angles. There are many softwares able to define this molecular initial structure. A quantum chemistry optimization is required at this step in order to adjust internal parameters. The best method must be chosen according to the system under study.

2. **Dihedral Pair Rotation:** The PDA-SA conformational search begins and the combination of the existing pairs of angles is taking account. Sometimes it is only possible to choose a dihedral increment with a less refined value. A rough PES is obtained in this case. The matrix to be analyzed consists of energy values from potential surfaces for angle combinations, and they are grouped according to Figure 7 for N angles. Appendix A shows the energy values for omprazole basic structure. The idea is to perform a cyclical permutation on the data, and this matrix form ensures that no information about the total PES is lost. The energy values obtained for each angle rotation as a function of the others allow the conformational space to be completely mapped. The major advantage is that the shape of these small portions can be visually observed, since we have a 3-D fitting. (see Figures 9 and 10)
3. **PCA application on data matrix:** After the energy matrix statement, PCA is performed on the data. The regions with minimum energy points on the grid search can be easily selected. The number of selected regions will depend on the nature of the studied system.
4. **Refinement with a short dihedral increment:** The regions initially obtained in step 3 can be refined with a small angle increment. It is important to emphasize that this step is not obligatory, since a small dihedral increment can be used in step 3, depending on the studied system. However, previous experience in this methodology (Bruni et al., 2002; Bruni & Ferreira, 2008) shows that this is the easiest procedure, *i.e.*, firstly make rotations with a large dihedral increment and subsequently refine the minimum energy regions selected by PCA with small dihedral increments.
5. **Optimization of the final structure:** the procedure described above provides angle values for the conformational search with a good level of accuracy. When these values are combined, we obtain all the possible minimum structures. Those structures constrained by the angle values obtained by PCA analysis are submitted to final optimization and the resulting structures are considered to be those of minimum energy.

In the study that introduced this method, the approach was successfully tested in the analysis of omeprazole and its derivatives, in which the results were in agreement with the experimental ones.(Bruni et al., 2002)

In a second study, the technique was used to find minimum energy conformations of omeprazole derivative molecules in a QSAR study. (Bruni & Ferreira, 2002) It was shown that conformational analysis is crucial when establishing SAR/QSAR models using theoretically calculated descriptors, and they are strongly dependent on the details of molecular structure. Though all minima conformation have similar energetic values, some calculated properties are very sensitive to the structural variation, which is understandable since electronic properties are intrinsically dependent on molecular conformation.(Bruni & Ferreira, 2002)

Omeprazole's racemization barrier and decomposition reaction was also studied. Quantum chemistry coupled to PDA-SA chemometric method was used to find all omeprazole minimum energy structures. To obtain the racemization barriers it was essential that the starting structure was in a global energy minimum. In that work, for all the studied structures, there was no change in the values of the racemization barriers, which confirmed the identification of the most stable structures for omeprazole.(Bruni & Ferreira, 2008)

$i\backslash j$	1	2	3	...	N
1	-	E_{12}	E_{13}	...	E_{1N}
2	E_{21}	-	E_{23}	...	E_{2N}
3	E_{31}	E_{32}	-	...	E_{3N}
$\vdots$	$\vdots$	$\vdots$	$\vdots$	$\ddots$	$\vdots$
N	E_{N1}	E_{N2}	E_{N3}	...	-

Fig. 7. Matrix scheme for N angles: the discrete energy values for each rotation angle must be evaluated. E_{ij} are the energy matrices with elements E_{ij}^{km}, in which k and m are the angle increment indices for the angles i and j, respectively.

This approach is straightforward and in principle would have no size limits for its application. However, it presents limitation due to some initial condition dependence. Given a system with N degrees of freedom, for each pair of angles there are N-2 parameters that can interfere in the method. For example, in Figure 4 the potential energy surface for first and last dihedral angle combinations depends on the dihedral angles conformation between them. When the dihedral angles are too far from each other along the chain of atoms, the method may not become feasible. In this case the method may need to be repeated with different initial conditions to improve the sampling of the configurational phase space, and moreover we cannot be sure that we have reached the global minimum. When the correlations between the pairs of angles do not depend strongly on these initial conditions the method is very useful. Such system corresponds to small molecules, not so flexible, in which there are few large potential basins, such as omeprazol and its derivatives.(Bruni et al., 2002; Bruni & Ferreira, 2002; Bruni & Ferreira, 2008)

The limit of validity for this method is under investigation. We are applying this method to study the IAN peptide, which is a tetrapeptide isobutyryl-(ala)3-NH-methyl. (Nascimento et al., 2009) This is the smallest polypeptide that can have secondary-like structure (an helix) (Becker & Karplus, 1997) and it has 11 free rotation bonds. For a flexible system, such as this, the initial condition dependence in the calculation of the minimum energy conformations is expected to increase with the size of the system. Since the system is more flexible and expected to be more rugged, we partially overcome this problem by using small angle increments steps, in order to probe all local minima of the system and compare them.

7. Numerical results

7.1 Study of basic structure for omeprazole and derivatives

Initially, the basic structure of omeprazole and derivatives was evaluated. This structure has three bonds with free rotation. To validate the proposed methodology, two different approaches were performed. In the first approach, pairs of angles were taken account ((1,2), (1,3) e (2,3) in Fig. 8) and the number of conformations is given according to Equation 5. The resulting matrix analyzed was composed by the energy values from the potential energy surface for each angle combination (see matrix example in Figure 7). A matrix with discrete energy values for the basic structures with 30° angle increment in Equation 2 is showed in Appendix A.

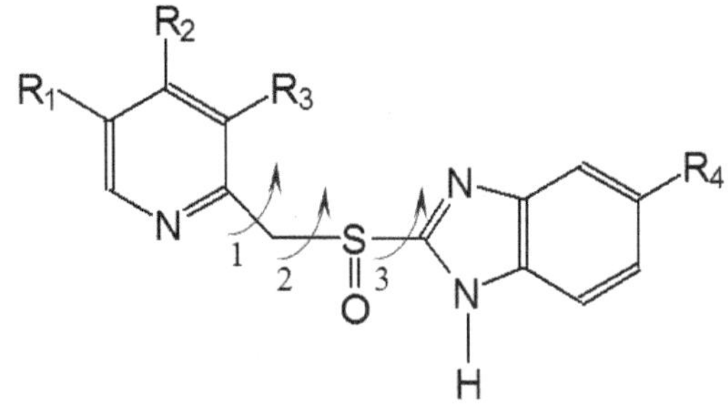

Fig. 8. Basic structure for omeprazole and derivatives.

Three PES were obtained for a 30° dihedral increment and are showed in Figures 9 and 10. Figure 9 shows the original energy values and Figure 10 shows the same surfaces, but with a 0,12 hartrees cut off for better visualization. PCA was performed on autoscaled original data and the results are shown in Figure 11. 64% of the whole information is cumulated in first and second Factors (or Principal Components-PCs). The convergence of the points for one region is observed. Figure 12 shows the PCA results for the leveled data in 0,12 hartrees. Factor 1 and Factor 2 now cumulate 73% of the entire information.

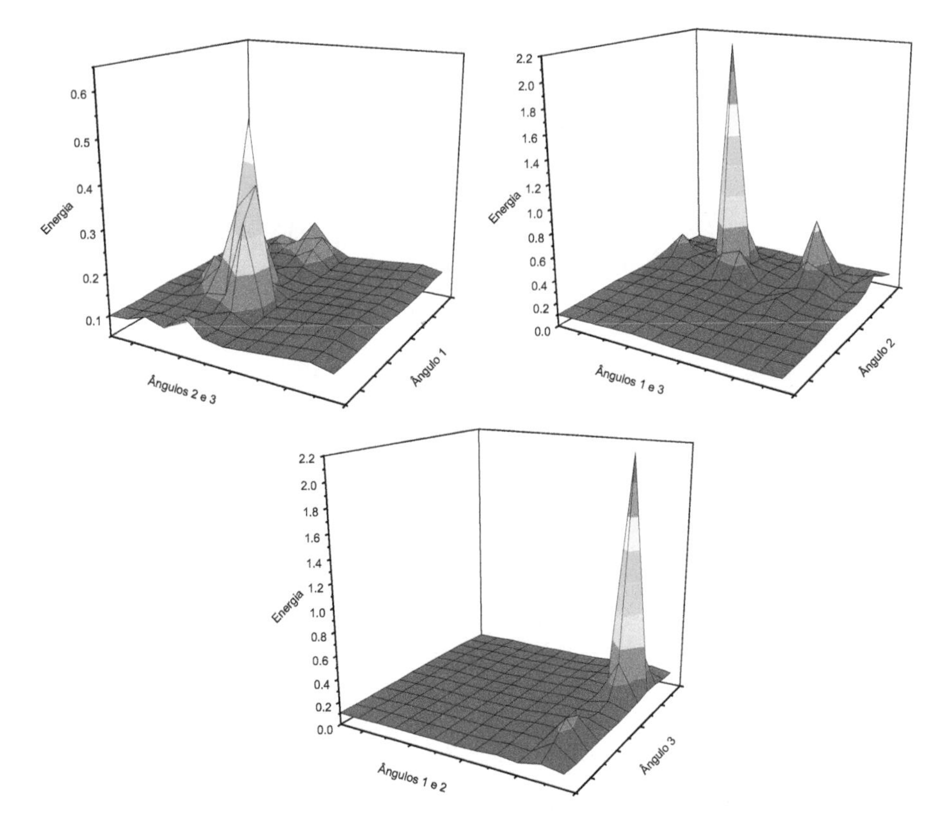

Fig. 9. Orignal PES obtained from PDA-SA method for structure from Fig. 8.

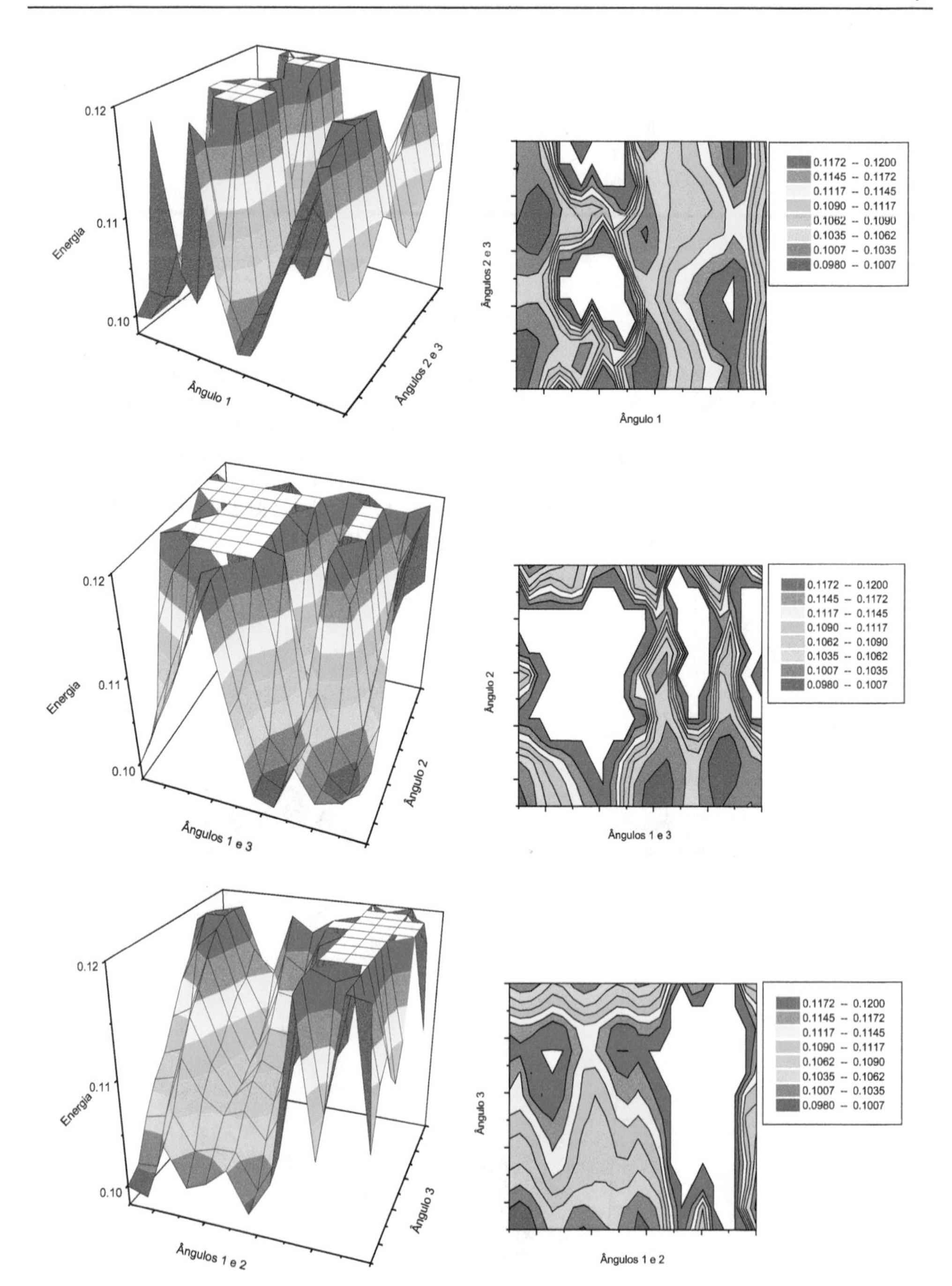

Fig. 10. PES obtained from PDA-SA method for structure from Fig. 8, with a 0,12 hartress cutoff.

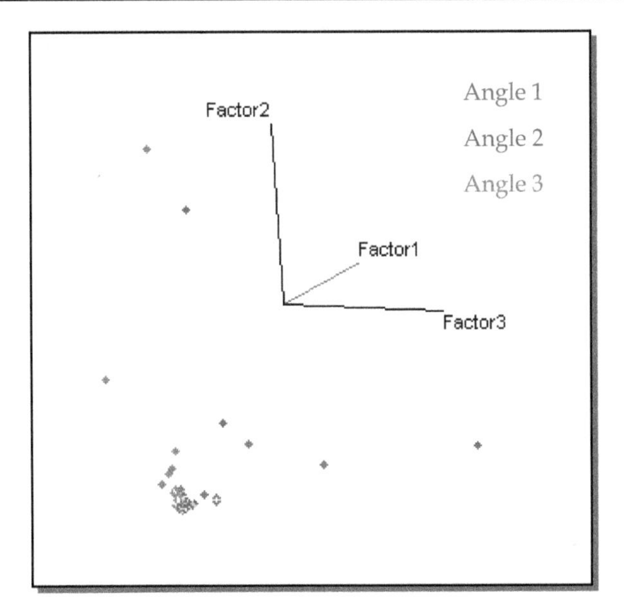

Fig. 11. PCA for data from original PES (Fig.9).

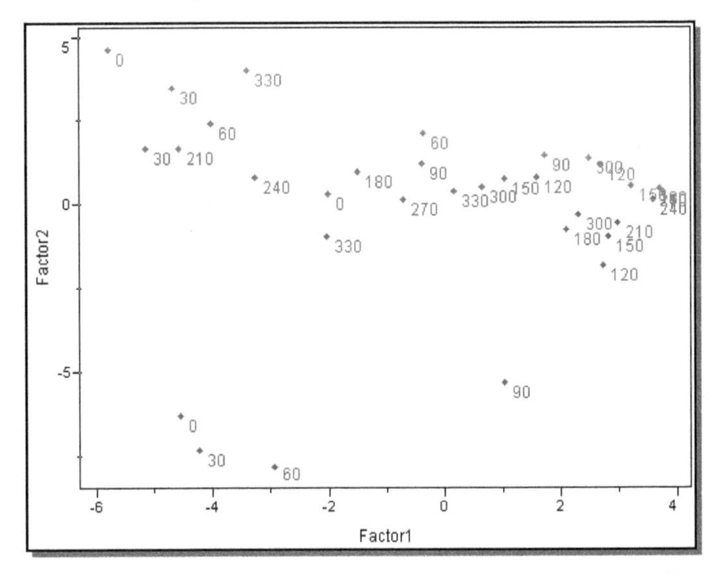

Fig. 12. Principal Component Analysis forPES from Fig. 10, with a 0,12 cutoff.

Factor 1 accounts to the minimum region in each case and Factor 2 accounts for the energy range for the different combinations. Table 1 shows the selected minimum energy for each angle. The first column shows that two different regions were chosen for Angle 1 and only one region for Angles 2 and 3. Second column shows the rotation over the initial angle value (third column) resulting in the fourth column.

Angle	Rotation	Initial Value	Value obtained by PCA
1 (a)	0° - 60°	48,48°	48,48° - 108,48°
1 (b)	180° - 240°	48,48°	228,48° - 288,48°
2	0° - 60°	209,79°	209,79° - 269,79°
3	330° - 30°	289,09°	259,09° - 319,09°

Table 1. Regions separated by PCA

Once minima energy regions were defined, a small angle increment (5°) was used on them. Results for PCA are in Figure 13. In all cases a parabolic behavior was observed. When data variation decreases, curves are more easily observed and the minimum point is detectable. The amount of information accounted for both first and second PC´s (Factors) is around 90%. Table 2 shows the final values for each angle. When these values are combined, two different geometries were obtained with similar energy values (Table 3). These conformations are shown in Figure 14.

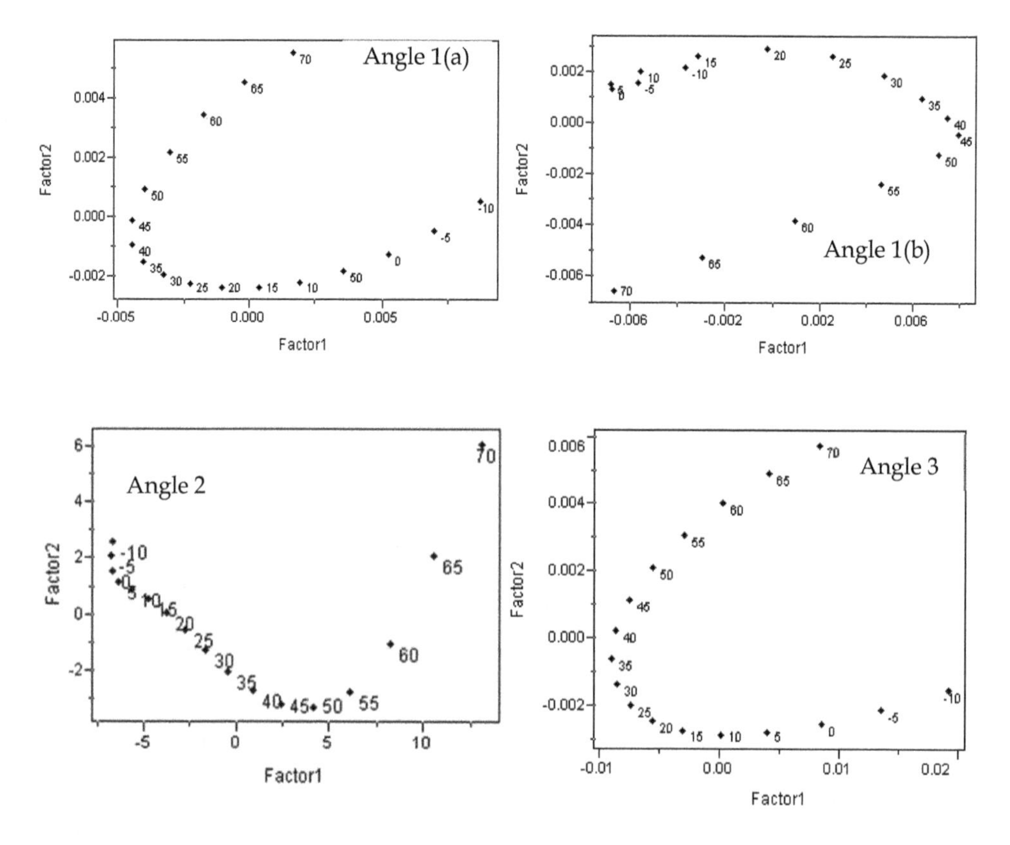

Fig. 13. PCA results for 5° angle increment refinement.

Angle	Rotation	Initial Value	Value obtained by PCA
1 (a)	45°	48,48° - 108,48°	93,48°
1 (b)	45°	228,48° - 288,48°	273,48°
2	45°	209,79° - 269,79°	254,79°
3	35°	259,09° - 319,09°	294,09°

Table 2. Regions obtained through PCA

Conformation	Angle	Obtained value	$\Delta H_{f\,(PM3)}$/kcal mol^{-1}	$E_{e(6\text{-}31G^{**})}$/hartree
A	1	92.29	54.96	-1134.33
	2	107.10		
	3	298.32		
B	1	266.45	54.71	-1134.35
	2	175.70		
	3	296.45		

Table 3. Minimum conformation characteristics (basic structure)

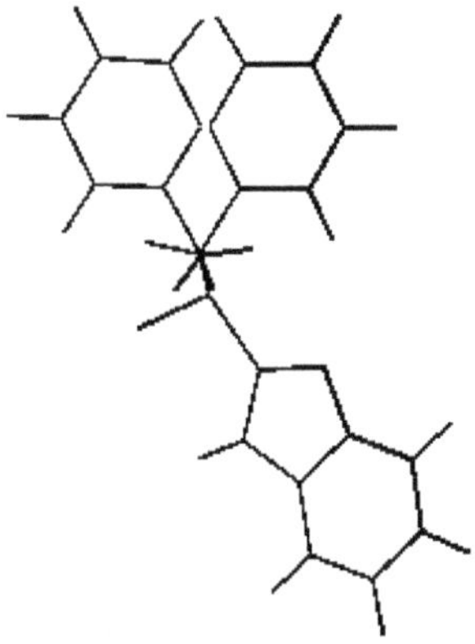

Fig. 14. Optimized superposed conformations for structure from Fig. 8.

In the second approach, the conformational analysis was made according to Equation 1 and took into account all possible conformations. PCA was performed on data matrix and minimum energy regions were selected. The next step a lower dihedral increment of 5° was used to refine those selected regions. PCA was performed again, and the same structures and energy, shown in Table 3, were obtained. This indicates that the two approaches are equivalent. The details of this complete systematic search can be found in (Bruni et al., 2002).

7.2 IAN preliminary studies

IAN (isobutyryl-Ala3-NH-methyl) tetrapeptide has also been studied to validate PDA-SA methodology. IAN has 11 consecutives dihedrals and its main characteristic is to be the shorter peptide able to make a complete helix turn. Figure 15 shows the IAN 2D structure

(Becker, 1998). Red arrows indicate the ψ, Φ e ω dihedrals. The dihedral angles ψ, ω and Φ are related to the rotations of single bonds between atoms in the main chain C (i)-C, OC-NH and N-C(i+1), respectively, where C (i) is the ith alpha carbon of the polypeptide chain. Angles ψ and Φ are connected to two arrays of functional protein chain: alpha-helix or beta-sheet.

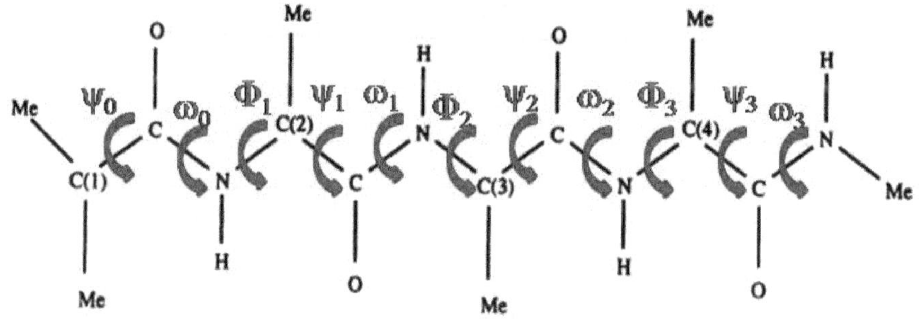

Fig. 15. 2D IAN peptide structure.

Ten random different starting conformations were studied. Table 4 shows the angles and energy values corresponding to these initial conformations. The red values indicate dihedrals that were changed in comparison to initial conformation number 1. The starting conformation 2 is close to an alpha-helix. Energy values correspond to single point AM1 semi-empirical calculation, in kcal mol^{-1}.

Number	Energy	ψ_0	ω_0	ϕ_0	ψ_1	ω_1	ϕ_1	ψ_2	ω_2	ϕ_2	ω_3	ϕ_3
1	-179.66	79.45	-169.01	-64.73	-44.75	171.78	-84.62	44.47	179.30	-144.22	-59.37	178.07
2	-122.24	-60.55	179.00	-64.73	-64.75	-180.00	-64.62	-65.55	180.00	-64.22	-59.37	178.07
3	-156.02	79.45	170.99	-64.73	-34.75	171.78	-84.62	74.47	179.30	-124.22	-59.37	178.07
4	195.46	79.45	-169.01	-14.73	-44.75	171.78	-134.62	44.47	179.30	-144.22	-39.37	178.07
5	-80.48	49.45	-169.01	-64.73	-44.75	151.78	-84.62	44.47	179.30	-144.22	-39.37	178.07
6	-169.24	79.45	-149.01	-64.73	-44.75	171.78	-84.62	44.47	159.30	-144.22	-79.37	178.07
7	-149.11	59.45	-169.01	-84.73	-44.75	-168.22	-84.62	74.47	179.30	-144.22	-59.37	178.07
8	-147.67	79.45	-169.01	-64.73	-24.75	171.78	-104.62	44.47	-160.70	-144.22	-39.37	178.07
9	-29.05	109.45	-169.01	-64.73	-44.75	171.78	-84.62	24.47	179.30	-174.22	-59.37	178.07
10	-167.76	79.45	-169.01	-44.73	-44.75	-178.22	-84.62	44.47	179.30	-144.22	-59.37	178.07

Table 4. Energy(kcal mol^{-1}) and dihedrals values (degrees) for each starting IAN structure.

IAN was analyzed using the PDA-SA procedure. The eleven dihedral angles provide 55 different conformations according to all possible combinations. Conformational analysis was performed with a 20° increment. PCA was carried out and Figure 16 shows that all points converge to specific regions of the phase space. Each selected region for each angle was refined with a 5° angle increment. PCA was performed again and the final structures characteristics are shown in Table 5.

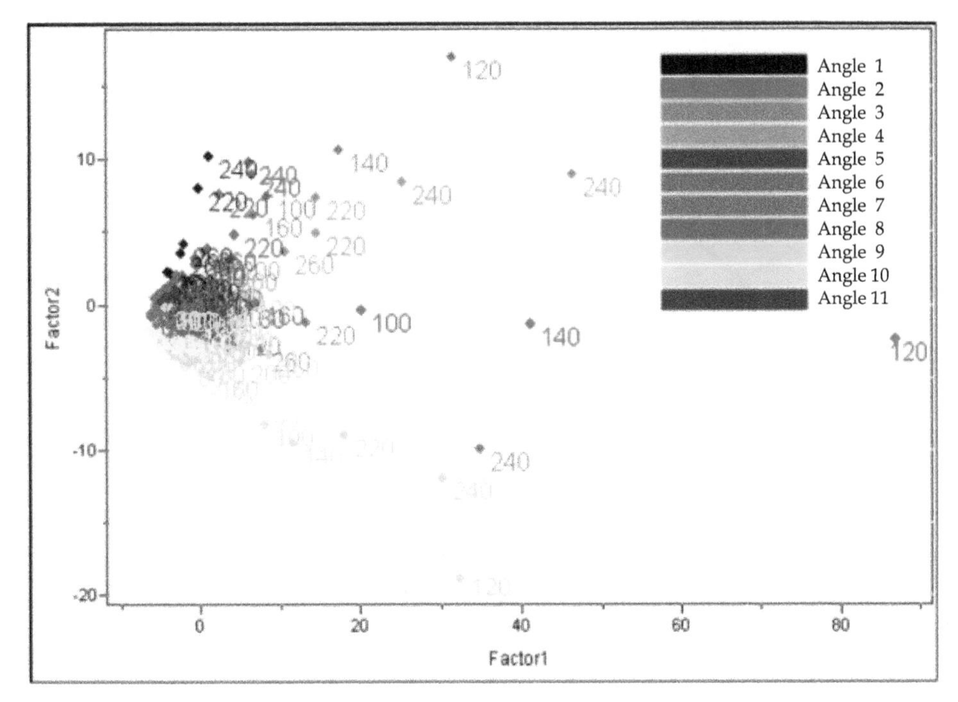

Fig. 16. PCA results for IAN peptide.

Table 5 shows the obtained energy values for the final structures and they indicate that some correspond to identical conformations. Three different groups were identified. Figure 17 shows the group that corresponds to structures 1, 5, 6, 9 and 10 superposed (blue ones in Table 5). Structure 9 shows a slightly different value on ψ_0 but it does not change the energy value. These five structures have two stabilizing hydrogen bonds which are indicated by the red circle and the resulting conformations for them resemble a beta-sheet.

Number	Energy	ψ_0	ω_0	ϕ_0	ψ_1	ω_1	ϕ_1	ψ_2	ω_2	ϕ_2	ω_3	ϕ_3
1	- 181.498	70.50	176.87	85.57	66.93	-177.14	-85.01	67.39	179.60	-112.24	-46.78	177.92
2	- 180.70	-61.83	-172.66	-78.26	25.05	157.24	-90.68	-29.19	177.15	-110.70	-45.10	178.62
3	- 179.661	79.45	-168.99	-64.80	-44.71	171.76	-84.61	44.32	179.26	143.99	-59.33	178.07
4	-179.276	72.93	-174.15	-104.24	-32.54	-179.9	-82.66	70.38	-179.69	-115.57	-51.50	179.42
5	-181.498	70.49	-176.89	-85.56	66.89	-177.17	-84.96	67.52	179.62	-112.34	-46.80	177.93
6	-181.498	70.49	-176.88	-85.57	66.93	-177.15	-85.01	67.40	179.61	-112.26	-46.78	177.92
7	-179.276	72.93	-174.15	-104.24	-32.52	-179.9	-82.65	70.39	-179.69	-115.58	-51.50	179.44
8	-179.276	72.94	-174.14	-104.27	-32.45	-179.89	-82.66	70.40	-179.67	-115.55	-51.48	179.36
9	-182.385	-62.43	-177.31	-84.72	67.46	-177.22	-84.95	67.38	179.58	-112.28	-46.85	177.94
10	-181.498	70.49	-176.89	-85.56	66.93	-177.15	-85.00	67.43	179.62	-112.32	-46.82	177.91

Table 5. Energy (kcal mol^{-1}) and dihedrals values (degrees) for each obtained IAN structure.

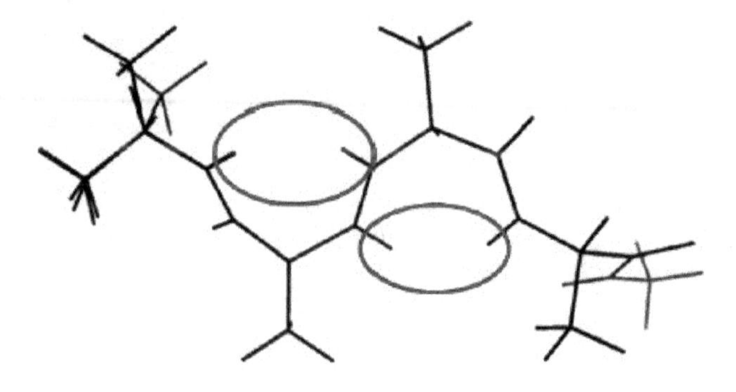

Fig. 17. Superposed final conformations for 1, 5, 6, 9 and 10 structures (blue ones in Table 5).

The second group is composed by final conformations 4, 7 and 8 (green ones in Table 5). The superposed conformations can be observed in Figure 18. These conformations are more open and have only one hydrogen-bond (red circle in Fig.18). The last group, the black ones in Table 5, are superposed in Figure 19. The resulting structures show an alpha-helix like behavior, with two stabilizing hydrogen bond (red circles, Fig. 19).

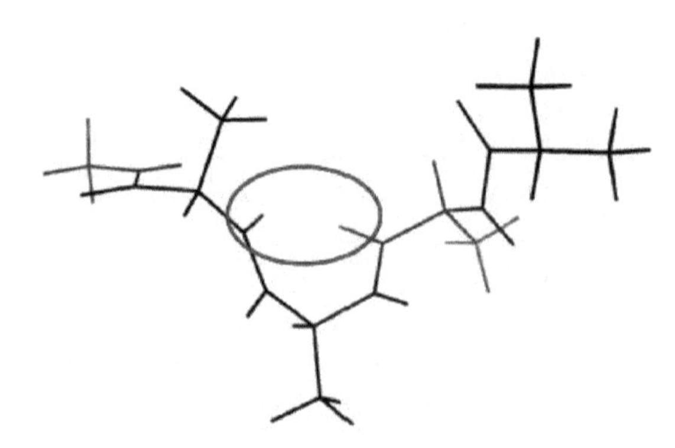

Fig. 18. Superposed final conformations for 4,7and 8 structures (green ones in Table 5).

Fig. 19. Superposed final conformations for structures 2 and 3 (black ones in Table 5).

Results presented for IAN peptide are partial and were only performed for one minimum region for each starting structure. Other minimum energy regions of this system are being investigated. A gradual increase in the size of the chain is also been explored.

8. Conclusion

The arrangement of atoms in a molecule or its structure determination has intrigued scientists through history. However only with recent experimental and computational advances the discussions on this theme became more effective and elucidative. The nature of PES is intrinsically multidimensional, usually has a very complex landscape. The global minima search, like the one encountered in the protein folding problem, is a NP-hard problem. This means that this task belongs to a large set of computational problems, assumed to be very hard ("conditionally intractable") (Fraenkel, 1993). The search for its relevant minima in molecular modeling has motivated the development of methods with very specific applications, as discussed in this chapter. For each particular problem one finds a variety of methods that allows feasible solutions, and most likely a combination of methods provides the optimum solution.

In this chapter, we discussed some aspects of conformational search that controls the combinatorial explosion. In particular, Principal Component Analysis was associated with a systematic search method to find structures with low energy in PES. The methodology can be useful to handle small- and medium-size molecules. The maximum size which the method can efficiently handle is being investigated (Nascimento et al., 2009). Due to the PCA dimension reduction, the method's efficiency is highly increased, allowing it to be of practical use in the study of more complex molecules.

9. Acknowledgment

We thank Prof. Márcia M.C. Ferreira (Unicamp) for the helpful discussions. We were supported by Fundação de Amparo à Pesquisa do Estado de São Paulo (FAPESP) and Conselho Nacional de Desenvolvimento Científico e Tecnológico (CNPq), Brazil. Computational resources were provided by Centro Nacional de Processamento e Alto Desempenho em São Paulo (CENAPAD-SP), Brazil.

10. Apendix A

Matrix with the discrete values for each rotation angle and its corresponding energy value for the first rotation for basic structure in Figure 8. Labels in bold were not used in PCA analysis, they are shown to help the matrix notation and visualization.

		2												3											
	Rotation	**0**	**30**	**60**	**90**	**120**	**150**	**180**	**210**	**240**	**270**	**300**	**330**	**0**	**30**	**60**	**90**	**120**	**150**	**180**	**210**	**240**	**270**	**300**	**330**
	0	0.0998	0.100	0.103	0.111	0.119	0.113	0.106	0.118	0.146	0.133	0.106	0.099	0.099	0.100	0.104	0.107	0.108	0.111	0.114	0.115	0.118	0.116	0.108	0.102
	30	0.0984	0.099	0.101	0.105	0.106	0.103	0.102	0.106	0.117	0.111	0.103	0.099	0.098	0.099	0.103	0.106	0.108	0.110	0.113	0.114	0.118	0.115	0.107	0.101
	60	0.0992	0.099	0.100	0.102	0.103	0.104	0.103	0.103	0.107	0.115	0.123	0.108	0.099	0.101	0.104	0.107	0.109	0.112	0.117	0.125	0.119	0.115	0.107	0.101
	90	0.1020	0.101	0.101	0.103	0.107	0.115	0.126	0.108	0.132	0.201	0.310	0.151	0.102	0.103	0.107	0.110	0.112	0.116	0.120	0.119	0.120	0.117	0.109	0.103
	120	0.1046	0.104	0.104	0.107	0.124	0.196	0.174	0.145	0.322	0.634	0.383	0.116	0.104	0.106	0.110	0.113	0.115	0.117	0.119	0.120	0.122	0.119	0.111	0.106
1	**150**	0.1037	0.103	0.104	0.109	0.173	0.186	0.136	0.252	0.516	0.459	0.137	0.105	0.103	0.105	0.110	0.112	0.113	0.115	0.117	0.118	0.120	0.118	0.110	0.105
	180	0.1010	0.100	0.100	0.103	0.116	0.116	0.121	0.166	0.243	0.145	0.107	0.100	0.101	0.103	0.107	0.109	0.110	0.112	0.113	0.114	0.116	0.114	0.107	0.102
	210	0.1003	0.099	0.098	0.100	0.104	0.104	0.104	0.112	0.119	0.111	0.104	0.101	0.100	0.103	0.106	0.107	0.108	0.109	0.110	0.111	0.114	0.112	0.106	0.101
	240	0.1019	0.099	0.098	0.099	0.102	0.105	0.107	0.107	0.109	0.114	0.118	0.108	0.101	0.104	0.106	0.106	0.107	0.108	0.109	0.110	0.114	0.113	0.106	0.102
	270	0.1033	0.101	0.100	0.101	0.107	0.118	0.121	0.114	0.120	0.150	0.146	0.117	0.103	0.104	0.106	0.107	0.107	0.108	0.110	0.113	0.117	0.115	0.109	0.104
	300	0.1036	0.102	0.102	0.106	0.121	0.132	0.127	0.121	0.152	0.249	0.163	0.111	0.103	0.104	0.106	0.108	0.109	0.111	0.114	0.117	0.120	0.118	0.111	0.106
	330	0.1023	0.102	0.103	0.112	0.128	0.128	0.117	0.126	0.180	0.216	0.126	0.103	0.102	0.103	0.106	0.108	0.110	0.112	0.115	0.117	0.120	0.118	0.111	0.105
Angle		3												1											
	Rotation	**0**	**30**	**60**	**90**	**120**	**150**	**180**	**210**	**240**	**270**	**300**	**330**	**0**	**30**	**60**	**90**	**120**	**150**	**180**	**210**	**240**	**270**	**300**	**330**
	0	0.0998	0.100	0.104	0.107	0.108	0.111	0.114	0.115	0.118	0.116	0.108	0.102	0.099	0.098	0.099	0.102	0.104	0.103	0.101	0.100	0.101	0.103	0.103	0.102
	30	0.1006	0.101	0.105	0.108	0.110	0.113	0.115	0.118	0.123	0.115	0.108	0.103	0.100	0.099	0.099	0.101	0.104	0.103	0.100	0.099	0.099	0.101	0.102	0.102
	60	0.1032	0.104	0.108	0.111	0.113	0.116	0.118	0.120	0.120	0.117	0.111	0.105	0.103	0.101	0.100	0.101	0.104	0.104	0.100	0.098	0.098	0.100	0.102	0.103
	90	0.1115	0.113	0.117	0.120	0.122	0.125	0.127	0.128	0.129	0.125	0.120	0.113	0.111	0.105	0.102	0.103	0.107	0.109	0.103	0.100	0.099	0.101	0.106	0.112
	120	0.1194	0.121	0.126	0.129	0.132	0.134	0.136	0.138	0.141	0.138	0.127	0.121	0.119	0.106	0.103	0.107	0.124	0.173	0.116	0.104	0.102	0.107	0.121	0.128
2	**150**	0.1136	0.116	0.122	0.127	0.130	0.131	0.132	0.134	0.139	0.129	0.120	0.114	0.113	0.103	0.104	0.115	0.196	0.186	0.116	0.104	0.105	0.118	0.132	0.128
	180	0.1062	0.109	0.119	0.134	0.141	0.133	0.125	0.123	0.127	0.150	0.171	0.116	0.106	0.102	0.103	0.126	0.174	0.136	0.121	0.104	0.107	0.121	0.127	0.117
	210	0.1182	0.133	0.172	0.238	0.189	0.147	0.133	0.147	0.319	1.241	0.376	0.140	0.118	0.106	0.103	0.108	0.145	0.252	0.166	0.112	0.107	0.114	0.121	0.126
	240	0.1467	0.225	0.339	0.279	0.174	0.146	0.167	0.434	2.168	0.552	0.181	0.135	0.146	0.117	0.107	0.132	0.322	0.516	0.243	0.119	0.109	0.120	0.152	0.180
	270	0.1467	0.133	0.166	0.167	0.140	0.121	0.126	0.178	0.533	0.368	0.167	0.120	0.133	0.111	0.115	0.201	0.634	0.459	0.145	0.111	0.114	0.150	0.249	0.216
	300	0.1196	0.106	0.111	0.113	0.110	0.108	0.111	0.123	0.146	0.140	0.117	0.108	0.106	0.103	0.123	0.310	0.383	0.137	0.107	0.104	0.118	0.146	0.163	0.126
	330	0.1047	0.099	0.101	0.104	0.105	0.107	0.109	0.113	0.115	0.116	0.112	0.107	0.099	0.099	0.108	0.151	0.116	0.105	0.100	0.101	0.108	0.117	0.111	0.103
Angle		1												2											
	Rotation	**0**	**30**	**60**	**90**	**120**	**150**	**180**	**210**	**240**	**270**	**300**	**330**	**0**	**30**	**60**	**90**	**120**	**150**	**180**	**210**	**240**	**270**	**300**	**330**
	0	0.0998	0.098	0.099	0.102	0.104	0.103	0.101	0.100	0.101	0.103	0.103	0.102	0.099	0.100	0.103	0.111	0.119	0.113	0.106	0.118	0.146	0.146	0.119	0.104
	30	0.1009	0.099	0.101	0.103	0.106	0.105	0.103	0.103	0.104	0.104	0.104	0.103	0.100	0.101	0.104	0.113	0.121	0.116	0.109	0.133	0.225	0.133	0.106	0.099
	60	0.1045	0.103	0.104	0.107	0.110	0.110	0.107	0.106	0.106	0.106	0.106	0.106	0.104	0.105	0.108	0.117	0.126	0.122	0.119	0.172	0.339	0.166	0.111	0.101
	90	0.1071	0.106	0.107	0.110	0.113	0.112	0.109	0.107	0.106	0.107	0.108	0.108	0.107	0.108	0.111	0.120	0.129	0.127	0.134	0.238	0.279	0.167	0.113	0.104
	120	0.1088	0.108	0.109	0.112	0.115	0.113	0.110	0.108	0.107	0.107	0.109	0.110	0.108	0.110	0.113	0.122	0.132	0.130	0.141	0.189	0.174	0.140	0.110	0.105
3	**150**	0.1115	0.110	0.112	0.116	0.117	0.115	0.112	0.109	0.108	0.108	0.111	0.112	0.111	0.113	0.116	0.125	0.134	0.131	0.133	0.147	0.146	0.121	0.108	0.107
	180	0.1140	0.113	0.117	0.120	0.119	0.117	0.113	0.110	0.109	0.110	0.114	0.115	0.114	0.115	0.118	0.127	0.136	0.132	0.125	0.133	0.167	0.126	0.111	0.109
	210	0.1157	0.114	0.125	0.119	0.120	0.118	0.114	0.111	0.110	0.113	0.117	0.117	0.115	0.118	0.120	0.128	0.138	0.134	0.123	0.147	0.434	0.178	0.123	0.113
	240	0.1187	0.118	0.119	0.120	0.122	0.120	0.116	0.114	0.114	0.117	0.120	0.120	0.118	0.123	0.120	0.129	0.141	0.139	0.127	0.319	2.168	0.533	0.146	0.115
	270	0.1166	0.115	0.115	0.117	0.119	0.118	0.114	0.112	0.113	0.115	0.118	0.118	0.116	0.115	0.117	0.125	0.138	0.129	0.150	1.241	0.552	0.368	0.140	0.116
	300	0.1086	0.107	0.107	0.109	0.111	0.110	0.107	0.106	0.106	0.109	0.111	0.111	0.108	0.108	0.111	0.120	0.127	0.120	0.171	0.376	0.181	0.167	0.117	0.112
	330	0.1025	0.101	0.101	0.103	0.106	0.105	0.102	0.101	0.102	0.104	0.106	0.105	0.102	0.103	0.105	0.113	0.121	0.114	0.116	0.140	0.135	0.120	0.108	0.107

11. References

Allen, F. H., Galek, P. T. a, & Wood, P. a. (2010). Energy matters! *Crystallography Reviews, 16*(3), 169-195. doi:10.1080/08893110903476919

Araujo-Andrade, C., Lopes, S., Fausto, R., & Gómez-Zavaglia, A. (2010). Conformational study of arbutin by quantum chemical calculations and multivariate analysis. *Journal of Molecular Structure, 975*(1-3), 100-109. doi:10.1016/j.molstruc.2010.04.002

Becker, O M. (1998). Principal coordinate maps of molecular potential energy surfaces. *Journal of Computational Chemistry, 19*(11), 1255-1267. 605 Third Ave, New York, NY 10158-0012 Usa: John Wiley & Sons Inc. doi:10.1002/(SICI)1096-987X(199808)19:11<1255::AID-JCC5>3.3.CO;2-H

Becker, Oren M., & Karplus, M. (1997). The topology of multidimensional potential energy surfaces: Theory and application to peptide structure and kinetics. *The Journal of Chemical Physics, 106*(4), 1495. doi:10.1063/1.473299

Beckers, M. L. M., Derks, E. P. P. A., Melssen, W. J., & Buydens, L. M. C. (1996). Pergamon oo!n-8485(%~6-0. *Science, 20*(4), 449-457.

Beckers, M. L., Buydens, L. M., Pikkemaat, J. a, & Altona, C. (1997). Application of a genetic algorithm in the conformational analysis of methylene-acetal-linked thymine dimers in DNA: comparison with distance geometry calculations. *Journal of biomolecular NMR, 9*(1), 25-34. Retrieved from http://www.ncbi.nlm.nih.gov/pubmed/9081542

Beebe, K. R. (1998). *Chemometrics: A Practical Guide* (p. 360). Wiley-Blackwell. Retrieved from http://www.amazon.co.uk/Chemometrics-Practical-Wiley-Interscience-Laboratory-Automation/dp/0471124516/ref=sr_1_1?ie=UTF8&qid=1320084260&sr=8-1

Beusen, D. D., Shands, E. F. B., Karasek, S. F., Marshall, G. R., & Dammkoehler, R. A. (1996). Systematic search in conformational analysis. *Theochem-Journal of Molecular Structure, 370*(2-3), 157-171. Po Box 211, 1000 Ae Amsterdam, Netherlands: Elsevier Science Bv.

Biarnés, X., Bongarzone, S., Vargiu, A. V., Carloni, P., & Ruggerone, P. (2011). Molecular motions in drug design: the coming age of the metadynamics method. *Journal of computer-aided molecular design, 25*(5), 395-402. doi:10.1007/s10822-011-9415-3

Brodmeier, T., & Pretsch, E. (1994). Application of genetic algorithms in molecular modeling. *Journal of Computational Chemistry, 15*(6), 588-595. doi:10.1002/jcc.540150604

Bruni, A. T., Leite, V. B. P., & Ferreira, M. M. C. (2002). Conformational analysis: A new approach by means of chemometrics. *Journal Of Computational Chemistry, 23*(2), 222-236. Commerce Place, 350 Main St, Malden 02148, MA USA: Wiley-Blackwell. doi:10.1002/jcc.10004

Bruni, A. T., & Ferreira, M. M. C. (2008). Theoretical study of omeprazole behavior: Racemization barrier and decomposition reaction. *International Journal of Quantum Chemistry, 108*(6), 1097-1106. doi:10.1002/qua.21597

Bruni, A. T., & Ferreira, M. M. C. (2002). Omeprazole and analogue compounds: a QSAR study of activity againstHelicobacter pylori using theoretical descriptors. *Journal of Chemometrics, 16*(8-10), 510-520. doi:10.1002/cem.737

Brush, S. G. (1999). Dynamics of Theory Change in Chemistry : Part 1 . The Benzene Problem 1865 - 1945. *Science, 30*(1), 21-79.

Cintas, P. (2007). Tracing the origins and evolution of chirality and handedness in chemical language. *Angewandte Chemie (International ed. in English), 46*(22), 4016-24. doi:10.1002/anie.200603714

Das, G., Gentile, F., Coluccio, M. L., Perri, a M., Nicastri, a, Mecarini, F., Cojoc, G., et al. (2011). Principal component analysis based methodology to distinguish protein SERS spectra. *Journal of Molecular Structure, 993*(1-3), 500-505. Elsevier B.V. doi:10.1016/j.molstruc.2010.12.044

Derks, E. P. P. A., & Buydens, L. M. C. (1996). E. P. P. A DERKS,* M. L. M. BECKER& W. J. MELSSEN and L. M. C. BUYDENS, *20*(4), 439-448.

Drayer, D. (1993). The Early History of Stereochemistry: From the Discovery of Molecular Asymmetry and the First Resolution of a Racemate by Pasteur to the Asymmetrical Chiral Carbon of van't Hoff and Le Bel. *Clinical Pharmacology-New York-Marcel Dekker Incorporated-, 18*(3), 1-1. Marcel Dekker Ag. Retrieved from http://scholar.google.com/scholar?hl=en&btnG=Search&q=intitle:The+Early+History+of+Stereochemistry+:+From+the+Discovery+of+Molecular+Asymmetry+and+the+First+Resolution+of+A+Racemate+By+Pasteur+To+The+Asymmetrical+Chiral+Carbon+Of+Van+?+T+Hoff+And+Le+Bel+*#0

Fraenkel, a S. (1993). Complexity of protein folding. *Bulletin of mathematical biology, 55*(6), 1199-210. Retrieved from http://www.pubmedcentral.nih.gov/articlerender.fcgi?artid=3042729&tool=pmcentrez&rendertype=abstract

Gal, J. (2007). Review Article Carl Friedrich Naumann and the Introduction of Enantio Terminology : A Review and Analysis on the 150th Anniversary. *Chirality, 98*(May 2006), 89-98. doi:10.1002/chir

Gal, J. (2011). Review Article Louis Pasteur , Language , and Molecular Chirality . I . Background and Dissymmetry. *Clinical Laboratory, 16*(March 2010), 1-16. doi:10.1002/chir

Geladi, P. (2003). Chemometrics in spectroscopy. Part 1. Classical chemometrics. *Spectrochimica Acta Part B Atomic Spectroscopy, 58*(5), 767-782. doi:10.1016/S0584-8547(03)00037-5

Golub, G. H., & Loan, C. F. van V. (1996). *Matrix Computations (Johns Hopkins Studies in Mathematical Sciences)(3rd Edition)* (p. 728). The Johns Hopkins University Press. Retrieved from http://www.amazon.com/Computations-Hopkins-Studies-Mathematical-Sciences/dp/0801854148

Grouleff, J., & Jensen, F. (2011). Searching Peptide Conformational Space. *Journal of Chemical Theory and Computation*, 1783-1790.

Hanselman, D., Littlefield, B., Inc., M., & Mathworks. (1997). *The Student Edition of Matlab Version 5 User's Guide* (p. 429). Prentice Hall College Div. Retrieved from http://www.amazon.com/Student-Matlab-Version-Users-Guide/dp/0132725509

Havel, T. F., Crippen, G. M., Kuntz, I. D., & Blaney, J. M. (1983). The combinatorial distance geometry method for the calculation of molecular conformation. II. Sample problems and computational statistics. *Journal of theoretical biology, 104*(3), 383-400. Retrieved from http://www.ncbi.nlm.nih.gov/pubmed/6197591

Havel, T. F., Kuntz, I. D., & Crippen, G. M. (1983). The combinatorial distance geometry method for the calculation of molecular conformation. I. A new approach to an old problem. *Journal of theoretical biology, 104*(3), 359-81. Retrieved from http://www.ncbi.nlm.nih.gov/pubmed/6656266

Hunger, J., & Huttner, G. (1999). Optimization and analysis of force field parameters by combination of genetic algorithms and neural networks. *Journal of Computational Chemistry, 20*(4), 455-471. doi:10.1002/(SICI)1096-987X(199903)20:4<455::AID-JCC6>3.0.CO;2-1

Izgorodina, E. I., Lin, C. Y., & Coote, M. L. (2007). Energy-directed tree search: an efficient systematic algorithm for finding the lowest energy conformation of molecules. *Physical chemistry chemical physics : PCCP, 9*(20), 2507-16. doi:10.1039/b700938k

Jordan, S. N., Leach, A. R., & Bradshaw, J. (1995). The Application of Neural Networks in Conformational Analysis. 1. Prediction of Minimum and Maximum Interatomic Distances. *Journal of Chemical Information and Modeling, 35*(3), 640-650. doi:10.1021/ci00025a035

Jorgensen, W. L., & TiradoRives, J. (1996). Monte Carlo vs molecular dynamics for conformational sampling. *Journal of Physical Chemistry, 100*(34), 14508-14513. 1155 16th St, Nw, Washington, Dc 20036: Amer Chemical Soc. doi:10.1021/jp960880x

Kiralj, R., Ferreira, M. C., Donate, P. M., & Silva, R. (2007). Combined Computational , Database Mining , NMR , and Chemometric Approaches. *Analysis*, 6316-6333.

Lucasius, C. B., & Kateman, G. (1994). Understanding and Using Genetic Algorithms.2. Representation, Configuration and Hybridization. *Chemometrics and Intelligent Laboratory Systems, 25*(2), 99-145. Po Box 211, 1000 Ae Amsterdam, Netherlands: Elsevier Science Bv. doi:10.1016/0169-7439(94)85038-0

Leach, A. (2001). *Molecular Modelling: Principles and Applications (2nd Edition)*. Prentice Hall. Retrieved from http://www.amazon.ca/exec/obidos/redirect?tag=citeulike09-20&path=ASIN/0582382106

Leach, A. R., & Smellie, A. S. (1992). A combined model-building and distance-geometry approach to automated conformational analysis and search. *Journal of Chemical Information and Modeling, 32*(4), 379-385. doi:10.1021/ci00008a019

Lucasius, C. (1993). Understanding and using genetic algorithms Part 1. Concepts, properties and context. *Chemometrics and Intelligent Laboratory Systems, 19*(1), 1-33. doi:10.1016/0169-7439(93)80079-W

Malinowski, E. R. (2003). *Factor Analysis in Chemistry. Technometrics* (Vol. 45, pp. 180-181). Wiley. doi:10.1198/tech.2003.s145

Li Manni, G., Barone, G., Duca, D., & Murzin, D. Y. (2009). Systematic conformational search analysis of the SRR and RRR epimers of 7-hydroxymatairesinol. *Journal of Physical Organic Chemistry*, (June 2009), n/a-n/a. doi:10.1002/poc.1595

Massart, D. L., Heyden, Y. V., & Brussel, V. U. (2004). What Can Chemometrics Do for Separation Science ? *Europe, 17*(9).

Mo, Y., & Gao, J. (2007). Theoretical analysis of the rotational barrier of ethane. *Accounts of chemical research, 40*(2), 113-9. doi:10.1021/ar068073w

Nair, N., & Goodman, J. M. (1998). Genetic Algorithms in Conformational Analysis. *Journal of Chemical Information and Modeling, 38*(2), 317-320. doi:10.1021/ci970433u

Nascimento, R. R., Bruni, A. T. , & Leite, V. B. P. (2009). Estudo conformacional do peptídeo IAN e seus fragmentos pelo método de análise sistemática reduzida. *07/10/09*. Retrieved November 1, 2011, from http://www.athena.biblioteca.unesp.br/exlibris/bd/brp/33004153068P9/2009/nascimento_rr_me_sjrp_parcial.pdf

Oblinsky, D. G., Vanschouwen, B. M. B., Gordon, H. L., & Rothstein, S. M. (2009). Procrustean rotation in concert with principal component analysis of molecular dynamics trajectories: Quantifying global and local differences between conformational samples. *The Journal of chemical physics, 131*(22), 225102. doi:10.1063/1.3268625

Pietropaolo, A., Branduardi, D., Bonomi, M., & Parrinello, M. (2011). A Chirality-Based Metrics for Free-Energy Calculations in Biomolecular Systems. *Journal of Computational Chemistry*. doi:10.1002/jcc

Ramberg, P. J., & Somsen, G. J. (2001). Annals of Science The Young J . H . van ' t Hoff : The Background to the Publication of his 1874 Pamphlet on the Tetrahedral Carbon Atom , Together with a New English Translation. *Annals of Science*, (September 2011), 51-74.

Ramos, L. S., Beebe, K. R., Carey, W. P., M, E. S., Erickson, B. C., Wilson, B. E., Wangen, L. E., et al. (1986). L. Scott Ramos, Kenneth R. Beebe, W. Patrick Carey, Eugenio Sfinchez M., Brice C. Erickson, Bruce E. Wilson, Lawrence E. Wangen,' and Bruce R. Kowalski* Laboratory for Chemometrics, Department. *Education*, (300), 31-49.

Seabra, G. D. M., Walker, R. C., & Roitberg, A. E. (2009). Are current semiempirical methods better than force fields? A study from the thermodynamics perspective. *The journal of physical chemistry. A, 113*(43), 11938-48. doi:10.1021/jp903474v

Silva, D.-A., Domínguez-Ramírez, L., Rojo-Domínguez, A., & Sosa-Peinado, A. (2011). Conformational dynamics of L-lysine, L-arginine, L-ornithine binding protein reveals ligand-dependent plasticity. *Proteins, 79*(7), 2097-108. doi:10.1002/prot.23030

Smellie, A., Stanton, R., Henne, R., & Teig, S. (2003). Conformational analysis by intersection: Conan. *Journal of Computational Chemistry, 24*(1), 10-20. 111 River St, Hoboken, Nj 07030 Usa: John Wiley & Sons Inc. doi:10.1002/jcc.10175

Spellmeyer, D. C., Wong, a K., Bower, M. J., & Blaney, J. M. (1997). Conformational analysis using distance geometry methods. *Journal of molecular graphics & modelling, 15*(1), 18-36. Retrieved from http://www.ncbi.nlm.nih.gov/pubmed/9346820

Wold, S., Esbensen, K., & Geladi, P. (1987). Principal Component Analysis. *Chemometrics And Intelligent Laboratory Systems, 2*(1-3), 37-52. Po Box 211, 1000 Ae Amsterdam, Netherlands: Elsevier Science Bv. doi:10.1016/0169-7439(87)80084-9

Wold, S. (1995). Chemometrics; what do we mean with it, and what do we want from it? *Chemometrics and Intelligent Laboratory Systems, 30*(1), 109-115. doi:10.1016/0169-7439(95)00042-9

Yanmaz, E., Sarıpınar, E., Şahin, K., Geçen, N., & Çopur, F. (2011). 4D-QSAR analysis and pharmacophore modeling: electron conformational-genetic algorithm approach for penicillins. *Bioorganic & medicinal chemistry, 19*(7), 2199-210. doi:10.1016/j.bmc.2011.02.035

4

Quantum Chemical Calculations for some Isatin Thiosemicarbazones

Fatma Kandemirli*,** et al.
Niğde University,
Turkey

1. Introduction

Derivatives of isatin are reported to be present in mammalian tissues and body fluids (Casas et al., 1996; Agrawal & Sartorelli, 1978; Casas et al., 1994; Medvedev et al., 1998; Boon, 1997; Pandeya & Dimmock, 1993; Rodríguez-Argüelles et al., 1999; Casas et al., 2000) and possess antibacterial (Daisley & Shah, 1984), antifungal (Piscopo et al., 1987), and anti-HIV (Pandeya et al., 1998, 1999) activities. *N*-methylisatin-β-4′, 4′ - diethylthiosemicarbazone were also reported to have activity against the viruses such as cytomegalo and moloney leukemia viruses (Sherman et al., 1980; Ronen et al., 1987). With the help of combinatorial method, the cytotoxicity and antiviral activities of isaitin-β-thiosemicarbazones against the vaccine virus and cowpox virus-infected human cells were evaluated (Pirrung et al., 2005).

Some 5-fluoroisatin, 5-fluoro-1-morpholino/piperidinomethyl, and 5-nitroisaitn synthesized. They are reported to have anti-TB activity. ETM Study has also been carried out on these compounds (Karali et al., 2007). Synthesis and quantum chemical calculations of 5-methoxyisatin-3-(N-cyclohexyl), its Zn (II) and Ni (II) complexes (Kandemirli et al., 2009a), and 5-methoxyisatin-3-(*N*-cyclohexyl)thiosemicarbazone (Kandemirli et al., 2009b) were studied. The thiosemicarbazones likely possess anti-HIV activity according to 3D pharmacophoric distance map analysis (Bal et al., 2005).

Isatin-thiosemicarbazones may coordinate through the deprotonated nitrogen atom, sulphur atom of thiosemicarbazone group, and carbonyl oxygen atom with the metal, depending on its nature. Zinc (II) and mercury (II) complexes of isatin-3-thiosemicarbazones were reported to be coordinated through imino nitrogen and thiolato sulfur atoms and was suggested to have tetrahedral structures (Akinchan et al., 2002).

It was reported that only amino nitrogen atom coordinates in the Cu (II) complex (Ivanov et al., 1988). Quantum chemical calculations and IR studies on Zn (II) and Ni (II) complexes of

* M. Iqbal Choudhary[2], Sadia Siddiq[2], Murat Saracoglu[3], Hakan Sayiner[4],
Taner Arslan[5], Ayşe Erbay[6] and Baybars Köksoy[6]
[2]*University of Karachi, Pakistan,*
[3]*Erciyes University, Turkey,*
[4]*Kahta State Hospital, Turkey,*
[5]*Osmangazi University, Turkey, Turkey,*
[6]*Kocaeli University, Turkey*
**Corresponding Author

5-fluoro-isatin -3-(*N*-benzylthiosemicarbazone) have recently been reported (Gunesdogdu-Sagdinc et al., 2009).

During the current study, we prepared [Zn(HICHT)$_2$], [Zn(HMIPT)$_2$], [Zn(HIPT)$_2$], [Zn(HICPT)$_2$], [Zn(HIBT)$_2$], [Ni(HMIPT)$_2$], [Ni(HIPT)$_2$], [Ni(HICPT)$_2$], [Ni(HIBT)$_2$], and [Ni(HICHT)$_2$] derivatives, and characterized them with elemental analysis, and IR, UV, and ^{1}H-NMR spectroscopic techniques.

In view of the reports about antimicrobial and antifungal activities of the isatin derivatives, we synthesized and screened compounds **1-16** (Table 1) for their antimicrobial effects in vitro against Bacillus subtilis, Escherichia coli, Stahpylococcus aureus, Shigella flexnari, Pseudomonas aeruginosa, and Salmonella typhi bacterial strains and Aspergillus flavus, Candida albicans, Microsporum canis, Fusarium solani, and Candida glabrata fungal strains. Compounds **1**, **14**, and **16** were found to be moderately active, compounds **2**, and **4** possess a good activity, while compound **13** exhibited a significant activity against Microsporum canis. Compounds **13**, **12**, and **4** exhibited moderate activities against Fusarium solani. Compound **10** showed a moderate activity against Candia albicans. Compound **5** was only moderately active against the Candida albicans.

Compound No	List of the Compounds
1	5-Methoxyisatin-3-(N-cyclohexyl) thiosemicarbazone (H_2MICT)
2	5-Methoxyisatin-3-(N-benzyl)thiosemicarbazone (H_2MIBT)
3	5-Methoxyisatin-3-(N-phenyl)thiosemicarbazone (H_2MIPT)
4	5-Methoxyisatin-3-(N-chlorophenyl)thiosemicarbazone (H_2MICPT)
5	Isatin-3-(N-cyclohexyl)thiosemicarbazone(H_2ICHT)
6	[Zn(HICHT)$_2$]
7	[Zn((HMICT)$_2$]
8	[Zn(HMIPT)$_2$]
9	[Zn(HIPT)$_2$]
10	[Zn(HICPT)$_2$]
11	[Zn(HIBT)$_2$]
12	[Ni((HMICHT)$_2$]
13	[Ni(HMIPT)$_2$]
14	[Ni(HIPT)$_2$]
15	[Ni(HICPT)$_2$]
16	[Ni(HIBT)$_2$]
17	[Ni(HICHT)$_2$]
18	Isatin-3-(N-benzyl)thiosemicarbazone (H_2IBT)
19	Isatin-3-(N-phenyl)thiosemicarbazone (H_2IPT)
20	Isatin-3-(N-chlorophenyl)thiosemicarbazone (H_2ICPT)

Table 1. Studied compounds

2. Experimental

Elemental analyses were performed by using a LECO CHN Elemental Analyzer. IR Spectra were recorded by Shimadzu FT-IR 8201 spectrometer with the KBr technique in the region of 4000-300 cm^{-1}, which was calibrated by polystyrene. There was no decomposition of the samples due to the effect of potassium bromide. The ^{1}H-NMR spectra were recorded in DMSO-d_6 on a BRUKER DPX-400 (400 MHz) spectrometer.

The ligands under study were obtained by refluxing an ethanolic solution of 4-cyclohexyl-3-thiosemicarbazide, 4-benzyl-3-thiosemicarbazide, 4-phenyl-3-thiosemicarbazide, and 4-(4-chlorophenyl)-3-thiosemicarbazide with isatin (1H-indole-2,3-dione) or 5-methoxyisatin (all were purchased from Aldrich Chemical Company USA and used without purification), as described in the literature (Karali et al., 2007; Kandemirli et al., 2009a, 2009b).

2.1 General procedure for synthesis of Ni and Zn complexes

1 mmol of appropriate ligand was dissolved in 20 mL of ethanol at 50-55 °C and then slowly added to ethanol solution (10 mL) of 0.5 mmol zinc acetate dihydrate or nickel acetate tetrahydrate. The mixture was refluxed for 2 h for nickel complex, and 6 h for zinc complex at approximately 75 °C. The zinc complex precipated at the end of the reflux, while the nickel complex precipated only after two days of stirring. The solid was filtered, washed with ethanol, and diethyl ether, and dried under vacuum.

2.1.1 [Zn(HICHT)$_2$] (6)

Yield: (80%). (M.p.: 288-290 °C)

^{1}H-NMR (DMSO-d_6, ppm): δ 1-2 (cyclohexyl C-H), 4.11 (m, cyclohexyl C-H), 6.90-8.3 (aromatic C-H), 8.95 (d, J= 7.8 Hz), NH 10.80 (s, indole-NH)

IR (cm^{-1}): 1688 (C=O), 1595 (C=N), 819 (C=S)

Calculated: % C: 53.92, % H: 5.128, % N: 16.77, % S: 9.60, found: % C: 53.49, % H: 5.644, % N: 16.06, % S: 9.58.

2.1.2 [Zn(HMIPT)$_2$] (8)

Yield: (64%). (M.p.: 320 °C)

^{1}H-NMR (DMSO-d_6, ppm): δ 3.35 (s, CH_3-methoxy), 6.91-7.72 (aromatic C-H), 10.76 (s, NH), 11.04 (s, indole-NH)

IR (cm^{-1}): 1697 (C=O), 1589 (C=N), 817 (C=S)

Calculated: % C: 53.67, % H: 3.66, % N: 15.64, % S: 8.95, found: % C: 53.64, % H: 3.65, % N: 15.67, % S: 9.08.

2.1.3 [Zn(HIPT)$_2$] (9)

Yield: (90%). (M.p.: 310 °C)

^{1}H-NMR (DMSO- d_6, ppm): δ 7.01-8.11 (aromatic C-H), 10.68 (s, NH), 11.01 (s, indole-NH)

IR (cm^{-1}): 1703 (C=O), 1595 (C=N), 802 (C=S)

Calculated: % C: 54.92, % H: 3.38, % N: 17.08, % S: 9.77, found: % C: 54.52, % H:3.31, % N: 16.91, % S: 10.10.

2.1.4 $[Zn(HICPT)_2]$ (10)

Yield: (82%) (M.p.: 321 °C)

^{1}H-NMR (DMSO-d_6, ppm): δ 7.03-8.08 (aromatic C-H), 10.73 (s, NH), 11.06 (s, indole-NH)

IR (cm^{-1}): 1695 (C=O), 1600 (C=N), 816 (C=S)

Calculated: % C: 48.69, % H: 2.78, % N: 15.45, % S: 8.84, found: % C: 48.43, % H: 3.03, % N: 14.96, % S: 8.81.

2.1.5 $[Zn(HIBT)_2]$ (11)

Yield: (79%) (M.p.: 318 °C)

^{1}H-NMR (DMSO-d_6, ppm): δ 4.80 (d, benzyl-CH_2), 6.95-7.40 (aromatic C-H), 9.45 (t, *J*=7.52 Hz), NH), 10.81 (s, indole-NH).

IR (cm^{-1}): 1690 (C=O), 1599 (C=N), 814 (C=S)

Calculated: % C: 56.18, % H: 3.83, % N: 16.38, % S: 9.37, found: % C: 55.72, % H: 3.79, % N: 16.24, % S: 9.58.

2.1.6 $[Ni(HMIPT)_2]$ (13)

Yield: (90%) (M.p.: 300 °C)

IR (cm^{-1}): 1670 (C=O), 1589 (C=N), 818 (C=S),

Calculated: % C: 54.17, % H: 3.69, % N: 15.79, % S: 9.04, found: % C: 54.00, % H: 3.73, % N: 15.71, % S: 9.07.

2.1.7 $[Ni(HIPT)_2]$ (14)

Yield: (81%) (M.p.: 296 °C)

IR (cm^{-1}): 1660 (C=O), 1595 (C=N), 802 (C=S),

Calculated: % C: 52.48, % H: 3.41, % N: 16.25, % S: 9.87, found: % C: 52.79, % H: 3.60, % N: 16.24, % S: 9.33.

2.1.8 $[Ni(HICPT)_2]$ (15)

Yield: (70%) (M.p.: 265-267 °C)

IR (cm^{-1}): 1672 (C=O), 1595 (C=N), 817 (C=S)

Calculated: % C: 51.16, % H: 2.80, % N: 15.59, % S: 8.92, found: % C: 51.37, % H: 3.08, % N: 15.71, % S: 8.63.

2.1.9 [Ni(HIBT)$_2$] (16)

Yield (84%) (M.p.: 265-267 °C)

IR (cm^{-1}): 1659 (C=O), 1595 (C=N), 818 (C=S)

Calculated: % C: 56.73, % H: 3.86, % N: 16.54, % S: 9.46, found: % C: 56.35, % H: 3.84, % N: 16.40, % S: 9.60.

2.1.10 [Ni(HICHT)$_2$] (17)

Yield: (90%). (M.p.: 265 °C)

IR (cm^{-1}): 1664 (C=O), 1595 (C=N), 823 (C=S)

Calculated: % C: 53.47, % H: 5.18, % N: 16.54, % S: 9.69, found: % C: 52.99, % H: 5.17, % N: 16.14, % S: 9.37.

2.2 Antibacterial activity

The antibacterial activities were determined by using the agar well diffusion method (Rahman et al., 2001). The wells were dugged in the media with a sterile borer and an eight-hour-old bacterial inoculum containing 0>ca. 104-106 colony forming units (CFU)/mL were spread on the surface of the nutrient agar. The recommended concentration of the test sample (2 mg/mL in DMSO) was introduced into the respective wells. Other wells containing DMSO and the reference antibacterial drug imipenum served as negative and positive controls, respectively. The plates were incubated immediately at 37 °C for 20 h. The activity was determined by measuring the diameter of the inhibition zone (in mm), showing complete inhibition. Growth inhibition was calculated with reference to the positive control.

2.3 Antifungal assay

To test for antifungal activity, the agar dilution method, a modification of the agar dilution method of Washington and Sutter (980), was employed (Ajaiyeoba et al., 1988). Test tubes having sterile SDA were inoculated with test samples (200 mg/mL), and kept in a slanting position at room temperature. Test fungal culture was inoculated on the slant and growth inhibitions were observed after an incubation period of 7 days at 27 °C. Control agar tubes were made in paralellel and treated smilarly, except for the presence of test sample. Growth inhibition was calculated with reference to positive control.

2.4 Theoretical and computational details

All the quantum chemical calculations on the compounds **18**, **5**, **19**, and **20** were performed with full geometrical optimizations by using standard Gaussian 03 and 09 software package (Frisch et al., 2004). Geometrical optimization were carried out with two different methods, *ab initio* methods at the Hartree-Fock (HF) level, and density functional theory (DFT) by using the B3LYP change-correlation corrected functional (Becke, 1993; Lee et al., 1988) with 6-31G(d,p), 6-311G(d,p), 6-311++G(d,p), LANL2DZ 6-31G(d,p) basis sets, and BP86/CEP-31G* hybrid functional with 30% HF exchange and Stevens-Basch-Krauss pseudo potentials with polarized split valence basis sets (CEP-Compact Effective Potentials -31G*) (Hill et al., 1992; Stevens et al., 1984).

Fukui functions, which are common descriptors of site reactivity and can be expressed by the following equations, were calculated by AOMix program (Gorelsky, 2009; Gorelsky & Lever, 2001).

$$f_k^+ = \rho_k(N+1) - \rho_k(N) \text{ (for nucleophilic attack)}$$

$$f_k^- = \rho_k(N) - \rho_k(N-1) \text{ (for electrophilic attack)}$$

$$f_k^o = \rho_k(N+1) - \rho_k(N-1)/2 \text{ (for radical attack)}$$

Where, k represents the sites (atoms/molecular fragments) for nucleophilic, electrophilic and radical agents and ρ_k are their gross electron populations. An elevated value of f_k implies a high reactivity of the site k.

3. Results and discussion

The optimized structures of the compounds **18**, **5**, **19**, and **20** ligands, and the optimized structures of their corresponding Ni(II) and Zn(II) complexes are shown in Figure 1.

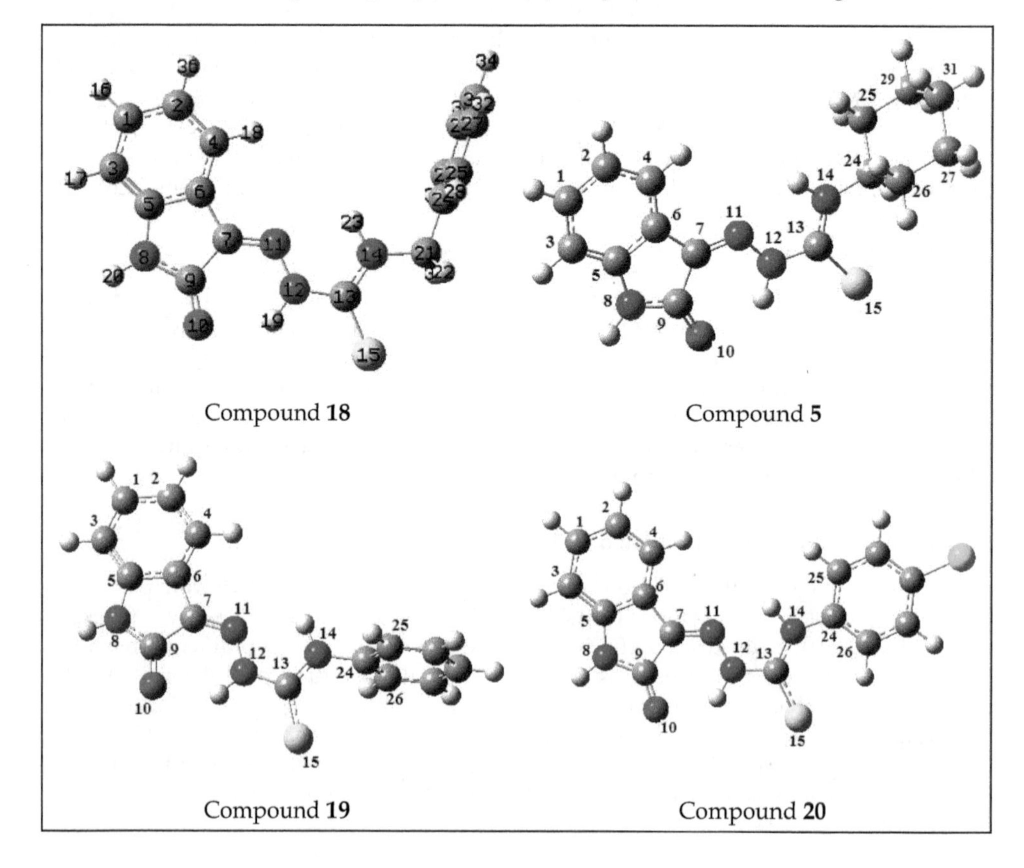

Compound **18** Compound **5**

Compound **19** Compound **20**

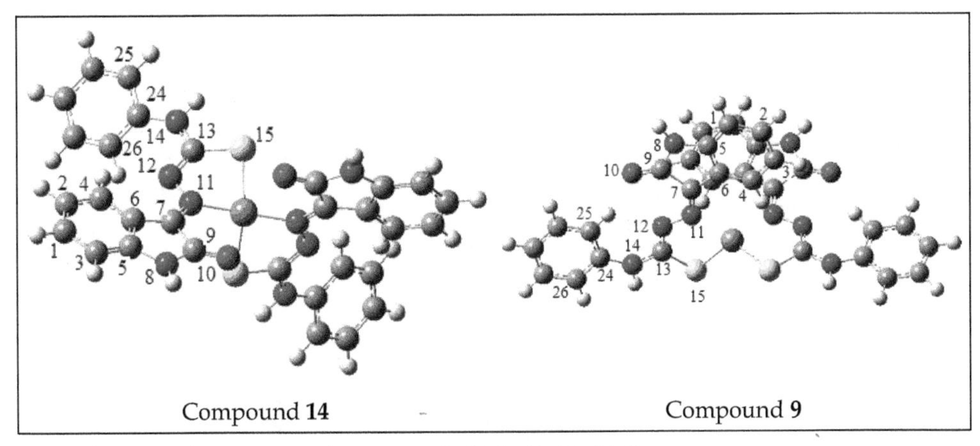

Fig. 1. Optimized structures of compounds **18, 5, 19, 20, 14, 9**

Values of the optimized geometrical parameters for all compounds are presented in Tables 2 and 3. The bond lengths, bond angles, and dihedral angles for compounds **18**, **5**, **19**, and **20** are almost the same as in isatin group. Mulliken charges of most atoms, except N and S, belong to thiosemicarbazone group are the same value. Therefore comparision among ligands, and deprotonated forms of ligand and complexes were made for compound **19**. In the deprotonated forms of **19**, as in N11-N12, C6-C7, and C13-C14, the bond lengths decrease, while C7-C9, C5-N8, and C4-C6 bond lengths increase. In Zn (II) complexes, C4-C6, C6-C7 bond lengths are similar to those of ligands. N11-N12 bond length in the deprotonated form of the ligands decreases, whereas in the complexes this bond length increases due to a transfer of charge from N atoms to metal. In the complex form, C-S bond length increases from 1.667 Å to 1.759 Å in Ni (II) complex, and 1.744 Å in Zn (II) complex.

The calculated bond lengths of the Zn-S and Zn-N bonds for compound **9** were found to be 2.323 and 2.107 Å. The bond lengths of the Ni-N and Ni-S, Ni-O bonds for compound **14**

	Compounds						
Atoms	**18**	**5**	**20**	**19**	**19**[a]	**9**	**14**
Bond distances (Å)							
C1-C2	1.396	1.396	1.396	1.396	1.408	1.396	1.395
C1-C3	1.399	1.399	1.399	1.399	1.396	1.398	1.399
C2-C4	1.396	1.396	1.395	1.395	1.382	1.393	1.396
C3-C5	1.385	1.385	1.385	1.385	1.385	1.386	1.385
C4-C6	1.390	1.390	1.390	1.390	1.406	1.393	1.394
C5-C6	1.411	1.411	1.411	1.411	1.424	1.413	1.417
C6-C7	1.456	1.456	1.455	1.456	1.418	1.455	1.450
C7-C9	1.502	1.502	1.503	1.502	1.548	1.525	1.502

C5-N8	1.403	1.403	1.403	1.403	1.391	1.392	1.394
N8-C9	1.380	1.380	1.378	1.379	1.380	1.387	1.388
C7-N11	1.296	1.296	1.295	1.296	1.299	1.310	1.307
N11-N12	1.332	1.332	1.333	1.333	1.283	1.347	1.350
N12-C13	1.391	1.393	1.395	1.391	1.389	1.326	1.331
C13-S15	1.673	1.676	1.669	1.667	1.661	1.759	1.744
C13-N14	1.339	1.340	1.353	1.348	1.325	1.360	1.348
N14-C21	1.463	1.462	1.407	1.432	1.425	1.412	1.348
C21-C24	1.511	-	-	-	-	-	-
C24-C25	1.398	1.538	1.404	1.393	1.399	1.405	1.534
C24-C26	1.398	1.535	1.399	1.393	1.398	1.400	1.539
S15-Zn	-	-	-	-	-	2.323	2.259
N11-Zn	-	-	-	-	-	2.107	1.903
O10-Ni	-	-	-	-	-	-	2.830
Bond angles (°)							
C5-C6-C7	106.79	106.80	106.76	106.78	106.83	107.33	106.42
C6-C7-C9	107.02	107.00	107.01	107.01	107.29	106.47	107.54
C5-N8-C9	111.81	111.82	111.79	111.81	111.65	112.60	111.87
N8-C9-O10	126.91	126.87	105.29	105.29	104.17	124.43	126.18
C7-C9-O10	127.80	127.84	127.64	127.77	126.50	131.09	129.00
C7-N11-N12	119.25	119.25	119.33	119.29	126.94	116.83	117.32
C9-C7-N11	126.46	126.46	126.29	126.46	123.14	126.33	120.51
N11-N12-C13	120.79	121.01	121.47	120.93	116.34	117.34	112.11
N12-C13-S15	33.46	118.35	117.54	27.38	102.33	128.18	112.11
S15-C13-N14	125.98	127.13	130.01	126.98	135.52	113.71	118.44
C13-N14-C21	123.87	125.55	132.89	125.18	127.50	133.38	125.44
N14-C21-C24	110.10	109.87	115.97	119.79	117.49	115.70	109.61
C13-Zn/Ni-N11	-	-	-	-	-	91.80	90.93
S15- Zn/Ni -N11	-	-	-	-	-	130.33	95.84
N11- Zn/Ni -S15	-	-	-	-	-	123.82	84.44
O10-Ni-S15	-	-	-	-	-	-	79.44
O10-Ni-O1	-	-	-	-	-	-	74.20
O10-Ni-N11	-	-	-	-	-	-	73.70

[a] Protonated form of 19

Table 2. Selected bond distances (Å) and bond angles charges calculate with B3LYP/6-311G(d,p) for compounds **18**, **5**, **20**, **19**, **9**, and **14**

were 2.259, 1.903, and 2.830 Å, respectively. Experimental data of Ni-N and Ni-S, Ni-O bonds for the crystallographic analyses of complex of isatin-β-thiosemicarbazone were 2.023, 2.368, and 2.226 Å, respectively. The calculated dihedral angles ∠C6-C7-N11-N12 and ∠N12-C13-N14-C21 for compound **19** were 179.98° and -180.00°, respectively, very close to the crystal values of 5-methoxyisatin-3-(*N*-cyclohexyl)thiosemicarbazone) (Kandemirli et al., 2009a).

	Compounds						
Atoms	18	5	20	19	19[a]	9	14
Dihedral angles (°)							
C3-C5-N8-C9	-180.00	179.96	-	-179.99	-176.17	-	-
C5-C6-C7-N11	-180.00	-179.87	180.00	179.99	-178.71	176.47	-178.31
C5-N8-C9-O10	180.00	179.98	180.00	-179.98	176.75	179.62	175.86
N8-C9-C7-N11	180.00	179.85	-180.00	-179.98	-155.39	-176.13	179.37
C6-C7-N11-N12	-180.00	179.96	180.00	-179.98	28.48	177.24	1.70
C9-C7-N11-N12	0.00	0.08	0.00	0.01	-4.73	-5.54	-174.73
O10-C9-C7-N11	0.00	-0.07	0.00	-0.023	-147.12	2.76	1.58
C7-N11-N12-C13	-180.00	-179.78	-180.00	-179.99	123.93	176.10	160.68
N11-N12-C13-S15	180.00	179.55	-180.00	-180.00	-172.93	-3.78	-3.65
N12-C13-N14-C21	180.00	177.56	-179.98	-180.00	8.06	-0.22	-6.96
S15-C13-N14-C21	0.00	-2.69	0.02	0.00	-38.29	-179.42	176.54
C13-N14-C21-C24	179.99	-	-	-	-	-	-
N14-C21-C24-C25	89.51	-92.75	179.96	-91.29	142.56	5.42	156.43
N14-C21-C24-C26	-89.48	143.32	-0.04	91.24	142.56	-175.30	-79.66
Mulliken charges (ē)							
C5	0.239	0.239	0.240	0.239	0.261	0.253	0.245
C6	-0.134	-0.133	-0.136	-0.134	-0.110	-0.157	-0.140
C7	0.081	0.082	0.094	0.084	0.142	0.203	0.205
N8	-0.489	-0.489	-0.489	-0.489	-0.483	-0.490	-0.489
C9	0.421	0.420	0.423	0.421	0.436	0.614	0.376
O10	-0.360	-0.359	-0.357	-0.359	-0.231	-0.412	-0.322
N11	-0.225	-0.228	-0.240	-0.231	-0.207	-0.605	-0.476
N12	-0.274	-0.273	-0.270	-0.270	-0.082	-0.352	-0.266
C13	0.226	0.228	0.229	0.215	0.145	0.329	0.205
S15	-0.237	-0.244	-0.208	-0.201	0.163	-0.572	-0.317
N14	-0.405	-0.392	-0.463	-0.437	-0.381	-0.537	-0.379
Zn or Ni	-	-	-	-	-	1.568	1.036

[a] Protonated form of 19

Table 3. Selected dihedral angles and Mulliken charges calculate with B3LYP/6-311G(d,p) for compounds **18**, **5**, **20**, **19**, **9**, and **14**

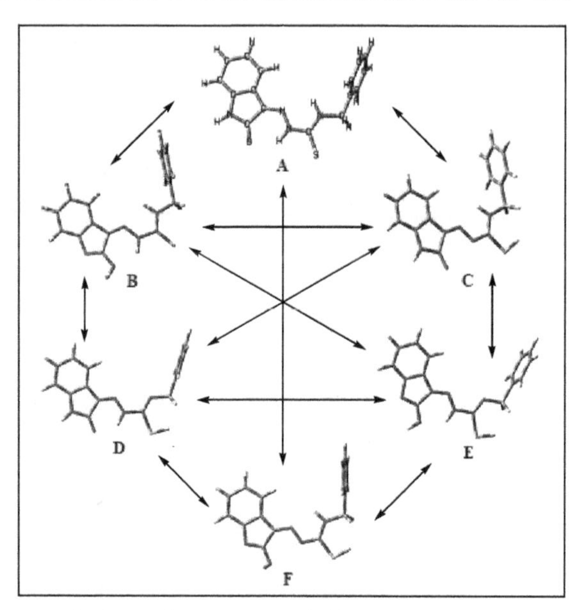

Fig. 2. Possible tautomeric forms for compound **18**

Calculated theoretical angle for **16** (∠C13-Ni-N11, ∠S15-Ni -N11, ∠N11-Ni -S15, ∠O10-Ni-S15, ∠O10-Ni-O1, ∠O10-Ni-N11 were 90.93°, 95.84°, 84.44°, 79.44°, and 74.20°, respectively), which indicated that the complex is in a distorted octahedral coordination. Two terdentate monodeprotonated thiosemicarbazone groups, each of which is attached to the metal with the sulfur, the nitrogen atom from the hydrazine chain, and the carbonylic oxygen of the isatin moiety. The calculated dihedral angles of ∠C3-C5-N8-C9, ∠C6-C7-N11-N12°, ∠N11-N12-C13-S15° were -179.99°, -179.98°, and -180.00° which showed that ligands are planar.

As presented in Table 3, Mulliken charges of C5, C6, N8, C9, and N12 were similar to each other in the neutral form. Most of the changes are on the N14 atom because of the changes in substituent attached to the N14 atom. The Mulliken charge of N14 atom was -0.405 ē for H_2IBT, -0.392 ē for H_2ICHT, -0.463 ē for H_2IPT, and -0.437 ē for H_2ICPT. Mulliken charges of atoms, both in the indole ring and thiosemicarbazone group, undergo a significant change, depending whether ligand is in the deprotonated form or in complexation process.

3.1 The possible tautomeric forms of ligands

The optimized structure for **18** ligand, calculated with B3LYP/6-311G(d,p), is shown in Figure 2. Electronic and zero point energies for compounds **18**, **5**, **19**, and **20** were given in Table 4. The most stable tautomeric form is A for studied ligands. In the form A, calculated electronic and zero point energy for compounds **18**, **5**, **19**, and **20** were -1310.525279 au, -1274.797093 au, -1271.228351 au, and 1730.859463 au, respectively. Therefore, all discussions are based on A form.

Tautomeric forms	Compounds			
	18	**5**	**19**	**20**
A	-1310.525279	-1274.797093	-1271.228351	-1730.859463
B	-1310.516899	-1274.770217	-1271.201356	-1730.830654
C	-1310.505963	-1274.758544	-1271.196548	-1730.826130
D	-1310.492187	-1274.764865	-1271.207983	-1730.837828
E	-1310.463311	-1274.737477	-1271.179113	-1730.808789
F	-1310.466475	-1274.743568	-1271.175689	-1730.804186

Table 4. Electronic and zero point energy calculated with B3LYP/6-311G(d,p) for compounds **18**, **5**, **19**, and **20** ligands

3.2 Fukui functions

Fukui functions values for compounds **18**, **5**, **19**, and **20** calculated with B3LYP/6-31G(d,p) were summarized in Table 5. The contribution of sulphur atom to the HOMO is 90.2% and 89.52% for compounds **18**, and **5**, whereas, for compounds **19**, and **20**, the contribution of S decreases to 48.32%, and 41.90%, respectively. The other contributions for compound **19** comes from N14 atom (12.63%) belonging to thiosemicarbazone and phenyl groups (C24: 5.41%, C25: 5.65%, C25: 15%, C29: 1.93%, C25: 9.18%), and isatin group (C1: 1.49%, C5: 1.14%, C6: 1.06%, C7: 2.08%). For compound **5** contribution involves mostly N14 (13.45%), belonging to thiosemicarbazone group and phenyl ring (C24: 6.96%, C25: 5.75%, C25: 4.78%, C29: 2.76%, C25: 10.14%). Proportions of contribution change were according to the groups attached to N14 atom. Attaching phenyl group instead of cyclohexyl to N14 atom decreases the contribution of S atom to the HOMO orbital to approximately half.

Tauto. Forms*	For electrophilic attack													
	C1	C5	C6	C7	O10	N12	C13	N14	S 15	C24	C25	C26	C29	C31
A	-	-	-	-	-	1.06	4.26	-	90.20	-	-	-	-	-
B	-	-	-	-	-	1.35	4.09	-	89.52	1.12	-	-	-	-
C	1.49	1.14	1.06	2.08	1.10	2.05	-	12.63	48.32	5.41	5.65	4.15	1.93	9.18
D	-	-	-	1.20	-	-	-	13.45	41.90	6.96	5.75	4.78	2.76	10.14
	For nucleophilic attack													
	C 1	C 4	C5	C6	C7	N8	C9	O10	N11	N12	C13	N14	S15	-
A	6.69	5.03	3.98	3.61	12.92	1.46	10.45	8.96	26.63	5.35	4.45	2.16	6.98	-
B	6.71	5.05	3.98	3.64	12.90	1.44	10.40	8.94	26.68	5.44	4.42	2.14	6.89	-
C	6.52	5.01	3.94	3.26	13.23	1.39	9.95	8.70	25.21	4.57	5.39	1.86	8.12	-
D	6.51	5.05	3.97	3.15	13.51	1.39	9.90	8.70	24.87	4.23	5.66	1.79	8.28	-

* Tautomeric forms

Table 5. Fukui functions for calculated with B3LYP/6-31G(d,p) for compounds **18**, **5**, **19**, and **20** ligands

The other molecular parameters, obtained through the theoretical calculations by using the level of B3LYP and RHF theory by using 6-31G(d,p), 6-311G(d,p), 6-311++G(d,p) basis sets, are the E_{HOMO} (highest occupied molecular orbital), and the E_{LUMO} (lowest unoccupied molecular orbital). The highest occupied and the lowest unoccupied molecular orbital energies (E_{HOMO} and E_{LUMO}), and the five molecular orbital energies for the nearest frontier orbitals of compounds **18**, **5**, **19**, and **20** were presented in Table 6. HOMO and LUMO electron densities for compounds **18**, **5**, **19**, and **20** which are calculated with B3LYP/6-311++G(d,p) and shown in Figure 3. E_{HOMO} values, obtained with B3LYP method, are higher than those of RHF method. Calculated E_{HOMO} at the level of RHF/6-311G(d,p) and B3LYP/6-311G(d,p) theory are -0.30438 au and -0.21286 au for **18**; 0.30266 au and -0.21701 au for **5**; -0.30386 au and -0.21656 au for **19**; and -0.30899 au and -0.22050 au for compound **20**. HOMO of compounds **5**, **18**, and **19** consist of mainly S atom belonging to the thiosemicarbazone group, and HOMO of compound **20** includes mainly S and chlorophenyl group. LUMO orbitals are distributed mainly over isatin and thiosemicarbazone groups for compounds **18**, **5**, **19**, and **20.**

Electron densities for compounds **18**, **5**, **19**, and **20** are also shown in Figure 3. As shown in Figure 3, electron rich regions shown as red are found in the vicinity of the double bonded O and S atoms attached to C which belongs to isatin moiety and thiosemicarbazone moiety, respectively. Electron poor regions shown as blue, are mainly consists of N-H atoms belonging to isatin group.

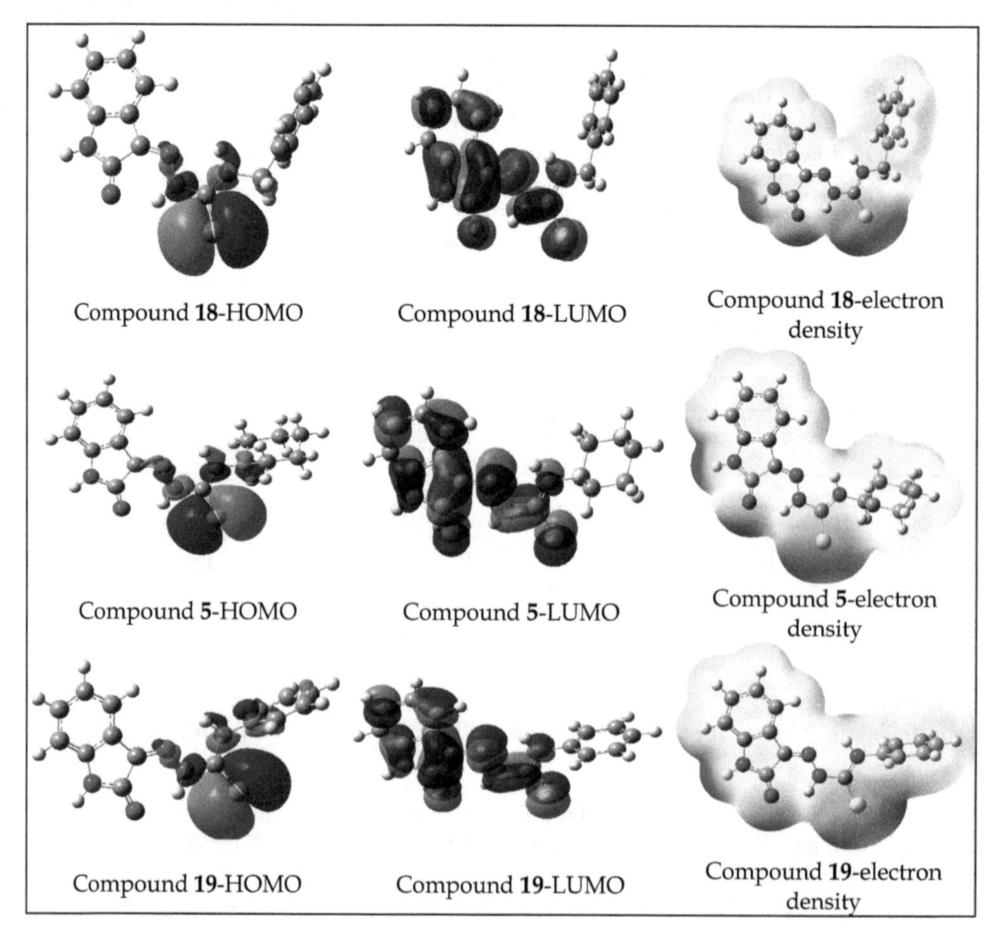

Compound **18**-HOMO | Compound **18**-LUMO | Compound **18**-electron density

Compound **5**-HOMO | Compound **5**-LUMO | Compound **5**-electron density

Compound **19**-HOMO | Compound **19**-LUMO | Compound **19**-electron density

Fig. 3. HOMO, LUMO and electron density calculated at the level of B3LYP/6-311++G(d,p) theory for compounds **18**, **5**, **19**, and **20**

3.3 IR spectra

Thiosemicarbazones can coordinate with metal ions as neutral ligands (HTSC) or as anionic species (TSC) upon deprotonation at the N(2′) (Beraldo & Tosi, 1983, 1986; Borges et al., 1997; Rejane et al., 1999).

Metal acetates with the ligands (H_2L) lead to isolation of complexes of formula $M(HL)_2$ (Rodriguez-Argiielles et al., 1999). As seen from the C, H, N analyses, the synthesized complexes are with the formula $M(HL)_2$ (if the ligand is written as H_2IPT, then complex has the formula of $M(HIPT)_2$) *and* form 2:1 ligand-to-metal complexes which two of the ligands were anionic.

Infrared absorptions in the range of 4000-400 cm^{-1} have been calculated for **19** with the method, B3LYP/6-31G(d,p), B3LYP/6-311G(d,p) and B3LYP/6-311++G(d,p) and for their Zn(II) and Ni (II) complexes with the method B3LYP/6-31G(d,p), B3LYP/6-311G(d,p), and B3LYP/LanL2DZ. Beside this, infrared frequencies for compounds **18**, **5**, and **20**, and their zinc (II) and nickel (II) complexes have been calculated. Calculated bands and their correspondence experimental values are presented in Tables 7 and 8. Loosing of N12 H band due to deprotonation is one of the main changes (Bresolin et al., 1997).

MO Energy	6-31G(d,p)		6-311G(d,p)		6-311G++(d,p)		LANL2DZ	
(a)	RHF	B3LYP	RHF	B3LYP	RHF	B3LYP	RHF	B3LYP
6′	0.18055	0.01472	0.14439	0.00491	0.05089	-0.03326	0.16053	0.00090
5′	0.15887	0.00827	0.13996	-0.00377	0.04569	-0.03286	0.13557	-0.00545
4′	0.13325	-0.01245	0.12068	-0.02526	0.04149	-0.03142	0.11213	-0.02372
3′	0.13102	-0.01563	0.12043	-0.02712	0.03267	-0.01705	0.11212	-0.02547
2′	0.12559	-0.01632	0.11615	-0.02911	0.03241	-0.00870	0.10372	-0.03112
1′	0.04691	-0.08870	0.04111	-0.09642	0.02804	-0.01309	0.02603	-0.10317
1	-0.29966	-0.20958	-0.30438	-0.21286	-0.30635	-0.22160	-0.30530	-0.21316
2	-0.31169	-0.21266	-0.31672	-0.21770	-0.31882	-0.22471	-0.31549	-0.21892
3	-0.32555	-0.21266	-0.32999	-0.24209	-0.33211	-0.24364	-0.33497	-0.24024
4	-0.34287	-0.24654	-0.34749	-0.25866	-0.34886	-0.25832	-0.35162	-0.25603
5	-0.34330	-0.25940	-0.34807	-0.26857	-0.34983	-0.27125	-0.35190	-0.26582
6	-0.34878	-0.26066	-0.35334	-0.26974	-0.35503	-0.27277	-0.35772	-0.26721

(b)								
6'	0.22036	0.07889	0.15398	0.03320	0.05283	-0.00555	0.22904	0.07491
5′	0.21615	0.07539	0.14812	0.02913	0.04826	-0.00929	0.21900	0.07219
4′	0.18048	0.01426	0.14059	0.00407	0.04213	-0.01325	0.16181	0.00122
3′	0.15787	0.00751	0.13494	-0.00274	0.03393	-0.01759	0.13511	-0.00560
2′	0.12486	-0.01671	0.11558	-0.02654	0.03241	-0.03341	0.10324	-0.03120
1′	0.04648	-0.08890	0.04082	-0.09696	0.02768	-0.10207	0.02635	-0.10269
1	-0.29829	-0.20836	-0.30266	-0.21701	-0.30407	-0.21948	-0.30214	-0.21091
2	-0.31042	-0.21108	-0.31499	-0.21922	-0.31631	-0.22241	-0.31288	-0.21568
3	-0.32484	-0.22993	-0.32907	-0.23819	-0.33081	-0.24172	-0.33388	-0.23854
4	-0.34721	-0.24596	-0.35142	-0.25421	-0.35263	-0.25725	-0.35609	-0.25495
5	-0.39928	-0.26914	-0.40313	-0.27761	-0.40410	-0.28176	-0.40838	-0.27693
6	-0.43098	-0.28528	-0.43466	-0.29358	-0.43699	-0.29626	-0.43871	-0.29228
(c)								
6′	0.18135	0.03242	0.14358	0.00235	0.05130	-0.01079	0.17626	0.01936
5′	0.15783	0.00696	0.13850	-0.00421	0.04653	-0.01753	0.13359	-0.00865
4′	0.14350	-0.00375	0.13029	-0.01837	0.04127	-0.02585	0.12023	-0.01588
3′	0.13463	-0.02009	0.12802	-0.02132	0.03540	-0.02825	0.10782	-0.03380
2′	0.12322	-0.02198	0.11434	-0.02779	0.02997	-0.03488	0.09968	-0.03534
1′	0.04323	-0.09503	0.03855	-0.09899	0.02732	-0.10450	0.01959	-0.10946
1	-0.29931	-0.21623	-0.30386	-0.21656	-0.30594	-0.21996	-0.30582	-0.22084
2	-0.30783	-0.21628	-0.31251	-0.22104	-0.31487	-0.22489	-0.32135	-0.22190
3	-0.32430	-0.22239	-0.33168	-0.24187	-0.33378	-0.24588	-0.32368	-0.23349
4	-0.33602	-0.24315	-0.34277	-0.25858	-0.34444	-0.26205	-0.34504	-0.25289
5	-0.34001	-0.25398	-0.34717	-0.26456	-0.34914	-0.26793	-0.34924	-0.26157
6	-0.35686	-0.27301	-0.35596	-0.26571	-0.35765	-0.26889	-0.37209	-0.28058
(d)								
6′	0.17471	0.02265	0.13955	0.01052	0.05007	-0.02179	0.15295	0.00965
5′	0.15375	0.00354	0.13592	-0.00666	0.04981	0.00440	0.12967	-0.01248
4′	0.12806	-0.01734	0.11543	-0.02957	0.04023	0.00291	0.10620	-0.03124
3′	0.12168	-0.02339	0.11423	-0.03312	0.03302	-0.01424	0.10191	-0.03896
2′	0.11948	-0.03026	0.11139	-0.04013	0.02637	-0.00571	0.09757	-0.04281
1′	0.03893	-0.09939	0.03459	-0.10727	0.02552	-0.01692	0.01736	-0.11453
1	-0.30480	-0.22050	-0.30899	-0.22895	-0.31067	-0.23154	-0.31149	-0.22767
2	-0.31382	-0.22213	-0.31803	-0.23052	-0.31983	-0.23281	-0.32058	-0.22812
3	-0.32819	-0.22582	-0.33599	-0.23414	-0.33779	-0.23730	-0.34034	-0.23818
4	-0.33930	-0.24652	-0.34810	-0.25459	-0.34911	-0.25774	-0.35297	-0.25692
5	-0.35484	-0.26768	-0.35724	-0.27708	-0.35800	-0.27925	-0.36610	-0.28266
6	-0.36077	-0.27427	-0.36142	-0.28272	-0.36254	-0.28521	-0.37048	-0.27736
HOMO↔LUMO: 1 ↔ 1′								

Table 6. Calculated values for the highest occupied and the lowest unoccupied molecular orbital energies HOMO and LUMO, and the five molecular orbital energy the nearest frontier orbitals for (a) **18**, (b) **5**, (c) **19**, and (d) **20**

The FT-IR spectra for **19** shows bands at 3298, 3244, 3177, 1703, and 1612 cm^{-1}, assigned to stretching vibration modes νN(8)H, N(14)H, νC=O and νC=N, respectively. Stretching vibration modes of νN(8)H, νN(14)H, νC=O, and νC=N were found at 3372, 3179, 1703, and 1612 cm^{-1} for its zinc (II) complex and at 3381, 3271, 1661, and 1614 cm^{-1} for nickel (II) complex, respectively. The IR peak of N(12)-H of thiosemicarbazide regions at 3283, 3244, 3242, and 3242 cm^{-1} in the spectra of compounds **18**, **5**, **19**, and **20** ligands are not present in their zinc (II), and nickel (II) complexes, due to proton dissociation, which is the main change (Bresolin et al., 1997). Infrared spectra of the ν(C=N) for compounds **5**, **18**, **19**, and **20** ligands were assigned as 1618, 1620, 1620, and 1622 cm^{-1}, respectively. In their nickel (II) and zinc (II) complexes, the position of these bands shifted to 1595 cm^{-1}. On complex formation with nickel (II), and zinc (II), C=S band shifted towards the lower side.

	6-31G(d,p)		6-311G(d,p)		6-311++G(d,p)		
Exp.	Freq. (cm^{-1})	Intensity	Freq. (cm^{-1})	Intensity	Freq. (cm^{-1})	Intensity	Assignment
Compound **19**							
3298	3661	75	3642	71	3640	75	$\nu(N_8H)_{indole}$
3244	3516	86	3540	96	3543	94	$\nu(N_{14}H)_{tiyo}$
3177	3432	91	3437	85	3433	84	$\nu(N_{12}H)_{tiyo}$
			3198	16	3198	13	$\nu(CH)_{ring\ A\ combination}$
3148	3254	12	3198	4	3197	4	$\nu(C_{26}\text{-}H)_{ring\ C\ combination}$
			3194	15	3193	12	$\nu(C_{25}\text{-}H)_{ring\ C}$, $\nu(C_{26}\text{-}H)_{ring\ C}$, $\nu(C_{27}\text{-}H)_{ring\ C}$, $\nu(C_{29}\text{-}H)_{ring\ C}$
3071	3216	18	3189	17	3189	13	$\nu(CH)_{ring\ A\ combination}$
3059	3210	27	3186	27	3185	23	$\nu(CH)_{ring\ C\ combination}$
	3207	18	3181	5	3181	5	$\nu(CH)_{ring\ A\ combination}$
	3200	5					$\nu(CH)_{ring\ A\ combination}$
3028	3196	22	3176	6	3175	5	$\nu(C_{25}\text{-}H)_{ring\ C}$, $\nu(C_{26}\text{-}H)_{ring\ C}$, $\nu(C_{27}\text{-}H)_{ring\ C}$, $\nu(C_{29}\text{-}H)_{ring\ C}$
	3168	12					$\nu(CH)_{ring\ C\ combination}$
1694	1790	222	1775	257	1758	297	ν(C=O), $\delta(N_8H)$, $\delta(N_{12}H)$
1620							
1594	1675	91	1661	98	1658	94	$\nu(CC)_{ring\ A\ combination}$, $\delta(N_8H)$
1539	1661	25	1644	14	1641	15	$\nu(CC)_{ring\ A\ combination}$, $\delta(N_{14}C_{13})$
1492	1652	124					$\nu(CC)_{ring\ C\ combination}$, $\delta(N_{14}H)$
			1636	21	1632	15	$\nu(CC)_{ring\ A\ combination}$
1477	1649	16					$\nu(CC)_{ring\ A\ combination}$
			1630	3	1627	3	$\nu(CC)_{ring\ C\ combination}$
1462	1637	175	1623	191	1618	216	$\nu(C_7\text{-}N_{11})$, $\delta(N_{12}H)$, $\nu(CC)_{ring\ C\ combination}$
1439	1598	750	1544	620	1539	542	$\delta(N_{14}H)$, $\delta(CH)_{ring\ C\ combination}$
1412	1541	74					$\delta(CH)_{ring\ C\ combination}$, $\delta(N_{12}H)$, $\delta(N_{14}H)$
			1525	190	1522	183	$\delta(CH)_{ring\ A\ combination}$, $\delta(N_8H)$, $\delta(N_{12}H)$

1377	1531	168					$\delta(CH)_{\text{ring A combination}}$, $\delta(N_8H)$, $\delta(N_{12}H)$
			1522	110	1519	139	$\delta(CH)_{\text{ring C combination}}$, $\delta(N_{14}H)$
1341	1522	54	1513	17	1510	17	isatin, $\delta(N_{12}H)$
1316	1505	143	1494	143	1492	138	$\delta(CH)_{\text{ring A combination}}$, $\delta(N_{12}H)$
1298	1487	64	1483	4	1480	4	$\delta(CH)_{\text{ring C combination}}$
1271	1429	148	1418	58	1415	56	$\delta(N_8H)$, $\delta(CH)_{\text{ring A combination}}$, $\delta(N_{14}H)$
	1407	212	1385	128	1380	136	$\delta(N_{14}H)$, $\nu(N_{12}C_{13})$
1248	1369	28					ω(ring A), $\delta(N_{14}H)$
1227	1365	39					ω(ring A, B, C)
			1355	72	1353	73	ω(isatin)
1207	1354	23	1312	16	1311	17	δ(NH), $\delta(CH)_{\text{ring A combination}}$
1148	1198	895	1189	935	1187	836	ν(N-N)+δ(ring A)+ν(C-N)
1136	1183	153	1179	212	1177	336	ν(C-N)+δ(ring A)+ν(C-N)
1101	1127	14	1122	24	1121	28	ω(ring A)
1028	1059	7	1048	9	1047	11	ω(phenyl)
982	1048	3	1021	5	1020	4	τ(phenyl ring)
964	1000	1	1000	4	999	6	τ(indole ring)
941	961	38	955	25	953	23	γ(phenyl ring)+ω(N-H)
903	940	1	932	15	932	15	τ(phenyl ring)
864	889	7	890	7	889	6	τ(indole ring)
793	817	100	823	80	822	81	γ(phenyl ring)+γ(indole ring)
760	797	45	806	32	796	29	$\gamma(N\text{-}H)_{thio}$+$\nu(C\text{-}C)_{\text{indole ring}}$+ τ(indole ring)
700	771	64	761	33	757	45	γ(indole ring)
679	763	21	738	54	733	44	γ(N-H)thio+γ(indole ring)
658	716	97	710	40	707	44	γ(phenyl ring)
644	697	9	691	18	690	16	τ(indole ring)+γ(phenyl ring)
590	673	18	659	19	658	19	τ(phenyl ring)+ν(C-S)
561	592	8	624	13	620	13	γ(C-S)
503	514	109	517	169	509	107	γ(N-H)indole+$\gamma(N\text{-}H)_{thio}$
480	490	24	458	17	456	19	τ(phenyl ring)
Compound **9**							
3372	3667	66	3649	68	3674	37	$\nu(N_8H)_{indole}$
	3667	70	3649	70	3674	105	$\nu(N_8H)_{indole}$
3179	3615	99	3601	106	3596	98	$\nu(N_{14}H)_{thio}$
	3615	105	3601	107	3596	142	$\nu(N_{14}H)_{thio}$
3150	3260	28	3242	26	3272	20	$\nu(C_{25}\text{-}H)_{\text{ring C}}$
			3242	105	3272	121	$\nu(C_{25}\text{-}H)_{\text{ring C}}$
3063	3260	116			3266	14	$\nu(CH)_{\text{ring A combination}}$
	3215	10	3198	7	3234	22	$\nu(CH)_{\text{ring A combination}}$
	3214	27					$\nu(CH)_{\text{ring A}}$+$\nu(CH)_{\text{ring C}}$

3023	3208	77	3198	23	3234	35	$\nu(CH)_{ring\ A\ combination}$
	3204	14			3203	4	$\nu(CH)_{ring\ A\ combination}$
	3204	11					$\nu(CH)_{ring\ A\ combination}$
	3195	29	3191	71			$\nu(CH)_{ring\ C\ combination}$
	3195	19			3198	3	$\nu(CH)_{ring\ C\ combination}$
	3190	4	3186	15			$\nu(CH)_{ring\ C\ combination}$
			3186	10			$\nu(CH)_{ring\ C\ combination}$
			3178	26			$\nu(C_{26}\text{-}H)_{ring\ C}$, $\nu(C_{27}\text{-}H)_{ring\ C}$, $\nu(C_{29}\text{-}H)_{ring\ C}$
			3178	19			$\nu(C_{26}\text{-}H)_{ring\ C}$, $\nu(C_{27}\text{-}H)_{ring\ C}$, $\nu(C_{29}\text{-}H)_{ring\ C}$
			3173	4			$\nu(CH)_{ring\ A\ combination}$
	3165	6	3150	6	3175	4	$\nu(C_{26}\text{-}H)_{ring\ C}$, $\nu(C_{29}\text{-}H)_{ring\ C}$
	3165	13	3150	11	3175	18	$\nu(C_{26}\text{-}H)_{ring\ C}$, $\nu(C_{29}\text{-}H)_{ring\ C}$
1703	1808	509	1794	559	1690	22	$\nu(C\text{-}O)$
	1808	10	1794	7	1690	436	$\nu(C\text{-}O)$
1612	1674	184	1652	192	1671	243	$\delta(N_8H)$, $\nu(CC)_{ring\ A\ combination}$
	1674	21	1661	19	1671	48	$\delta(N_8H)$, $\nu(CC)_{ring\ A\ combination}$
	1656	10	1642	6	1649	6	$\delta(N_{14}H)$, $\nu(CC)_{ring\ C\ combination}$
	1648	4	1642	6	1649	5	$\delta(N_{14}H)$, $\nu(CC)_{ring\ C\ combination}$
1595	1648	54	1636	73	1643	48	$\delta(N_{14}H)$, $\nu(CC)_{ring\ C\ combination}$
	1642	16	1631	18	1639	28	$\delta(N_8H)$, $\nu(CC)_{ring\ A\ combination}$
1508	1603	51	1593	42	1582	24	$\nu(N_{11}C_7)$
	1602	12	1592	13	1582	66	$\nu(N_{11}C_7)$
	1582	5	1576	10			$\delta(N_{14}H)$, $\nu(CC)_{ring\ C\ combination}$
1456	1578	635	1572	591	1573	423	$\delta(N_{14}H)$, $\nu(CC)_{ring\ C\ combination}$
	1537	181	1525	183	1525	10	$\delta(N_{14}H)$, $\nu(CC)_{ring\ C\ combination}$
	1526	12	1516	13	1524	273	$\delta(N_8H)$, $\nu(CC)_{ring\ A\ combination}$
	1510	5	1500	9	1501	15	$\nu(CC)_{ring\ A\ combination}$
1396	1510	160	1500	179	1501	218	$\nu(CC)_{ring\ A\ combination}$
	1489	128	1479	90	1478	55	$\delta(N_{14}H)$, $\nu(CC)_{ring\ C\ combination}$
	1488	100	1478	70	1477	76	$\delta(N_{14}H)$, $\nu(CC)_{ring\ C\ combination}$
1338	1464	1142	1446	1189	1451	837	$\nu(N_{11}C_7)$, $\nu(N_{12}C_{13})$, $\nu(CH)_{ring\ C}$
	1462	2109	1444	2160	1451	1948	$\nu(N_{11}C_7)$, $\nu(N_{12}C_{13})$, $\nu(CH)_{ring\ C}$
1277	1428	89	1420	129	1431	74	$\delta(N_8H)$, $\nu(CC)_{ring\ A\ combination}$
	1371	41	1363	60	1386	40	$\delta(N_{14}H)$, $\nu(CC)_{ring\ C\ combination}$
1242	1368	81	1359	138	1378	101	$\nu(NH)$, $\nu(CH)$
	1366	101	1358	119	1377	127	$\delta(N_8H)$, $\nu(CC)_{ring\ A\ combination}$
			1348	17	1373	5	$\delta(N_{14}H)$, $\nu(CC)_{ring\ C\ combination}$
	1361	164	1347	155	1373	95	$\delta(N_{14}H)$, $\nu(CC)_{ring\ C\ combination}$
1178	1324	32	1312	24	1333	45	ν(isatin)
1144	1272	143	1265	138	1300	221	ρ(indole ring), $\rho(NH)_{semicarbazone}$
1094	1204	228	1192	222	1273	39	$\sigma(CH)_{indole}+\nu(C\text{-}S)$

1043	1176	443	1166	511	1222	124	ν(N-N)+σ(CH)$_{indole}$
1007	1122	13	1125	111	1208	123	ω(indole ring)
966	998	64	1116	22	1187	347	ω(phenyl ring)
883	924	5	1000	25	1127	135	τ(indole ring)
833	858	9	995	54	1111	84	τ(indole ring)+ρ(C-N)
802	807	154	855	8	983	92	γ(phenyl ring)
783	795	62	824	28	974	10	γ(C-N+N-H)
750	771	46	799	216	894	9	ν(C-S)+ν(C-C)$_{indole}$
690	705	19	768	54	846	18	δ(phenyl ring)
658	696	16	764	19	814	33	δ(indole ring)
611	630	30	706	25	800	121	δ(phenyl ring)+δ(N-H)
594	605	74	640	19	785	226	δ(N-H)+ω(phenyl ring)
550	537	19	637	34	725	60	δ(N-H)
Compound **14**							
3381	3671	103	3653	103	3680	97	ν(N_8H)$_{indole}$
	3670	49	3653	48	3679	57	ν(N_8H)$_{indole}$
3271	3624	40	3610	40	3606	66	ν($N_{14}H$)$_{thio}$
	3624	92	3610	97	3606	81	ν($N_{14}H$)$_{thio}$
					3265	5	ν(C_4H)$_{ring\ A}$
3122	3210	19	3192	17	3230	24	ν(CH)$_{ring\ C\ combination}$
	3210	44	3192	40	3230	62	ν(CH)$_{ring\ C\ combination}$
	3209	10	3192	8	3230	20	ν(CH)$_{ring\ A\ combination}$
3065	3209	57	3192	49	3229	66	ν(CH)$_{ring\ A\ combination}$
	3200	15	3182	16	3214	13	ν(CH)$_{ring\ A\ combination}$
	3200	13	3182	12	3214	18	ν(CH)$_{ring\ A\ combination}$
	3195	43	3178	40	3210	52	ν(CH)$_{ring\ C\ combination}$
	3173	14	3156	12	3210	4	ν(CH)$_{ring\ C\ combination}$
					3182	19	ν(CH)$_{ring\ C\ combination}$
1661	1803	11	1801	24	1686	2	ν(isatin), ν($N_{11}C_7$), ν(C-O)
	1792	540	1792	620	1678	251	ν(isatin), ν($N_{11}C_7$), ν(C-O)
					1668	6	ν(isatin), ν($N_{11}C_7$), ν(C-O)
					1666	534	ν(isatin), ν($N_{11}C_7$), ν(C-O)
1614	1670	216	1658	240			ν(CC)$_{ring\ A\ combination,}$ δ(N_8H)
	1660	32	1645	27	1653	25	ν(CC)$_{ring\ C\ combination\ ,}$ δ($N_{14}H$)
1595	1659	65	1645	61	1652	87	ν(CC)$_{ring\ C\ combination\ ,}$ δ($N_{14}H$)
1535	1648	16	1635	25	1643	39	ν(CC)$_{ring\ C\ combination\ ,}$ δ($N_{14}H$)
	1647	60	1634	80	1642	89	ν(CC)$_{ring\ C\ combination\ ,}$ δ($N_{14}H$)
	1643	22	1630	16	1640	11	ν(CC)$_{ring\ A\ combination,}$ δ(N_8H), ν($N_{11}C_7$)
	1642	13	1629	19	1638	74	ν(CC)$_{ring\ A\ combination,}$ δ(N_8H), ν($N_{11}C_7$)
	1626	4	1612	9	1614	27	ν(CC)$_{ring\ A\ combination,}$ δ(N_8H), ν($N_{11}C_7$)
1518	1620	48	1566	184	1604	24	ν(CC)$_{ring\ A\ combination,}$ δ(N_8H), ν($N_{11}C_7$)

1496	1572	265			1571	139	$\nu(CC)_{ring\ C\ combination}$, $\delta(N_{14}H)$
	1569	645	1562	611	1568	539	$\nu(CC)_{ring\ C\ combination}$, $\delta(N_{14}H)$
1456	1539	61	1527	69	1527	172	$\nu(CH)_{ring\ C\ combination}$, $\delta(N_{14}H)$
	1538	130	1527	133	1526	311	$\nu(CH)_{ring\ C\ combination}$, $\delta(N_{14}H)$
	1525	86	1514	81	1512	16	ν(isatin), ν(N-C-N)
	1525	68	1514	90	1511	26	ν(isatin), ν(N-C-N)
1408	1501	259	1492	303	1490	8	$\nu(CC)_{ring\ A\ combination}$, $\delta(N_8H)$
	1501	172	1492	333	1489	145	$\nu(CC)_{ring\ A\ combination}$, $\delta(N_8H)$
	1498	220	1488	153	1480	65	$\nu(CC)_{ring\ A}$, $\nu(CC)_{ring\ C}$, ν(CH), ν(NH)
	1496	360	1487	228	1480	128	$\nu(CC)_{ring\ A}$, $\nu(CC)_{ring\ C}$, $\nu(N_{12}C_{13})$, $\nu(N_{11}C_7)$
1348	1478	790	1465	841	1456	1179	$\nu(CC)_{ring\ C}$, ν(N-C-N)
	1475	687	1463	691	1453	1020	$\nu(CC)_{ring\ C}$, ν(N-C-N)
1315	1422	80	1414	62	1428	147	$\nu(CC)_{ring\ A\ combination}$, $\delta(N_8H)$
	1421	87	1414	74	1428	145	$\nu(CC)_{ring\ A\ combination}$, $\delta(N_8H)$
	1369	29	1357	31	1385	10	$\nu(CC)_{ring\ C\ combination}$, $\delta(N_{14}H)$
1275	1369	54	1356	309	1372	276	$\nu(CC)_{ring\ A}$, $\nu(CC)_{ring\ C}$, ν(NH)
	1364	223	1355	187	1371	141	$\nu(CC)_{ring\ A}$, $\nu(CC)_{ring\ C}$, ν(NH)
	1364	193					$\nu(CC)_{ring\ A}$, $\nu(CC)_{ring\ C}$, ν(NH)
	1356	5			1365	27	$\nu(CH)_{ring\ C\ combination}$
1229	1356	238	1344	278	1365	169	$\nu(CH)_{ring\ C\ combination}$
1165	1309	91	1310	38	1338	43	ρ(indole ring), $\rho(NH)_{semicarbazone}$
1143	1272	143	1286	71	1304	96	$\nu(NN)+\sigma(CH)_{indole}$
1094	1244	10	1265	195	1265	177	ω(indole ring)+ω(N-H)
1045	1204	228	1237	45	1257	128	$\nu(C_{phenyl}\text{-}N)+\rho(NH)_{semicarbazone}$
1009	1202	253	1194	693	1229	261	ω(phenyl ring)
937	1176	443	1148	343	1206	45	ν(NN)+σ(phenyl)
903	1129	79	1116	107	1189	680	ρ(indole ring)
881	1055	3	1029	35	1148	215	τ(indole ring)
839	998	64	915	12	1059	28	γ(indole ring)+ρ(C-N)
802	943	2	895	13	1026	55	γ(C-N+N-H)
777	858	9	851	46	951	17	δ(indole ring)
748	807	154	819	19	841	43	ν(C-S)+ν(Ni-N)
694	795	62	765	45	792	132	τ(phenyl ring)
657	705	19	663	17	631	29	δ(N-H)+ω(phenyl ring)
421	575	25	572	25	562	25	ν(Ni-N)
327	319	8	362	21	269	5	ν(Ni-S)

Table 7. The experimental and calculated infrared frequencies at the level of B3LYP/6-31G(d,p), B3LYP/6-311G(d,p) and B3LYP/6-311++G(d,p) theory for **19**, at the level of B3LYP/6-31G(d,p), B3LYP/6-311G(d,p), and B3LYP /LANL2DZ theory for its Zn(II) and Ni (II) complexes with their tentative assignment

	Theoretical IR			Experimental IR		
Compound	$v(N(8)H)$	$v(N(14)H)$	$v(N(12)H)$	$v(N(8)H)$	$v(N(14)H)$	$v(N(12)H)$
5	3919	3784	3843	3366	3283	3231
6	3924	-	3885	3379	-	3169
17	3925	-	3882	3296	-	3194
19	3919	3785	3843	3298	3244	3176
9	3922	-	3874	3372	-	3179
14	3924	-	3880	3381	-	3155
18	3919	3787	3858	3366	3242	3166
11	3924	-	3897	3169	-	3402
16	3925	-	3874	3167	-	3394
20	3918	3783	3841	3309	3242	3184
10	3920	-	3873	3348	-	3177
15	3922	-	3879	3383	-	3179
Compound	$v(C=0)$	$v(C=N)$	$v(C=S)$	$v(C=0)$	$v(C=N)$	$v(C=S)$
5	1887	1842	1181	1684	1618	864
6	1890	1817	1124	1688	1595	819
17	1834	1826	1138	1664	1595	823
19	1888	1845	1230	1693	1620	863
9	1898	1828	1124	1703	1595	802
14	1849	1838	1146	1660	1595	802
18	1888	1843	1209	1693	1620	862
11	1894	1828	1121	1690	1599	814
16				1659	1595	818
20	1890	1848	1229	1693	1622	864
10	1899	1830	1124	1695	1600	816
15	1852	1841	1146	1672	1595	817

Table 8. Some important experimental and theoretical IR assignments

The frequency of the **v(**C=O) vibration shifts to lower energy by (**v** = 20-30 cm^{-1}) upon coordination as the anion ligands (the frequency of the v(C=O) vibration for compounds **5**, **19**, **18**, and **20** are 1684, 1693, 1693, 1693 cm^{-1}, whereas for their nickel (II) complexes are 1664, 1660, 1659, and 1672 cm^{-1}, which may be due to the transfer of a charge from the oxygen to the nickel (Mulliken charge on O10 atom for **19**, its zinc (II) and nickel (II) complexes are 0.359, 0.412, and -0.322, respectively). Thus, it was concluded that nickel (II) complexex co-ordinate through the oxygen of (C=O) in the indole ring, and nitrogen, and sulphur atoms, belonging to thiosemicarbazone and C=S groups. In the zinc (II) complexes, no shifting for v(C=O) vibration occurred indicated that only thiosemicarbazone moiety of the compounds **5**, **18**, **19**, and **20** ligands is coordinated in a bidentate way through N and S. In the far-infrared region for zinc (II) complexes, Zn-N, Zn-S vibrations were observed at 422, and 318 cm^{-1} for compound **6**; 430, and 329 for compound **10**; 420, and 333 cm^{-1} for compound **11**; 434, and 325 cm^{-1} for compound **9**; in nickel (II) complexes, Ni-N, Ni-O, and Ni-S were assigned at 420, 374, 316, for compound **17**; 434, 352, 20^{-1} for compound **15**; 424, 345, 329 cm^{-1} for compound **16**; 420, 346, 327 cm^{-1} for compound **14**.

3.4 Electronic spectrum

Compounds **5**, **18**, **19**, and **20** were optimized by B3LYP method with 6-31G(d,p), 6-311G(d,p), 6-311++G(d,p), and with density functional theory by using the BP86 hybrid functional with 30% HF exchange (B3P86-30%), and Stevens-Basch-Krauss pseudo potentials with polarized split valence basis sets (CEP-31G*) (Stevens et al., 1984) and theoretical calculated wave numbers of electronic transitions for compounds **5**, **18**, **19**, and **20** are reported in Tables 9 and 10. Very weak features with oscillator strengths below 0.25 were omitted.

			UV-Visible spectrum data (nm)						
Comp. 18		Experimental	252	268	278	288	296		372
	B3LYP	6-31G(d,p)	219	233	249	264	326		399
		6-311G(d,p)	209	222	236	266	326		401
		6-311++G(d,p)	214	223	227	270	332		407
	BP86	CEP-31G*	210	225	238	255	309		378-349
Comp. 5		Experimental	250	268	278	288	296	364	373
	B3LYP	6-31G(d,p)	210	236	252	267	327		406
		6-311G(d,p)	212	228 -237	254	268	328		408
		6-311++G(d,p)	216	242	259	272	334		415
	BP86	CEP-31G*		227	240	257	312		382
Comp. 19		Experimental		258		296		364	370
	B3LYP	6-31G(d,p)	207	249		254-256		355	400
		6-311G(d,p)	210	224-236		264		321	405
		6-311++G(d,p)	213	231-240-256		269		327	412
	BP86	CEP-31G*	216	253-238-226		305		352	382
Comp. 20		Experimental	250	260				368	370
	B3LYP	6-31G(d,p)	208	256-257-262				357	429-402
		6-311G(d,p)	209	251-258-266				358	427-403
		6-311++G(d,p)	209	256-261-269				363	436-409
	BP86	CEP-31G*	236	256-254-245				338	390-382

Table 9. Experimental and theoretical UV assignments for compounds **18**, **5**, **19**, and **20** calculated by using TDB3LYP with the 6-31G(d,p), 6-311G(d,p), 6-311++G(d,p), BP86-CEP-31G* basis set

The UV visible spectral values, excitation energies, and ossillator strengths were calculated at TDB3LYP levels by using 6-31G(d,p), 6-311Gs(d,p), 6-311++G(d,p) basis sets and TDBP86-CEP-31G* level for compounds **5**, **18**, **19**, and **20**. There was an agreement between the UV calculations at the level of TDBP86-CEP-31G* for ligands, and experimental results.

The C=N transitions due to $n-\pi$ for isatin-3-thiosemicarbazones were assigned as 2070 cm^{-1} (483 nm) of (Akinchan et al., 2002). This band was found to be 372, 373, 370, 370 nm for compounds **5**, **18**, **19**, and **20**, respectively. At the level of TDBP86-CEP-31G*, this band for compounds **5**, **18**, **19**, and **20** was found at 382 nm, which arises from 3-1′, 2-1′, 3-1′, and 3-1′ transitions. The bands due to $\pi-\pi^*$ transitions of semicarbazone group for isatin-3-thiosemicarbazone were between the range of 400-286 nm (Akinchan et al., 2002). Band at

305 nm at the level of TDBP86-CEP-31G* was predicted for compounds **5**, **18**, **19**, and **20**, which arises mainly from 4-1′, transitions.

The UV transitions and their excitation energies and ossillator strengths for zinc (II) and nickel (II) complexes of compounds **5**, **18**, **19**, and **20** were calculated by using TDB3LYP with the 6-31G(d,p), 6-311G(d,p), BP86-CEP-31G basis sets. The transitions for nickel (II) complexes were calculated with both multiplicity states, 1 and 3, with two unpaired electrons. The results are summarized in Tables 11-14. The UV spectra, calculated with two unpaired electrons for nickel (II) complexes, were not in agreement with the experimental results. The band assigned at around 372, due to C=N transitions, were observed at 432, and 450 nm in the UV spectra of zinc (II) and nickel (II) complexes, respectively. Band at 434 nm (experimental) for compound **18** was found to be at 436 nm in the calculated of spectra of BP86-CEP-31G. Bands located around 360 nm in zinc (II) complexes and 374 nm in nickel (II) complexes due to ligand LMCT transition, suggested a metal–sulfur bond formation (Akinchan et al., 2002). Nickel (II) complexes showed three transitions around 450, belonging to transition $^3A_{2g}\rightarrow{}^3T_{2g}$ (F), 635 including transition $A_{2g}\rightarrow{}^3T_{2g}$ (F), and 842 in transition $^3A_{2g}\rightarrow{}^3T_{1g}$ (P), these transitions are characteristic of hexacoordinated nickel (II) complexes.

Compound **18**						
6-31G(d,p)	3.11(0.28)E2	3.80(0.35)E5	4.68(0.12)E9	4.98(0.07)E13	5.30(0.07)E16	5.66(0.09)E23
6-311G(d,p)	3.09(0.30)E2	3.80(0.35)E5	4.66(0.11)E9	5.26(0.07)E16	5.58(0.12)E22	5.91(0.10)E26
6-311++G(d,p)	3.04(0.30)E2	3.73(0.32)E5	4.59(0.09)E9	5.47(0.09)E26	5.54(0.07)E28	5.80(0.08)E39
BP86-CEP-31G*	3.28(0.36)E2	3.55(0.08)E3 4.01(0.26)E5	4.86(0.14)E9	5.22(0.10)E11	5.51(0.06)E17	5.88(0.16)E21
Compound **5**						
6-31G(d,p)	3.05(0.26)E2	3.78(0.41)E5	4.65(0.11)E6	4.91(0.05)E9	5.26(0.11)E10	5.88(0.10)E18
6-311G(d,p)	3.04(0.28)E2	3.78(0.41)E5	4.63(0.11)E6	4.88(0.04)E9	5.21(0.11)E10	5.43(0.11)E11 5.84(0.09)E18
6-311++G(d,p)	2.99(0.29)E2	3.71(0.39)E5	4.56(0.10)E6	4.79(0.04)E9	5.11(0.11)E12	5.75(0.08)E31
BP86-CEP-31G*	3.24(0.40)E2	3.97(0.34)E5	4.83(0.13)E6	5.16(0.08)E8	5.46(0.10)E10	5.75(0.08)E31
Compound **19**						
6-31G(d,p)	3.10(0.27)E3	3.48(0.46)E4	4.88(0.22)E9	4.88(0.18)E10	4.97(0.06)E12	5.11(0.10)E15 5.98(0.08)E26
6-311G(d,p)	3.06(0.34)E2	3.86(0.37)E5	4.69(0.16)E10	5.26(0.08)E16	5.51(0.06)E21	5.90(0.14)E25
6-311++G(d,p)	3.01(0.34)E2	3.79 0.34)E5	4.61(0.14)E10	4.84(0.05)E14	5.16(0.06)E18	5.37(0.04)E23 5.80(0.16)E35
BP86-CEP-31G*	3.24(0.41)E2	3.52 0.07)E3	4.06(0.29)E5	4.88(0.20)E9	5.21(0.08)E14	5.50(0.08)E17 5.74(0.06)E18
Compound **20**						
6-31G(d,p)	2.80(0.06)E2	3.08(0.26)E3	3.47(0.49)E4	4.72(0.29)E9	4.81(0.08)E10	4.83(0.17)E11 5.96(0.10)E28
6-311G(d,p)	2.90(0.06)E2	3.07(0.28)E3	3.47(0.48)E4	4.65(0.29)E9	4.80(024)E12	4.93(0.06)E13 5.92(0.09)E28
6-311++G(d,p)	2.84(0.05)E2	3.03(0.28)E3	3.41(0.46)E4	4.60(0.28)E9	4.74(018)E12	4.90(0.10)E14 5.93(0.10)E41
BP86-CEP-31G*	3.17(0.12)E2	3.24(0.32)E3	3.66(0.42)E4	4.83(0.24)E9	4.88(013)E10	5.05(0.25)E11 5.24(0.10)E13

Table 10. TDB3LYP method with the 6-31G(d,p), 6-311G(d,p), 6-311++G(d,p) basis sets and TDBP86-CEP-31G* excitation energies in eV for compounds **18**, **5**, **19**, and **20**, and Oscillator Strengths (in Parenthesis)

Compound 11	-	254	268	288	296	360	434
6-31G(d,p)			249	361	364	436-441	457
6-311G(d,p)			251	363	365	436-439	453
CEP-31G			239		356-352	422-420	436
Compound 6	-	254	268	290	296	364	434
6-31G(d,p)		211	250	359	365	437	444
6-311G(d,p)		213-212	252	361	366	439-436	450
CEP-31G			241		355-350	435-423-420	
Compound 9	250	-	258	-	302	308	444
6-31G(d,p)	228		243-264		401-403	450	461-463
6-311G(d,p)	245		250-265		400-401	450-462	474-478
CEP-31G	240-234		254		388	438-428	461-457-453
Compound 10	-	-	260		300	312	444
6-31G(d,p)	249-253-256		270		406-406	455	475-479
6-311G(d,p)	252-261		271		404-404	465	481-484
CEP-31G	260-248-241-236				435-390	448-443	461-460

Table 11. Experimental and theoretical UV assignments for compounds **11**, **6**, **9**, and **10** calculated by using TDB3LYP with the 6-31G(d,p), 6-311G(d,p), BP86-CEP-31G basis set

Comp. 16	254	262	296	374	450	635	842
6-31G(d,p)	251	-	346	363-374	408-419	659	729
311G(d,p)	255	266	342-308	365-355	423-412	657-617	-
CEP-31G	248-244	293-261	345-344-300	387-384	473-445	606	-
6-311G(d,p)(1)	-	-	346-346-344	386-381	490-434-432	-	-
CEP-31G(1)	-	-	330-330	367-366	415-411	476	-
Comp. 17	244	264	292	378	450	636	832
6-31G(d,p)	247-231	252	344	375-361	459-418-406	659	731
6-311G(d,p)	229-247	265	340-353-364	409-421	456	620-658	-
CEP-31G	246-241-237	301-265	343-342-332	386-382	475-447	614	-
6-311G(d,p)(1)	-	-	386-382	403-392	486-418	-	-
CEP-31(1)	253-248-242	263	366-363-359-328-327	-	416-412	-	-
Comp. 14	-	260	300	414	435	636	826
6-31G(d,p)	-	250-252-263	280-310	374-379-392	421-430	604-652	743
6-311G(d,p)	-	253-262-270-288-305	306-311-368	368-392	432-421	622	663
CEP-31G	253-248-248-239-236	298-287-265	370-321	413-410-397	486-448	614	-
6-311G(d,p)(1)	-	-	406-342-341	447-420-419	510	-	-
CEP-31G(1)	-	288	354	425-405-401	494	-	-
[Ni(HICPT)$_2$]	-	264	300	410	-	637	883
6-31G(d,p)		286-310	374	433	-	-	747
6-311G(d,p)	309	310	374	384			666

CEP-31G	256-255- 248-239	321-294- 289-268	394-371	413-409	485	-	613
6-311G(d,p)(1)	-	-	406-374-370	448-425-422	511	-	-
CEP-31G(1)	-	288	405-401-354	425	493	-	-

(1) triplet state

Table 12. Experimental and theoretical UV assignments for compounds **16**, **17**, **14**, and **15** calculated by using TDB3LYP with the 6-31G(d,p), 6-311G(d,p), BP86-CEP-31G basis set

Compound **11**						
6-31G(d,p)	3.11(0.28)E2	3.80(0.35)E5	4.68(0.12)E9	4.98(0.07)E13	5.30(0.07)E16	5.66(0.09)E23
6-311G(d,p)	3.09(0.30)E2	3.80(0.35)E5	4.66(0.11)E9	5.26(0.07)E16	5.58(0.12)E22	5.91(0.10)E26
6-311++G(d,p)	3.04(0.30)E2	3.73(0.32)E5	4.59(0.09)E9	5.47(0.09)E26	5.54(0.07)E28	5.80(0.08)E39
CEP-31G	3.24(0.41)E2	4.06(0.29)E5	4.88(019)E9	5.20(0.08)E14	5.50(0.08)E17	5.74(0.06)E18
Compound **6**						
6-31G(d,p)	3.05(0.26)E2	3.78(0.41)E5	4.65(0.11)E6	4.91(0.05)E9	5.26(0.11)E10	5.88(0.10)E18
6-311G(d,p)	3.04(0.28)E2	3.78(0.41)E5	4.63(0.11)E6	4.88(0.04)E9	5.21(0.11)E10	5.43(0.11)E11 5.84(0.09)E18
6-311++G(d,p)	2.99(0.29)E2	3.71(0.39)E5	4.56(0.10)E6	4.79(0.04)E9	5.11(0.11)E12	5.75(0.08)E31
CEP-31G	3.24(0.40)E2	3.97(0.34)E5	4.83(0.13)E6	5.16(0.08)E8	5.46(0.10)E10	5.75(0.08)E31
Compound **9**						
6-31G(d,p)	3.10(0.27)E3	3.48(0.46)E4	4.88(0.22)E9	4.88(0.18)E10	4.97(0.06)E12	5.11(0.10)E15 5.98(0.08)E26
6-311G(d,p)	3.06(0.34)E2	3.86(0.37)E5	4.69(0.16)E10	5.26(0.08)E16	5.51(0.06)E21	5.90(0.14)E25
6-311++G(d,p)	3.01(0.34)E2	3.79 0.34)E5	4.61(0.14)E10	4.84(0.05)E14	5.16(0.06)E18	5.37(0.04)E23 5.80(0.16)E35
CEP-31G	3.24(0.41)E2	3.52 0.07)E3	4.06(0.29)E5	4.88(0.20)E9	5.21(0.08)E14	5.50(0.08)E17 5.74(0.06)E18
Compound **10**						
6-31G(d,p)	2.88(0.06)E2	3.08(0.26)E3	3.47(0.49)E4	4.72(0.29)E9	4.81(0.08)E10	4.83(0.17)E11 5.96(0.10)E28
6-311G(d,p)	2.90(0.06)E2	3.07(0.28)E3	3.47(0.48)E4	4.65(0.29)E9	4.80(024)E12	4.93(0.06)E13 5.92(0.09)E28
6-311++G(d,p)	2.84(0.05)E2	3.03(0.28)E3	3.41(0.46)E4	4.60(0.28)E9	4.74(018)E12	4.90(0.10)E14 5.93(0.10)E41
CEP-31G	3.17(0.12)E2	3.24(0.32)E3	3.66(0.42)E4	4.83(0.24)E9	4.88(013)E10	5.05(0.25)E11 5.24(0.10)E13

Table 13. TDB3LYP method with the 6-31G(d,p), 6-311G(d,p), 6-311++G(d,p) basis sets and TDBP86-CEP-31G excitation energies in eV for compounds **11**, **6**, **9**, **10**, and Oscillator Strengths (in Parenthesis)

Compound **16**						
6-31G(d,p)	1.70(0.06)E3 1.88(0.02)E4	2.95(0.12)E11	3.31(015)E13	3.40(0.14)E16 3.41(0.24)E19	3.58(0.10)E22	4.92(0.21)E76
311G(d,p)	1.88(0.03)E3 2.01(0.07)E4	2.93(0.11)E12	3.01(011)E13	3.39(0.26)E18 3.49(0.09)E21	3.62(0.11)E24 4.03(0.08)E33	4.65(0.26)E60 4.92(0.21)E76
CEP-31	2.04(0.08)E5	2.62(0.06)E8 2.79(0.07)E9	3.20(013)E12 3.22(012)E13	3.58(0.25)E17 3.60(0.18)E18	4.12(0.08)E27 4.23(0.10)E29	4.74(0.08)E45 4.98(0.18)E53 5.07(0.30)E58
6-311G(d,p)(1)	2.53(0.06)E13	2.86(0.15)E19	2.87(007)E20 3.21(0.12)E28 3.24(0.07)E30	3.58(0.19)E36 3.58(0.08)E37	3.60(0.10)E39	3.82(0.10)E51

CEP-31[1]	2.60(0.10)E10	2.98(0.14)E19	3.01(0.08)E20	3.37(0.18)E27	3.38(0.06)E28	3.75(0.15)E39 3.76(0.20)E40
Compound **17**						
6-31G(d,p)	1.70(0.06)E3 1.88(0.02)E4	2.70(0.06)E9 2.96(0.13)E11	3.05(0.09)E13 3.29(0.15)E16 3.60(0.18)E22	4.91(0.23)E54	5.00(0.18)E57	5.34(0.17)E67
311G(d,p)	1.88(0.04)E3 1.99(0.06)E4	2.71(0.06)E9 3.02(013)E11	3.03(0.10)E13 3.38(0.17)E18 3.54(0.12)E21	3.64(0.11)E24	4.67(0.18)E47	5.02(0.27)E57 5.41(0.10)E69
CEP-31	2.02(0.08)E5	2.61(0.06)E8 2.77(0.06)E9	3.21(0.14)E12 3.24(0.10)E13 3.61(0.20)E17	3.62(0.18)E18 3.73(0.09)E20 4.11(0.08)E27	4.67(0.09)E37	5.03(0.42)E46 5.14(0.13)E49 5.21(0.14)E52
6-311G(d,p)[1]	2.55(0.07)E15	2.96(0.06)E26	3.07(0.08)E30 3.14(0.08)E32	3.27(0.05)E36 3.20(0.06)E38	3.62(0.22)E47	3.64(0.77)E48
CEP-31[1]	3.24(014)E19 3.19(008)E20	3.19(0.11)E25 3.41(0.05)E28	3.18(0.10)E30 3.06(0.19)E37 3.0(0.18)E38	3.20(0.06)E62 3.06(0.17)E67	4.99(0.06)E71	5.10(0.17)E76
Compound **14**						
6-31G(d,p)	1.67(0.08)E3 1.90(0.02)E5 2.05(0.01)E7	2.88(0.21)E11 2.95(0.09)E13	3.16(0.13)E17 3.26(0.12)E19 3.31(0.11)E21	4.00(0.11)E33 4.42(0.12)E50 4.70(0.09)E61	4.91(0.12)E72	4.97(0.19)E75
311G(d,p)	1.87(0.07)E3 1.99(0.04)E4	2.72(0.06)E9 2.88(0.19)E12	2.99(0.10)E13 3.15(0.16)E18 3.37(0.22)E23	3.98(0.09)E33 4.05(0.09)E37	4.29(0.11)E44	4.59(0.10)E57 4.72(0.08)E62 4.90(0.33)E73
CEP-31	2.02(0.10)E5 2.55(0.07)E8	2.77(0.05)E9 2.99(0.16)E11 3.02(0.32)E12	3.12(0.12)E15 3.35(0.07)E16	3.85(0.13)E25 4.15(0.11)E29 4.30(0.14)E33	4.67(0.10)E43 4.89(0.30)E51 4.99(0.12)E56 4.99(0.17)E57	5.13(0.08)E61 5.16(0.16)E62 5.23(0.35)E68
6-311G(d,p)[1]	2.43(0.10)E13	3.26(0.10)E19	2.95(0.25)E27	2.96(0.29)E28	3.05(0.12)E29	
CEP-31[1]	2.51(0.09)E10	2.91(0.09)E21	3.06(0.36)E23	3.09(0.38)E24	3.50(0.06)E36	4.31(0.11)E58
Compound **15**						
6-31G(d,p)	1.66(0.08)E3	2.86(0.22)E11	3.30(0.16)E21	3.99(0.13)E34	4.32(0.13)E48	4.66(0.10)E63 4.70(0.12)E67
311G(d,p)	1.86(0.06)E5	3.22(0.22)E18	3.30(0.16)E21	3.57(0.10)E23	3.99(0.13)E34	4.00(0.34)E37
CEP-31	2.02(0.10)E5	2.56(0.08)E8 3.00(0.17)E11 3.02(0.34)E12	3.14(0.08)E15 3.33(0.10)E16 3.86(0.16)E25	4.15(0.11)E29 4.28(0.14)E31 4.61(0.07)E41	4.83(0.06)E49 4.85(0.12)E51 4.87(0.36)E52	4.95(0.10)E57 4.99(0.11)E59 5.17(0.31)E65
6-311G(d,p)[1]	2.42(0.10)E13	3.27(0.11)E19	2.91(0.24)E27	3.42(0.34)E28	3.04(0.08)E29	3.31(0.09)E33
CEP-31[1]	2.51(0.09)E10	3.51(0.09)E10	2.91(0.09)E21	3.06(0.36)E23	3.08(0.38)E24	3.04(0.06)E36

[1] triplet state

Table 14. TDB3LYP method with the 6-31G(d,p), 6-311G(d,p), 6-311++G(d,p) basis sets and TDBP86-CEP-31G excitation energies in eV for compounds **16**, **17**, **14**, **15**, and Oscillator Strengths (in Parenthesis)

3.5 NMR spectra

Experimental and calculated theoretical NMR parameters for compounds **5**, **18**, **19**, and **20** were shown in Table 15. Geometrical optimization of studied ligands were performed at the level of B3LYP/6-311G(d,p), and B3LYP/6-311++G(d,p). NMR Chemical shifts were predicted from Gauge-Independent Atomic Orbital (GIAO) calculations at the same level of theory at DMSO solution with both gas phase optimization and DMSO optimization. Theoretically calculated ^{1}H-NMR parameters are generally in agreement with experimental results, except deviation in N8-H of isatin ring, and N14-H of thiosemicarbazone group.

The correlations between theoretical, and experimental NMR results, calculated with B3LYP/6-311, B3LYP/6-311++ level of theory are 97.6%, and 98.2% and at gas phase; 96.9%, 96.4%, in DMSO solution (but optimization was in gas phase), and at the B3LYP/6-311++ level of theory 96.9% in DMSO solution (and also optimisation in DMSO solution) for compound **18**. Although the experimental proton signal of N12-H of isatin group for compounds **5**, **18**, **19**, and **20** was 12.64, 12.57, 12.81 and 12.85, and theoretical one was δ 12.37, 12.21, 12.55, 12.51 in the DMSO solution at the level of B3LYP/6-311++G/dip) theory.

		6-311-G(d,p)		6-311++G(d,p)		6-311++G(d,p)	Exp,
	Atoms	gas	DMSO[a]	gas	DMSO[a]	DMSO	
Compound 18	C1-H	7.30	7.53	7.24	7.66	7.55	7.01
	C2-H	6.99	7.13	7.01	7.31	7.17	7.31
	C3-H	6.76	7.08	6.78	7.30	7.11	6.91
	C4-H	7.38	7.49	7.53	7.63	7.53	7.63
	C21-H	4.53	4.51	4.67	4.71	4.52	4.86
	C21-H	4.53	4.51	4.67	4.71	4.52	
	C25-H	7.66	7.78	7.75	8.06	7.80	7.32-7.36
	C26-H	7.66	7.78	7.75	8.05	7.80	7.32-7.36
	C27-H	7.62	7.76	7.75	8.05	7.77	7.32-7.36
	C29-H	7.62	7.76	7.72	8.05	7.77	7.32-7.36
	C31-H	7.58	7.74	7.54	7.88	7.75	7.23
	N8-H	6.16	6.80	6.43	7.86	6.88	9.77
	N12-H	12.27	12.17	12.62	12.46	12.37	12.64
	N14-H	6.84	7.23	7.17	7.44	7.16	11.18
Compound 5	C1-H	7.37	7.60	7.40	7.82	7.62	7.10
	C2-H	7.14	7.30	7.18	7.50	7.33	7.36
	C3-H	6.80	7.11	6.77	7.27	7.14	6.93
	C4-H	7.66	7.81	7.74	7.98	7.85	7.74
	C24-H	4.42	4.22	4.50	4.41	4.35	4.19
	C26-H	2.36	2.12	2.39	2.18	1.98	1.92
	C26-H	0.98	1.23	1.00	1.27	1.31	1.54-1.44
	C27-H	1.53	1.50	1.53	1.52	1.53	1.35-1.26
	C27-H	1.77	1.81	1.79	1.85	1.84	1.77
	C31-H	1.25	1.34	1.29	1.39	1.33	1.14
	C31-H	1.69	1.68	1.70	1.71	1.70	1.64
	C29-H	1.54	1.52	1.55	1.56	1.54	1.35-1.26
	C29-H	1.83	1.88	1.82	1.88	1.88	1.77
	C25-H	1.98	1.96	2.03	2.04	1.92	1.92
	C25-H	1.31	1.55	1.36	1.60	1.45	1.54-1.44
	N8-H	6.18	6.81	6.47	7.90	6.89	8.85
	N12-H	12.19	12.05	12.41	12.23	12.21	12.57
	N14-H	7.06	7.55	7.28	7.70	7.44	11.20

Compound 19	C1-H	7.39	7.63	7.46	7.87	7.64	7.12
	C2-H	7.15	7.29	7.21	7.51	7.33	7.38
	C3-H	6.83	7.14	6.78	7.29	7.18	6.95
	C4-H	7.69	7.84	7.71	8.03	7.88	7.78
	C25-H	7.47	7.56	7.58	7.86	7.58	7.63
	C26-H	7.47	7.56	7.58	7.86	7.81	7.63
	C27-H	7.61	7.80	7.67	8.04	7.80	7.44
	C29-H	7.61	7.80	7.67	8.04	7.81	7.44
	C31-H	7.58	7.79	7.56	7.93	7.58	7.28
	N8-H	6.22	6.86	6.46	7.89	6.94	10.83
	N12-H	12.40	12.32	12.61	12.45	12.55	12.81
	N14-H	8.65	9.08	8.80	9.45	9.03	11.26
Compound 20	C1-H	7.43	7.63	7.34	7.73	7.65	7.12
	C2-H	9.42	7.33	7.22	7.52	7.37	7.38
	C3-H	6.85	7.14	6.81	7.30	7.17	6.95
	C4-H	7.78	7.95	7.88	8.11	7.98	7.77
	C26-H	6.80	7.24	6.90	7.44	7.24	7.67
	C29-H	7.33	7.55	7.43	7.87	7.55	7.49
	C27-H	7.46	7.54	7.55	7.85	7.57	7.49
	C25-H	9.91	9.75	9.98	9.84	9.77	7.67
	N8-H	6.25	6.86	6.54	7.94	6.94	10.86
	N12-H	12.40	12.33	12.71	12.58	12.51	12.85
	N14-H	9.42	9.77	9.76	10.10	9.74	11.27

[a] Optimisation were calculated in the gas phase and Theoretical NMR shifting were calculated in the DMSO solution

Table 15. Experimental and theoretical NMR shifting for compounds **18**, **5**, **19**, and **20** calculated by using B3LYP with the 6-311G(d,p), 6-311++G(d,p) basis set for gas and DMSO phase

The proton signal of N8-H group seen at δ 9.77, 8.85, 10.83, 10.88, respectively, in the ^{1}H-NMR spectrum for compounds **5**, **18**, **19**, and **20,** also appeared in the spectrum of their zinc (II) complexes at δ 10.81, 10.80, 11.01, and 11.06, respectively. The upfield shift was due to the lack of intermolecular hydrogen bonding in its corresponding complex (Akinchan et al., 2002). A stronger hydrogen-bond interaction will shorten the O-H distance, will elongate the N–H distance, and will also cause a significant deshielding of the proton leading to a further downfield NMR signal (De Silva et al., 2007). The peak due to N12-H in the ^{1}H-NMR spectra of compounds **5**, **18**, **19**, and **20** disappeared in their corresponding complexes.

The experimental and theoretical NMR results of compounds **11**, **6**, **9**, and **10** are shown in Table 16. Geometrical optimizations of above mentioned studied complexes were performed by using the B3LYP method, and 6-31G(d,p), 6-311G(d,p) standard basis sets at the gas phase, and DMSO solution. The calculated peak, due to N12-H in the ^{1}H-NMR spectra of compounds **5**, **18**, **19**, and **20** disappears in their corresponding zinc (II) complexes, as the calculation was performed by considering the corresponding ligand, deprotonated from N12 belong to thiosemicarbazone group in its complex form.

		6-311G(d,p)		6-31G(d,p)	6-311G(d,p)	Experimental
	Atoms	gas	DMSO[a]	gas	DMSO	
Compound 11	C1-H	7.63	7.21	7.79	-	6.95-7.40
	C2-H	6.97	6.84	7.23	-	
	C3-H	7.11	6.55	7.27	-	
	C4-H	7.53	7.68	8.02	-	
	C25-H	7.80	7.77	7.97	-	
	C29-H	7.95	7.66	8.09	-	
	C31-H	7.84	7.53	7.99	-	
	C27-H	7.86	7.54	8.01	-	
	C22-H	7.78	7.47	7.88	-	
	C21-H	6.15	6.41	6.25	-	4.76
	C21-H	4.17	4.00	4.30	-	
	N8-H	7.38	5.98	8.11	-	10.81
	N14-H	6.33	5.33	6.91	-	9.26
Compound 6	C1-H	7.14	7.14	7.73	7.38	6.90-8.30
	C2-H	6.73	6.73	7.17	6.85	
	C3-H	6.51	6.51	7.23	6.89	
	C4-H	7.43	7.43	7.82	7.43	
	C24-H	4.80	4.80	4.98	4.64	1.15-2.34
	C26-H	2.41	2.41	2.34	2.08	
	C26-H	1.05	1.05	1.52	1.26	
	C27-H	1.69	1.69	1.96	1.76	
	C27-H	1.72	1.72	1.81	1.51	
	C31-H	1.69	1.69	1.87	1.69	
	C31-H	1.25	1.25	1.59	1.31	
	C29-H	1.82	1.82	2.06	1.90	
	C29-H	1.64	1.64	1.85	1.59	
	C25-H	1.24	1.24	1.69	1.40	
	C25-H	2.03	2.03	2.21	2.06	
	N8-H	5.97	5.97	8.06	6.69	10.80
	N14-H	5.30	5.30	6.83	5.65	8.77
Compound 9	C1-H	7.61	7.79	7.43	7.43	7.01-8.11
	C2-H	6.93	7.22	6.84	6.84	
	C3-H	7.13	7.31	6.98	6.98	
	C4-H	7.60	8.08	7.60	7.60	
	C26-H	7.49	7.51	7.24	7.24	
	C29-H	7.83	7.97	7.67	7.67	
	C31-H	7.56	7.76	7.43	7.43	
	C27-H	7.94	8.11	7.83	7.83	
	C25-H	10.10	10.22	10.14	10.14	
	N8-H	8.26	8.63	7.63	7.63	11.01
	N14-H	7.55	8.27	6.90	6.90	10.68

Compound 10	C1-H	7.61	7.22	7.81	7.43	7.03-8.08
	C2-H	6.92	6.74	7.21	6.82	
	C3-H	7.13	6.59	7.32	6.97	
	C4-H	7.54	7.70	8.01	7.64	
	C25-H	7.43	6.77	7.51	7.15	
	C29-H	7.67	7.29	7.83	7.48	
	C27-H	7.79	7.63	7.97	7.65	
	C26-H	10.09	10.21	10.28	10.13	
	N8-H	8.20	7.25	8.59	7.55	11.06
	N14-H	7.55	6.15	8.28	6.88	10.74

[a] Optimisation were calculated in the gas phase and Theoretical NMR shifting were calculated in the DMSO solution

Table 16. Experimental and theoretical NMR shifting for compounds **11**, **6**, **9**, and **10** calculated by using B3LYP with the 6-31G(d,p), 6-311G(d,p) basis set for gas and DMSO phase

3.6 Antibacterial activity

All the compounds **1-16** were tested against the two Gram-positive bacterial strains i.e. *Staphlococcus aureus* and *Bacillus subtilus,* and four Gram-negative bacterial strains i.e. *Escherichia coli. Shigella flexnari, Pseudomonas aeruginosa,* and *Salmonella typhi,* according to the literature protocol (Becke, 1993). The results were compared with the standard drug imipenem and shown in Table 17.

Compound Name	Escherichia coli	Bacillus subtilus	Shigella flexnari	Staphlococcus aureus	Pseudomonas aeruginosa	Salmonella typhi
1	00	00	00	00	00	00
2	00	00	00	00	00	00
3	00	00	00	00	00	00
4	00	00	00	00	00	00
5	00	00	00	00	00	00
6	00	00	00	00	00	00
7	00	00	00	00	00	00
8	00	00	00	00	00	00
9	00	00	00	00	00	00
10	00	00	00	00	00	00
11	00	00	00	00	00	00
12	00	00	00	00	00	00
13	00	00	00	00	00	00
14	00	00	00	00	00	00
15	00	00	00	00	00	00
16	00	00	00	00	00	00
SD* Imipenem	30	33	27	33	24	25

Table 17. Antibacterial activity for Escherichia coli, Shigella flexnari, Pseudomonas aeruginosa, and Salmonella typhi

3.7 Antifungal activity

All the compounds **1-16** were also tested against *Candida albicans, Aspergillus falvus, Microsporum canis, Fusarium solani,* and *Candida galbrata,* according to the literature protocol (Lee et al., 1988) and found that compounds showed a varying degree of percentage inhibition. These results were compared with the standard drugs miconazole and amphotericin B, as shown in Table 18.

Compound Name	**1**	**2**	**3**	**4**	**5**	**6**	**7**	**8**	SD
C. albicans	0	0	0	0	0	0	0	0	110
A. flavus	20	0	20	0	40	0	0	0	20
M. canis	50	60	00	60	30	0	0	80	98
F. Solani	20	0	30	40	10	0	25	30	73
C. glabrata	0	0	0	0	0	0	0	0	101
Compound Name	**9**	**10**	**11**	**12**	**13**	**14**	**15**	**16**	SD
C. albicans	0	40	0	0	0	0	0	0	110
A. flavus	30	0	20	0	20	0	0	0	20
M. canis	20	0	20	20	70	50	20	50	98
F. Solani	05	0	0	40	40	20	0	0	73
C. glabrata	0	0	0	0	0	0	0	0	101

Table 18. Antifungal activity for Candida albicans, Aspergillus falvus, Microsporum canis, Fusarium solani, and Candida galbrata

Compounds **1**, **14**, **16**, **8**, and **13** exhibited activity against *Microsporum canis* with 50, 50, 50, 80, and 70% inhibition, respectively. Compounds **4**, **12**, and **13** showed a moderate inhibition against *Fusarium solani* with a percentage inhibition of 40, 40, and 40%, respectively. Compound **5** demonstrated activity an against *Aspergillus flavus* (40% inhibition). Compound **10** was found to be active against Candia albicans with 40% inhibition.

Table 18 shows results of antifungal assay on compounds **1-16** (concentration used 200 μg/mL of DMSO, and Percentage Inhibition). All the compounds in this series showed non-significant activity against all the bacteria.

4. Acknowledgement

The financial support for this study was provided by the TUBİTAK Fund (Project Number: 108T974), and the High Education Commission of Pakistan.

5. References

Agrawal, K.C. & Sartorelli, A.C. (1978). The chemistry and biological activity of α-(N)-heterocyclic carboxaldehyde Thiosemicarbazones. *Prog. Med. Chem.*, Vol. 15, pp. (321-356), ISSN: 0079-6468.

Ajaiyeoba, E.O.; Rahman, A. & Choudhary, M.I. (1988). Preliminary antifungal and cytotoxicity studies of extracts of Ritchiea capparoides var. Longipedicellata. *J. Ethnopharm.*, Vol: 62, No. 3, pp. (243-246), ISSN: 0378-8741.

Akinchan, N.T.; Drozdzewski, P.M. & Holzer, W. (2002). Syntheses and spectroscopic studies on zinc(II) and mercury(II) complexes of isatin-3-thiosemicarbazone. *J. Mol. Struct.*, Vol. 641, No. 1, pp. (17-22), ISSN: 0022-2860.

Bal, T.R.; Anand, B.; Yogeeswari, P. & Sriram, D. (2005). Synthesis and evaluation of anti-HIV activity of isatin β-thiosemicarbazone derivatives. *Bioorg. Med. Chem. Lett.*, Vol. 15, No. 20, pp. (4451-4455), ISSN: 0960-894X.

Becke, A.D. (1993). Density-functional thermochemistry. III. The role of exact exchange. *J. Chem. Phys.*, Vol. 98, No. 7, pp. (5648-5652), ISSN: 0021-9606.

Beraldo, H. & Tosi, L. (1983). Spectroscopic studies of metal complexes containing п-delocalized sulfur ligands. The resonance raman spectra of the iron(II) and iron (III) complexes of the antitumor agent 2-formylpyridin thiosemicarbozone. *Inorg. Chim. Acta.*, Vol. 75, pp. (249-257), ISSN: 0020-1693.

Beraldo, H. & Tosi, L. (1986). Spectroscopic studies of metal complexes containing п-delocalized sulfur ligands. The pre-resonance Raman spectra of the antitumor agent 2-formylpyridine thiosemicarbazone and its Cu(II) and Zn(II) complexes. *Inorg. Chim. Acta.*, Vol. 125, No. 3, pp. (173-182), ISSN: 0020-1693.

Borges, R.H.U.; Paniago, E.; Beraldo, H. (1997). Equilibrium and kinetic studies of iron(II) and iron(III) complexes of some α(N)-heterocyclic thiosemicarbazones. Reduction of the iron(III) complexes of 2-formylpyridine thiosemicarbazone and 2-acetylpyridine thiosemicarbazone by cellular thiol-like reducing agents. *J. Inorg. Biochem.*, Vol. 65, No. 4, pp. (267-275), ISSN: 0162-0134.

Bresolin, L.; Burrow, R.A.; Hörner, M.; Bermejo, E. & Castineiras A. (1997). Synthesis and crystal structure of di [(μ-acetato)(2-acetylpyridine[4] N-ethylthiosemicarbazonato)zinc(II)]. *Polyhedron,* Vol. 16, No. 23, pp. (3947-3951), ISSN: 0277-5387.

Boon, R. (1997). Antiviral treatment: from concept to reality. *Antiviral Chem. Chemother.*, Vol: 8, pp. (5-10), ISSN: 0956-3202.

Casas, J.S.; Castineiras, A.; Sanche,z A.; Sordo, J.; Vazquez-Lopez, A.; Rodriguez- Argiuelles, M.C. & Russo U. (1994). Synthesis and spectroscopic properties of diorganotin(IV) derivatives of 2,6-diacetylpyridine bis(thiosemicarbazone). Crystal structure of diphenyl{2,6-diacetylpyridine bis(thiosemicarbazonato)}tin(IV) bis(dimethylformamide) solvate. *Inorg. Chem. Acta.*, Vol. 221, no. 1-2, pp (61-68), ISSN: 0020-1693.

Casas, J.S.; García-Tasende, M.S.; Maichle-Mössmer, C.; Rodríguez-Argüelles, M.C.; Sánchez, A.; Sordo, J.; Vázquez-López, A.; Pinelli, S.; Lunghi, P. & Albertini, R. (1996). Synthesis, structure, and spectroscopic properties of acetato (dimethyl) (pyridine-2-carbaldehydethiosemicarbazonato)tin(IV) acetic acid solvate, [$SnMe_2$ (PyTSC)(OAc)].HOAc. Comparison of its biological activity with that of some structurally related diorganotin(IV) bis(thiosemicarbazonates). *J. Inorg. Biochem.*, Vol. 62, No. 1, pp. (41-55), ISSN: 0162-0134.

Casas, J.S.; Castiñeiras, A.; Rodríguez-Argüelles, M.C.; Sánchez, A.; Sordo, J.; Vázquez-López, A. & Vázquez-López, E.M. (2000). Reactions of diorganotin(IV) oxides with isatin 3- and 2-thiosemicarbazones and with isatin 2,3-bis(thiosemicarbazone): influence of diphenyldithiophosphinic acid (isatin = 1*H*-indole-2,3-dione). *J. Chem. Soc. Dalton Trans.*, Vol. 2000, No. 22, pp. (4056-4063), ISSN: 1477-9226.

Daisley, R.W. & Shah, V.K. (1984). Synthesis and antibacterial activity of some 5-Nitro-3-phenyliminoindol-2(3*H*)-ones and their *N*-mannich bases. *J. Pharm. Sci.*, Vol. 73, No. 3, pp. (407-409), ISSN: 1520-6017.

De Silva, N.W.S.V.N. & Albu, T.V. (2007). A theoretical investigation on the isomerism and the NMR properties of thiosemicarbazones. *Central Eur. J. Chem.*, Vol. 5, No. 2, pp. (396-419), ISSN: 1644-3624.

Frisch, M.J.; Trucks, G.W.; Schlegel, H.B.; Scuseria, G.E.; Robb, M.A.; Cheeseman, J.R.; Montgomery, J.A.; Vreven, T.; Kudin, K.N.; Burant, J.C.; Millam, J.M.; Iyengar, S.S.; Tomasi, J.; Barone, V.; Mennucci, B.; Cossi, M.; Scalmani, G.; Rega, N.; Petersson, G.A.; Nakatsuji, H.; Hada, M.; Ehara, M.; Toyota, K.; Fukuda, R.; Hasegawa, J.; Ishida, M.; Nakajima, T.; Honda, Y.; Kitao, O.; Nakai, H.; Klene, M.; Li, X.; Knox, J.E.; Hratchian, H.P.; Cross, J.B.; Bakken, V.; Adamo, C.; Jaramillo, J.; Gomperts, R.; Stratmann, R.E.; Yazyev, O.; Austin, A.J.; Cammi, R.; Pomelli, C.; Ochterski, J.W.; Ayala, P.Y.; Morokuma, K.; Voth, G.A.; Salvador, P.; Dannenberg, J.J.; Zakrzewski, V.G.; Dapprich, S.; Daniels, A.D.; Strain, M.C.; Farkas, O.; Malick, D.K.; Rabuck, A.D.; Raghavachari, K.; Foresman, J.B.; Ortiz, J.V.; Cui, Q.; Baboul, A.G.; Clifford, S.; Cioslowski, J.; Stefanov, B.B.; Liu, G.; Liashenko, A.; Piskorz, P.; Komaromi, I.; Martin, R.L.; Fox, D.J.; Keith, T.; Al-Laham, M.A.; Peng, C.Y.; Nanayakkara, A.; Challacombe, M.; Gill, P.M.W.; Johnson, B.; Chen, W.; Wong, M.W.; Gonzalez, C. & Pople, J.A. (2004). Gaussian 03; Revision B.05, *Gaussian; Inc.*, Wallingford CT.

Gunesdogdu-Sagdinc, S.; Köksoy, B.; Kandemirli, F. & Bayari, S.H. (2009). Theoretical and spectroscopic studies of 5-fluoro-isatin-3-(*N*-benzylthiosemicarbazone) and its zinc(II) complex. *J. Mol. Struct.*, Vol. 917, No. 2-3, pp. (63-70), ISSN: 0022-2860.

Hill, M.G.; Mann, K.R.; Miller, L.L. & Penneau, J.F. (1992). Oligothiophene cation radical dimers. An alternative to bipolarons in oxidized polythiophene. *J. Am. Chem. Soc.*, Vol. 114, No. 7, pp. (2728-2730), ISSN: 0002-7863.

Gorelsky, S.I. & Lever, A.B.P. (2001). Electronic structure and spectra of ruthenium diimine complexes by density functional theory and INDO/S. Comparison of the two methods. *J. Organomet. Chem.*, Vol. 635, No. 1-2, pp. (187-196), ISSN: 0022-328X.

Gorelsky, S.I. (2009). AOMix: Program for Molecular Orbital Analysis, *University of Ottawa*, http://www.sg-chem.net/.

Ivanov, V.E.; Tihomirova, N.G. & Tomchin, A.B. (1988). *Zh. Obshch. Khim.*, Vol. 58, No. , pp. (2737-2743), ISSN: 1070-3632.

Kandemirli, F.; Arslan, T.; Karadayı, N.; Ebenso, E.E. & Köksoy, B. (2009). Synthesis and theoretical study of 5-methoxyisatin-3-(N-cyclohexyl)thiosemicarbazone and its Ni(II) and Zn(II) complexes. *J. Mol. Struct.*, Vol. 938, No. 1-3, pp. (89-96), ISSN: 0022-2860.

Kandemirli, F.; Arslan, T.; Koksoy, B. & Yılmaz, M. (2009). Synthesis, Characterization and Theoretical Calculations of 5-Methoxyisatin-3-thiosemicarbazone Derivatives. *J. Chem. Soc. Pakistan*, Vol. 31, No. , pp. (498-504), ISSN: 0253-5106.

Karali, N.; Gürsoy, A.; Kandemirli, F.; Shvets, N.; Kaynak, F.B.; Özbey, S.; Kovalishyn, V. & Dimoglo. A (2007). Synthesis and structure–antituberculosis activity relationship of 1*H*-indole-2,3-dione derivatives. *Bioorg. Med. Chem.*, Vol. 15, No. 17, pp. (5888-5904), ISSN: 0968-0896.

Lee, C.; Yang, W. & Parr, R.G. (1988). Development of the Colle-Salvetti correlation-energy formula into a functional of the electron density. *Phys. Rev. B*, Vol. 37, No. 2, pp. (785-789), ISSN: 1098-0121.

Medvedev, A.E.; Sandler, M. & Glover, V. (1998). Interaction of isatin with type-A natriuretic peptide receptor: possible mechanism. *Life Sci.*, Vol. 62, No. 26, pp. (2391-2398), ISSN: 0024-3205.

Pandeya, S.N. & Dimmock, J.R. (1993). Recent evaluations of thiosemicarbazones and semicarbazones and related compounds for antineoplastic and anticonvulsant activities. *Pharmazie*, Vol. 48, No. 9, pp. (659-666), ISSN: 0031-7144.

Pandeya, S.N.; Sriram, D.; De Clercq, E.; Pannecouque, C. & Witvrouw, M. (1998). Anti-HIV activity of some mannich bases of Lsatin derivatives. *Ind. J. Pharm. Sci.*, Vol. 60, No. 4, pp. (207-212), ISSN: 0250-474X.

Pandeya, S.N.; Sriram D; Nath G; De Clercq E (1999). Synthesis, antibacterial, antifungal and anti-HIV activities of Schiff and Mannich bases derived from isatin derivatives and *N*-[4-(4'-chlorophenyl)thiazol-2-yl] thiosemicarbazide. *Eur. J. Pharm. Sci.*, Vol. 9, No. 1, pp. (25-31), ISSN: 0928-0987.

Pirrung, M.C.; Pansare, S.V.; Sarma, K.D.; Keith, K.A. & Kern, E.R. (2005). Combinatorial Optimization of Isatin-β-Thiosemicarbazones as Anti-poxvirus Agents. *J. Med. Chem.*, Vol. 48, No. 8, pp. (3045-3050), ISSN: 0022-2623.

Piscopo, B.; Diumo, M.V.; Godliardi, R.; Cucciniello, M. & Veneruso, G, (1987). Studies on heterocyclic compounds Indole-2,3-dione derivatives variously substituted hydrazones with antimicrobial activity. *Boll. Soc. Ital. Biol. Sper.*, Vol. 63, No. , pp. (827-830), ISSN: 0037-8771.

Rahman, A.; Choudhary, M.I. & Thomsen, W.J. (2001). Antibacterial Assays, *Bioassay Techniques for Drug Development*, Rahman, A., pp. (14-22), Harwood Academic Publishers, ISBN: 90-5823-051-1, The Netherlands.

Rejane LL; Teixeira LRS; Carneiro TMG; Beraldo H (1999) Nickel(II), Copper(I) and Copper(II) Complexes of Bidentate Heterocyclic Thiosemicarbazones. *J. Braz. Chem. Soc.*, Vol. 10, No. 3, pp. (184-188), ISSN: 0103-5053.

Rodríguez-Argüelles, M.C.; Sánchez, A.; Ferrari, M.B.; Fava, G.G.; Pelizzi, C.; Pelosi, G.; Albertini, R.; Lunghi, P. & Pinelli, S. (1999). Transition-metal complexes of isatin-β-thiosemicarbazone. X-ray crystal structure of two nickel complexes. *J. Inorg. Biochem.*, Vol. 73, No. 1-2, pp. (7-15), ISSN: 0162-0134.

Ronen, D.; Sherman, L.; Bar-Nun, S. & Teitz, Y. (1987). N-methylisatin-beta-4',4'-diethylthiosemicarbazone, an inhibitor of Moloney leukemia virus protein production: characterization and in vitro translation of viral mRNA. *Antimicrob. Agents Ch.*, Vol. 31, No. 11, pp. (1798-1802), ISSN: 0066-4804.

Sherman, L.; Edelstein, F.; Shtacher, G.; Avramoff, M. & Teitz, Y. (1980). Inhibition of Moloney Leukaemia Virus Production by *N*-methylisatin-β-4':4'-diethylthiosemicarbazone. *J. Gen. Virol.*, Vol. 46, No. 1, pp. (195-203), ISSN: 0022-1317.

Stevens, W.J.; Basch, H. & Krauss, M. (1984). Compact effective potentials and efficient shared-exponent basis sets for the first- and second-row atoms. *J. Chem. Phys.*, Vol. 81, No. 12, pp. (6026-6033), ISSN: 0021-9606.

Elementary Molecular Mechanisms of the Spontaneous Point Mutations in DNA: A Novel Quantum-Chemical Insight into the Classical Understanding

Ol'ha O. Brovarets'[1,2,3], Iryna M. Kolomiets'[1] and Dmytro M. Hovorun[1,2,3]

[1]Department of Molecular and Quantum Biophysics,
Institute of Molecular Biology and Genetics,
National Academy of Sciences of Ukraine, Kyiv,
[2]Research and Educational Center
"State Key Laboratory of Molecular and Cell Biology", Kyiv,
[3]Department of Molecular Biology, Biotechnology and Biophysics,
Institute of High Technologies, Taras Shevchenko National University of Kyiv, Kyiv,
Ukraine

1. Introduction

DNA replication is an amazing biological phenomenon that is essential to the continuation of life (Kornberg & Baker, 1992). Faithful replication of DNA molecules by DNA polymerases is essential for genome integrity and stable transmission of genetic information in all living organisms. Although DNA replicates with immensely high fidelity, upon assembly of millions of nucleotides a DNA polymerase can make mistakes that are a major source of DNA mismatches. The overall accuracy and error spectrum of a DNA polymerase are determined mainly by three parameters: the nucleotide selectivity of its active site, its mismatch extension capacity, and its proofreading ability (Beard & Wilson, 1998, 2003; Joyce & Benkovic, 2004). Yet, natural and exogenous sources of DNA damage result in a variety of DNA modifications, the most common including nucleobase oxidation (Nakabeppu et al., 2007), alkylation (Drabløs et al., 2004) and deamination (Ehrlich et al., 1986; Kow, 2002; Labet et al., 2008).

Depending on the type of mismatch and the biological context of its occurrence, cells must apply appropriate strategies of postreplication repair to avoid mutation (Kunz et al., 2009). However, some replication errors make it past these mechanisms, thus becoming permanent mutations after the next cell division.

Mutations are stable, heritable alterations of the genetic material, namely DNA (Friedberg et al., 2006). They are an important contributor to human aging, metabolic and degenerative disorders, cancer, and cause heritable diseases, at the same time they are the kindling factor for biological evolution of living things. Beyond the individual level, perhaps the most dramatic effect of mutation relates to its role in evolution; indeed, without mutation, evolution would not be possible. The point mutations caused by the substitution of one nucleotide base for another are divided into *transitions* (replacement of a purine with another purine or replacement of a pyrimidine with another pyrimidine, i.e. purine-pyrimidine mismatches) and *transversions* (replacement of a purine with a pyrimidine or *vice versa,* i.e. purine-purine and pyrimidine-pyrimidine mispairs). Therefore, to maintain a stable genome, it is essential for cells to monitor the state of base pairing in their genomes and to correct mismatches that will occasionally occur.

Spontaneous mutations are generally occurring due to endogenous factors: endogenous chemical lesions generated during normal cell metabolism, errors in normal cellular processes and others.

It has been suggested that there are two major approaches to the origin of mutations arising during DNA replication:

1. *replication errors,* that occur due to mispair formation in the DNA double helix as a result of changing the coding property (for example, tautomeric) of DNA base in the template strand;
2. *incorporation errors,* that occur due to mispair formation in the DNA double helix as a result of changing the coding property (for example, tautomeric) of DNA base in the incoming deoxyribonucleoside triphosphate.

There is a natural – albeit low – error rate that occurs during DNA replication. So, the average frequency of spontaneous errors in DNA replication is in the range of $10^{-8}\div10^{-11}$ per base pair replicated per one cell division (Drake, 1991; Fersht & Knill-Jones, 1983; Loeb, 2001).

Nowadays the occurrence of the spontaneous point mutations can be explained by several physico-chemical mechanisms.

Today, scientists generally consider that most DNA replication errors are caused by mispairings with "correct" geometry formed either by the protonated or deprotonated bases (i.e., bases with an excess or missing proton, respectively) (Sowers et al., 1986, 1987; Yu et al., 1993), which generation and existence under physiological conditions remains disputable, because it was claimed that the methods used by researchers to determine ionized base pairing involve conditions different from those actually obtained during DNA replication. So, Bebenek et al. (Bebenek et al., 2011) demonstrated that wild-type DNA polymerase λ and its derivative polymerase λ DL misinsert dGTP opposite template Thy at substantially higher efficiencies in reactions performed at pH 9.0 as compared to those at physiological pH (7.0). These pH dependencies of enzymatic catalysis are in agreement with the results of Yu et al. (Yu et al., 1993) and are also consistent with the possible involvement of an ionized base pair. However, in our recent work (Brovarets' et al., 2010e), it was demonstrated that the ionization mechanism of spontaneous transitions appearance does not imply any advantages in comparison with other mechanisms described in literature. Moreover, we revealed that the protonation/deprotonation of base in any canonical nucleoside significantly perturbs its DNA-like conformations (Brovarets' et al., 2010e).

It is also generally accepted in the literature that wobble base pairs (Gua·Thy and Ade·Cyt) (Brown et al., 1985; Crick, 1966; Hunter et al., 1986; Kennard, 1985; Padermshoke et al., 2008; Patel et al., 1982a, 1982b, 1984a, 1984b) formed by bases in their canonical tautomeric forms and positioned in sheared relative to the Watson-Crick configuration represent erroneous occurrences leading to the substitution mutations. The wobble mispairings were observed in X-ray (Brown et al., 1985; Hunter et al., 1986; Kennard, 1985) and NMR (Patel et al., 1982a, 1982b, 1984a, 1984b) model experiments (in the absence of DNA polymerases) on co-crystallization of complementary oligonucleotides containing a single mismatched base pair. But such experimental conditions do not properly reflect those required for enzymatic DNA replication (Kornberg & Baker, 1992). The Gua·Thy and Ade·Cyt mismatches adopt a relatively stable and well-fitting wobble configurations, supporting intrahelical base pair stacking and affecting the DNA helical structure only marginally (Brown et al., 1985; Kunz et al., 2009). By structural considerations, mispairings that cause little distortion to the canonical Watson-Crick geometry are more likely to be tolerated by the polymerase active site and, therefore, to escape proofreading. This fact was demonstrated in structural and biochemical studies of DNA polymerases (Echols & Goodman, 1991; Kool, 2002). However, enzymes, involved in postreplication repair, can easily recognize and correct structural imperfections between such improperly paired nucleotides (Kunz et al., 2009).

Another mechanism of the spontaneously arising point mutations in DNA was originally proposed by James Watson and Francis Crick (Watson & Crick, 1953a, 1953b) and further elaborated by Topal and Fresco (Topal & Fresco, 1976) as the "rare tautomer hypothesis" which suggested that "spontaneous mutation may be due to a base occasionally occurring in one of its less likely tautomeric forms". Both the purine and pyrimidine bases in DNA exist in different chemical forms, so-called isomers or tautomers, in which the protons occupy different positions in the molecule. Tautomers of DNA bases – Ade, Gua, Thy and Cyt - can cause genetic mutations by pairing incorrectly with wrong complementary bases. Watson and Crick suggested two possible transition mispairs, Gua·Thy and Ade·Cyt, involving the enol form of guanine or thymine and the imino form of adenine or cytosine, respectively – Gua*·Thy, Gua·Thy*, Ade*·Cyt and Ade·Cyt* (herein and after mutagenic tautomeric forms of bases are marked by an asterisk). These mispairs fit well within the dimensions of the DNA double helix to preserve the geometry of a correct canonical base pair in such a way supporting the Watson and Crick's original idea that spontaneous base substitutions, namely transition mutations, may result from mismatches shaped like correct base pairs, which were experimentally confirmed by Bebenek et al. for DNA polymerase λ (Bebenek et al., 2011) and by Wang et al. for DNA polymerase I (W. Wang et al., 2011). However, it remains out of eyeshot whether these rare (or mutagenic) tautomers are dynamically stable and their lifetimes are long enough to cause mutations or they are short-lived structures unable to yield irreversible errors in DNA and finally induce genomic alterations. The actual lifetime was estimated only for mutagenic tautomer of Cyt, with a value being about 600 years (Zhao et al., 2006). But evidence for these types of tautomeric shifts remains sparse, because the limited sensitivity of the experimental methods prevents an accurate detection of the relative amount of the rare tautomers including mutagenic. Among all rare tautomers, only the imino tautomers of Cyt (Brown et al., 1989b; Dreyfus et al., 1976; Feyer et al., 2010; Szczesniak et al., 1988) and enol tautomers of Gua (Choi & Miller, 2006; Sheina et al., 1987; Plekan et al., 2009; Szczepaniak & Szczesniak, 1987) were experimentally detected. The lack of the experimental data on the rare tautomers of Ade (Brown et al., 1989a) and Thy can be explained by the high value of their relative energy (~12÷14 kcal/mol at 298.15 K) estimated by theoretical investigations (Basu et al., 2005; Brovarets' & Hovorun, 2010a; Fonseca Guerra et al., 2006; Mejía-Mazariegos & Hernández-Trujillo, 2009; Samijlenko et al., 2000, 2004).

Unusual tautomeric forms of modified bases have been found in damaged DNA duplex, indicating that the transition to such altered forms is indeed feasible (Chatake et al., 1999; Robinson et al., 1998). It is therefore likely that analogues of DNA bases have a propensity to adopt the rare, namely mutagenic tautomeric forms (Brovarets' & Hovorun, 2010b, 2011a).

The molecular nature of formation of mutagenic tautomers is not quite clear yet. Several alternative mechanisms of the rare tautomers formation have been discussed in the literature: i) intramolecular proton transfer in DNA bases (Basu et al., 2005; Brovarets' & Hovorun, 2010a, 2010d, 2011a; Gorb et al., 2005; Zhao et al., 2006), ii) proton transfer in a single base assisted by bulk aqueous solution, by micro-hydration or by a single interacting water molecule (Fogarasi, 2008; Furmanchuk et al., 2011; Gorb & Leszczynski, 1998a, 1998b; H.-S. Kim et al., 2007; Michalkova et al., 2008); iii) Löwdin's mechanism of tautomerisation involving double proton transfer (DPT) along two intermolecular hydrogen (H) bonds of complementary DNA base pairs (Löwdin, 1963, 1965, 1966).

On the basis of the Watson-Crick's model Löwdin (Löwdin, 1963, 1965, 1966) suggested that spontaneous mutagenesis causing aging and cancer could be induced by tautomerisation of Ade · Thy and Gua · Cyt Watson-Crick base pairs through DPT along neighbouring intermolecular H-bonds joining bases in pairs. Following the pioneering Löwdin's work the DNA base pairs have been extensively studied using a wide range of theoretical approaches, essentially in the gas phase (Cerón-Carrasco et al., 2011a; Cerón-Carrasco & Jacquemin, 2011b; Gorb et al., 2004; Florian et al., 1995; Florian & Leszczynski, 1996; Villani, 2005, 2006, 2010).

After a comprehensive literature review we came to a conclusion that although it is widely accepted that mutations *in vivo* play a very important role in cell functioning, elementary physico-chemical mechanisms of this process remain poorly understood.

The questions of existence of different tautomeric forms of nucleic acid bases and their possible role as mutagenic factors are under intense scrutiny. The understanding of the tautomeric behavior of the purine and pyrimidine bases of the nucleic acids is of fundamental importance not only for quantitative concepts of chemical bonding and physical chemistry, but also for molecular biology and the presumed role of the rare tautomers in mutagenesis.

The structural requirements for tautomeric shifts in the base pairs that may initiate mutations have been formulated in literature (Basu et al., 2005): (i) the bases open out during replication phase in their unusual tautomeric condition and (ii) the unusual tautomers form stable base pairs with isosteric Watson-Crick geometry with their wrong suite. Another group of researchers (Dąbkowska et al., 2005) based on the conclusions earlier reported by Florian et al. (Florian et al., 1994) established that tautomerisation reactions have to fulfill not only thermodynamic but also certain kinetic limits to be relevant to spontaneous DNA mutations. First, the lifetime of the canonical base should be shorter than the reproduction period of a given species. Second, the mutagenic tautomer needs to remain stable during the time period from the occurrence of tautomerisation until the replication process is completed. These conditions impose constraints on barriers for the forward and reverse reactions of DNA bases tautomerisation.

Our purpose in this study is to carefully analyse the molecular mechanisms of spontaneously arising point mutations proposed in literature, to offer truly new ideas for molecular and structural approaches to the nature of spontaneous DNA mutations caused by prototropic tautomerism of nucleotide bases and to provide a novel quantum-chemical insight into the classical understanding of this biologically important problem.

2. Computational methods

The *ab initio* methods were used to investigate the tautomerisation of the DNA bases and mispairs involving mutagenic tautomers. All quantum-chemical calculations were performed using the Gaussian 03 program package (Frisch et al., 2003).

Geometries and harmonic vibrational frequencies of molecules and complexes were obtained using Becke's three-parameter exchange functional (B3) (Becke, 1993) combined with Lee, Yang, and Parr's (LYP) correlation functional (Lee et al., 1988) implemented in Gaussian 03 that has good performance for calculating barrier heights, thermo-chemical kinetics or intra- and intermolecular H-bonds in the systems recently studied (Brovarets', 2010; Brovarets' & Hovorun, 2010a, 2010b, 2010d, 2010f, 2011a, 2011b; Brovarets' et al., 2010c, 2010e) and 6-311++G(d,p) basis set. The absence of imaginary vibrational frequencies proved that energy-minimized structures perfectly correspond to the local minima of the potential energy landscape.

To consider electronic correlation effects as accurately as possible, we performed single point energy calculations at the MP2/6-311++G(2df,pd) level of theory for the B3LYP/6-311++G(d,p) geometries.

As for the transition states (TS) of tautomerisation of the isolated bases or their complexes, they were located by means of Synchronous Transit-guided Quasi-Newton (STQN) method (Peng & Schlegel, 1993; Peng et al., 1996) using the Berny algorithm and proved to contain one and only one imaginary frequency corresponding to the reaction coordinate. Afterwards the reaction pathway of proton transfer was followed by performing an intrinsic reaction coordinate calculation in order to make sure that transition state really connects the expected reactants and products (Gonzalez & Schlegel, 1989). We applied the standard transition state theory (Atkins, 1998) to estimate barriers for tautomerisation reactions.

The equilibrium constants of tautomerisation were calculated using the standard equation $K=\exp(-\Delta G/RT)$, where ΔG is the relative Gibbs free energy of the reactant or product, T is the absolute temperature, and R is the universal gas constant.

The time $\tau_{99.9\%}$ necessary to reach 99.9% of the equilibrium concentration of the mutagenic tautomer in the system of reversible first-order forward (k_f) and reverse (k_r) reactions (canonical $\leftrightarrow$ mutagenic tautomer transitions) can be estimated from the equation (Atkins, 1998)

$$\tau_{99.9\%} = \frac{ln 10^3}{k_f + k_r} \quad (1)$$

and the lifetime τ and the half-lifetime $\tau_{1/2}$ of the complexes are given by $1/k$ and $\ln(2)/k$, respectively. We applied the standard transition state theory (Atkins, 1998) in which quantum tunneling effects are accounted by the Wigner's tunnelling correction (Wigner, 1932).

$$\Gamma = 1 + \frac{1}{24}\left(\frac{h\nu_i}{k_B T}\right)^2 \tag{2}$$

that is adequate for proton transfer reactions (Brovarets' & Hovorun, 2010a, 2010b, 2011a; Cerón-Carrasco & Jacquemin, 2011b) to estimate the values of rate constants k_f and k_r

$$k_{f,r} = \Gamma \cdot \frac{k_B T}{h} e^{-\frac{\Delta\Delta G_{f,r}}{RT}} \tag{3}$$

where k_B - the Boltzmann's constant, h – the Planck's constant, $\Delta\Delta G_{f,r}$ – the Gibbs free energy of activation for the proton transfer reaction, ν_i – the magnitude of the imaginary frequency associated with the vibrational mode at the transition state that connects reactants and products.

The electronic interaction energies have been computed at the MP2/6-311++G(2df,pd) level of theory for the B3LYP/6-311++G(d,p) geometries. In each case the interaction energy was corrected for the basis set superposition error (BSSE) (Boys & Bernardi, 1970; Gutowski et al., 1986) through the counterpoise procedure (Sordo et al., 1988; Sordo, 2001) implemented in the Gaussian 03 package (Frisch et al., 2003).

The topology of the electron density was analysed using program package AIMAll (AIMAll, 2010) with all the default options. The presence of a bond critical point (BCP), namely the so-called (3,-1) point, and a bond path between hydrogen donor and acceptor, as well as the positive value of the Laplacian at this bond critical point, were considered as the necessary conditions for H-bond formation. Wave functions were obtained at the level of theory used for geometry optimization.

3. DNA bases with amino group: Planar or nonplanar?

The amino group $-NH_2$ in DNA bases, namely, Gua, Cyt and Ade, plays a key role in formation of H-bonds in nucleic acids and in other molecular systems. Thus, the structure of this group is of fundamental importance in the molecular recognition phenomena. The DNA bases were believed to be planar for many years, until the nonplanarity of their amino groups has been predicted in the 1990s (Aamouche et al., 1997; Hobza & Šponer, 1999; Hovorun et al., 1995a, 1995b, 1999; Hovorun & Kondratyuk, 1996; Komarov & Polozov, 1990; Komarov et al., 1992; Šponer & Hobza, 1994; Šponer et al., 2001). Direct experimental results for the nucleic acid bases amino moieties are not available, but indirect experimental evidence does exist. The first indirect experimental evidence was connected with the excellent agreement between the theoretical anharmonic (Bludský et al., 1996) and experimental inversion-torsion (Kydd & Krueger, 1977, 1978; Larsen et al., 1976) vibrational frequencies that provided evidence concerning the nature of the predicted aniline potential energy surface, consistent with a strong nonplanarity of the amino group (Lister et al., 1974; Sinclair & Pratt, 1996; Quack & Stockburger, 1972).

Although a noticeable inertial defect of Ade was observed in a microwave study (Brown et al., 1989a), its source was not directly related to the nonplanarity of this base. Indirect experimental evidence was associated with the vibrational transition moment angles of Ade reported by Choi et al. (Choi et al., 2008). The mismatched $Gua_{anti}{\cdot}Ade_{anti}$ base pair (Privé et al., 1987) is an example exhibiting the strong out-of-plane H-bond character related to the nonplanar guanine amino group.

The internal nonplanarity of the amino group originates from the partial sp^3 hybridization of the amino group nitrogen atom (Govorun et al., 1992; Hovorun et al., 1995a, 1995b, 1999; Hovorun & Kondratyuk, 1996; Gorb & Leszczynski, 1998a, 1998b; Hobza & Šponer, 1999; Šponer & Hobza, 1994).

At least one conclusion that may be drawn from these investigations is that the amines could be much more flexible than previously expected because of the low values of the inversion and rotation barriers of the amino group. The inversion dynamics of the amino group have been investigated by *ab initio* methods with and without inclusion of correlation energy utilizing medium and extended basis sets (Bludský et al., 1996) and the barriers for inversion or internal rotation of the amino group in a quasi-classical approximation have been calculated (Y. Wang et al., 1993).

We present herein a more comprehensive analysis of the $\geq C\text{-}NH_2$ fragment interconversion in DNA bases - its plane inversion and anisotropic internal rotation of the amino group and its influence on the structural relaxation of the molecular ring. Summary of our findings makes it possible to describe a complex mechanism of the amino group motion which includes tunneling (only for rotations) and large amplitude motion above the barrier of planarization. Of particular interest, in this context, is the phenomenon of pyramidalization.

The nitrogenous bases with exocyclic amine fragment $\geq C\text{-}NH_2$ are known to have nonrigid structures (for details see (Bludský et al., 1996; Florian et al., 1995; Hovorun & Kondratyuk, 1996; Hovorun et al., 1999)). Their interconversion, i.e. conformational (without breaking chemical bonds) transitions within a molecule, is accomplished in three topologically and energetically distinct ways - plane inversion of the $\geq C\text{-}NH_2$ fragment and two, clockwise or counterclockwise, rotations of the amino group around exocyclic C-N bond *via* plane symmetrical transition states with substantially pyramidalized amine fragment. It should be mentioned that in the planar transition state (TS_1) of the $\geq C\text{-}NH_2$ fragment inversion the exocyclic C-N bond is shortened and the N-H bonds are elongated as compared to those in the nonplanar equilibrium configuration, the valence angle H-N-H becomes close to 120°. In the plane-symmetric transition states of the amino group rotations TS_2 and TS_3 the C-N bond becomes elongated, the N-H bonds become shortened and the valence angle H-N-H distinctly deviates from 120°, at that the amine fragment $\geq C\text{-}NH_2$ is highly pyramidalized as compared to the equilibrium configuration. All these results clearly demonstrate that the structural nonrigidity of nitrogenous bases is determined by intramolecular quantum-chemical effect - p-π-conjugation of a lone electron pair (LEP) of the nitrogen atom of the amine fragment $\geq C\text{-}NH_2$ with the π-electronic system of the ring (Dolinnaya & Gromova, 1983; Dolinnaya & Gryaznova, 1989).

3.1 Pyramidalization of the amine fragment of the Ade

So, we demonstrated that Ade (N1C6N6H=0.013°; C5C6N6H=-0.014°) is an effectively planar molecule (effective symmetry C_s) (Hovorun et al., 1995a, 1995b, 1999; Hovorun & Kondratyuk, 1996). Its interconversion is accomplished *via* two plane-symmetric transition states with Gibbs free energy of 14.34 and 14.57 kcal/mol and also through the planar transition state with the activation energy of 0.12 kcal/mol[1] (Table 1). MP2 complete basis set limit method with the aug-cc-pVTZ → aug-cc-pVQZ (aTZ → aQZ) extrapolation scheme has predicted very small planarization barrier of the Ade amino group, 0.015 kcal/mol (Zierkiewicz et al., 2008), which is in very good agreement with the MP2-predicted planarization barrier of 0.020 kcal/mol reported by Wang and Schaefer III (S. Wang & Schaefer III, 2006). Similar results were calculated using coupled cluster CCSD(T) complete basis set method - 0.125 kcal/mol (Zierkiewicz et al., 2008). Thus, the literature review highlights that the amino group in isolated Ade, in the gas phase, is very flexible with a small degree of nonplanarity.

Base	Plane inversion (TS_1)				Rotation (TS_2)				Rotation (TS_3)			
	ΔΔG	ΔΔE		ν,	ΔΔG	ΔΔE		ν,	ΔΔG	ΔΔE		ν,
		kcal/mol	cm^{-1}	cm^{-1}		kcal/mol	cm^{-1}	cm^{-1}		kcal/mol	cm^{-1}	cm^{-1}
Ade	-0.06*	0.12* 0.02#	42.3* 7.0#	309.4*	14.34	13.30	4652.7	539.5	14.57	13.51	4727.3	539.5
Gua	0.37	0.91 0.74#	318.6 258.9#	542.6	9.14	9.48	3316.4	327.9	5.40	5.35	1872.1	327.9
Cyt	0.06	0.08 0.03#	28.9 10.5#	212.2	11.9	11.74	4105.3	524.6	15.85	16.11	5633.7	524.6

* - values obtained at the MP2/6-311++G(2df,pd)//B3LYP/cc-pVDZ level of theory (Brovarets' & Hovorun, 2010b);
# - values obtained at the MP2/aug-cc-pVQZ level of theory (S. Wang & Schaefer III, 2006);
TS_2 - transition state of the amino group rotation toward the N1 atom for Ade, Gua or the N3 atom for Cyt;
TS_3 - transition state of the amino group rotation toward the N7 atom for Ade, the N3 atom for Gua or the C5-H group for Cyt

Table 1. Relative values of Gibbs free energy (ΔΔG) (T=298.15 K) and electronic energy (ΔΔE) (in kcal/mol) for the Ade, Gua, and Cyt transition states of amino group interconversion (plane inversion TS_1 and anisotropic rotations TS_2, TS_3) and corresponding vibrational modes (in cm^{-1}) obtained at the MP2/6-311++G(2df,pd)//B3LYP/6-311++G(d,p) level of theory in vacuum

We obtained that the deviations from the main geometric parameters of ≥C6-N6H_2 amine fragment of Ade are the following: the length of the C6-H6 bond is increased by 0.072 and 0.074 Å, the lengths of the N6-H are decreased on average by 0.011 Å, and the valence angle H-N6-H is decreased from 120.4° up to 105.8° and 105.9° at the transition states TS_2 and TS_3, respectively, as compared to those in the nonplanar equilibrium configuration of Ade (Brovarets' and Hovorun, 2010b). In the planar transition state TS_1 of the ≥C6-N6H_2 fragment inversion the exocyclic C6-N6 bond is shortened by 0.005 Å, the N6-H bonds are elongated by 0.002 Å as compared to those in the nonplanar equilibrium configuration, and the valence angle H-N6-H becomes close to 120° and is equal to 120.9° comparatively with the equilibrium state (118.7°).

3.2 Pyramidalization of the amine fragment of the Gua

It is commonly thought that exactly due to the presence of the neighbouring N1-H group, the pyramidalization of the amino group in guanine is higher than in canonical cytosine and

[1] The result obtained at the MP2/6-311++G(2df,pd)//B3LYP/cc-pVDZ level of theory.

adenine, which have no proton at the nitrogen atom located in the neighbourhood of the amino group. In guanine, one of the amino group hydrogen atoms oriented toward the N1-H bond is more bent down than the second amino group hydrogen atom oriented opposite to this bond. The amine fragment ≥C2-N2H_2 (N1C2N2H=-31.1°; N3C2N2H=12.2°) of Gua can not be considered to be pyramidalized even at T=0 K, since the zero-point vibrational energy associated with competent normal mode (542.6 cm^{-1}), which frequency becomes imaginary (371.1 i cm^{-1}) in the transition state of plane inversion, is higher than the planarization electronic energy barrier (0.91 kcal/mol or 318.6 cm^{-1}).

The Gibbs free energies of activation of Gua interconversion *via* the plane-symmetric transition states TS_2 and TS_3 of the amino group rotation (5.40 and 9.14 kcal/mol) from its *trans*- and *cis*-orientation relative to the N1-C2 bond differ markedly from each other. Such a difference in Gibbs free energies of activation can be explained by the fact that the transition state TS_2 is stabilized by electrostatic interactions of the LEP of the N2 atom with the hydrogen atom of the N1-H group and the amino group hydrogen atoms with the LEP of the N3 atom, while in the transition state TS_3 these electrostatic interactions are displaced by repulsion of LEP of the N2 and N3 atom and the amino group hydrogen atoms from the N1-H group hydrogen atom that leads to destabilization of this transition state (Brovarets' and Hovorun, 2010b).

In the Gua* mutagenic tautomer (ΔG=0.13 kcal/mol) which can mispair with Thy (Dąbkowska et al., 2005; Danilov et al., 2005; Mejía-Mazariegos & Hernández-Trujillo, 2009) the hydroxyl group O6-H is *cis*-oriented relatively to the N1-C6 bond. The barrier of planar inversion for Gua* is significantly lower than that for Gua (Brovarets' & Hovorun, 2010b).

3.3 Pyramidalization of the amine fragment of the Cyt

We also demonstrated that Cyt is a structurally nonrigid molecule. Its interconversion occurs through three topologically and energetically distinct ways - plane inversion of the amine fragment ≥C4-N4H_2 (N3C4N4H=7.2°; C5C4N4H=-11.7°) *via* the transition state TS_1 and two anisotropic (clockwise and counterclockwise) rotations of the amino group around the exocyclic C4-N4 bond *via* the transition states TS_2 and TS_3, respectively. The planarization barrier of Cyt amino group is not large enough (28.9 cm^{-1}) (Table 1) to allow the arrangement at least one vibrational level (n=0) of competent mode (212.1 cm^{-1}), which frequency becomes imaginary (154.6 i cm^{-1}) in the transition state TS_1 of planarization of the Cyt amino group. The calculated low planarization barrier of Cyt leads to large amplitude anharmonic vibration of the amino group of Cyt over the barrier (Brovarets' and Hovorun, 2011a).

The Gibbs free energy of activation for rotation of the amino group about the C4-N4 bond when the LEP of the N4 atom is oriented to the hydrogen atom of the C5-H group (N3C4N4H_1=56.6°; N3C4N4H_2=-56.5°; HN4H=104.8°) is found to be notably lower (11.85 kcal/mol) than in the case when the LEP of the N4 atom is oriented to the N3 atom (N3C4N4H_1=120.6°; N3C4N4H_2=-120.6°; HN4H=107.4°) - 15.85 kcal/mol. This can be explained by the fact that the attractive interactions in the first case (the LEP of the N4 atom with the C5-H and amino protons with the LEP of the N3 atom) are replaced by repulsive ones (between the LEPs of the N4 and N3 atoms and between the amino protons and the hydrogen atom of the C5-H group).

So, extremely low planarization barrier implies that Ade, Cyt and Gua require very little energy to conform the structure of the amino group for formation of the complementary H-bonds with other molecules. This fact is very important for base pairing in nucleic acids or other polymers containing Ade, Gua and Cyt residues.

3.4 Planarity or nonplanarity of DNA bases

The thorough analysis of our results and also interpretation of the data reported in literature (Bludský et al., 1996; Hobza & Šponer, 1999; Hovorun et al., 1995a, 1995b, 1999; Hovorun & Kondratyuk, 1996; Larsen et al., 1976; Lister et al., 1974; Šponer & Hobza, 1994; S. Wang & Schaefer III, 2006; Zierkiewicz et al., 2008) allow us to offer the following conclusions. The nucleobases with amino group are effectively planar structures with effective symmetry C_s. This is due to the fact that zero-point vibrational level of inverse out-of-plane vibration of their ≥C-NH_2 amine fragment is located above the barrier of its plane inversion, and the maximum of the quadrate of the ψ-function for this vibration coincides with the barrier of the inversion (Fig. 1). In other words, the above-mentioned inversion oscillator has an essentially quantum behavior and can not be appropriately described in the framework of classical mechanics. "Equilibrium", "static" characteristics of the ≥C-NH_2 amine fragment, namely the valence and dihedral angles, which are commonly interpreted by investigators as geometric parameters of equilibrium "nonplanarity" of amine fragment of Ade, Cyt and Gua, should be considered rather as dynamic characteristics of vibration mode of amine fragment inversion and no more than this.

At the same time, the two other nucleobases, Ura and Thy, are undoubtedly planar structures with point symmetry C_s (S. Wang & Schaefer III, 2006): the maximum of the quadrate of the ψ-function for low-frequency out-of-plane vibrations of pyrimidine ring coincides with the minimum of the potential energy that meets the planar structure (Fig. 1).

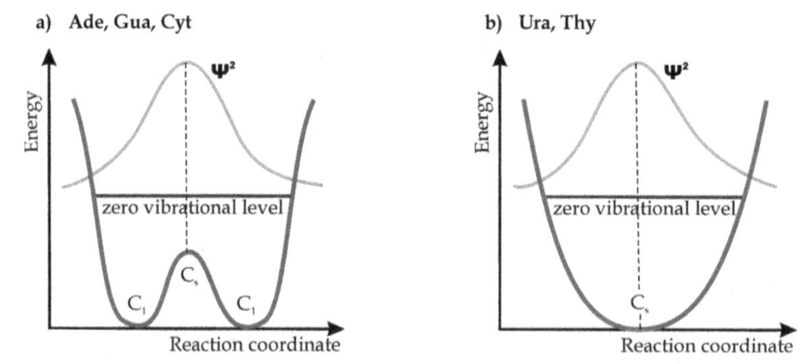

Fig. 1. Qualitative representation of potential energy profile (red line) of the amino group planarization in the case of Ade, Cyt and Gua (a) and out-of-plane ring deformation in the case of Ura and Thy (b) and quadrate of the ψ-function (blue line) for corresponding zero-point vibrational levels

All canonical DNA bases are rather "soft" structures taking into account nonplanar out-of-plane deformation. Therefore, their static, equilibrium nonplanarity, which is observed, particularly in crystal state and in isolated nucleosides (Yurenko et al., 2007a, 2007b, 2007c, 2008, 2009; Zhurakivsky & Hovorun, 2006, 2007a, 2007b) or nucleotides (Nikolaienko et al.,

2011a, 2011b, 2011c), is induced by anisotropic forces of crystal packaging and intramolecular interactions within nucleosides or nucleotides, respectively.

The amine fragment ≥C-NH_2 of DNA bases indeed determines their structural nonrigidity, which is in turn conditioned by quantum intramolecular effect, namely p-π-conjugation of a LEP of amino nitrogen atom with π-electron system of the ring. This specific phenomenon of conjugation is purely quantum and has no classical analogue.

Exactly the structural nonrigidity of the polar amine fragment in DNA bases is a reason to adequately explain a static nonplanarity of amine fragment induced by an external electrical field which deforms it so that the projection of the induced dipole moment on the field direction is maximal and coincides with vector of field strength (Brauer et al., 2011; Choi et al., 2005, 2008; Dong & Miller, 2002).

4. Mutagenic tautomers of DNA bases and possible molecular mechanisms of their formation

4.1 Mutagenic tautomers of DNA bases

For structural chemists, rare tautomers of DNA bases are of special interest because they exert strong mutational pressures on the genome(Friedberg et al., 2006; Harris et al., 2003; Kwiatkowski & Pullman, 1975). That's why the tautomerism of DNA bases and their biologically active modifications (Kondratyuk et al., 2000; Samijlenko et al., 2001) have been the subject of a great number of theoretical and experimental investigations due to their biochemical significance.

Numerous experimental and theoretical efforts have been directed towards the elucidation of qualitative and quantitative aspects of Cyt tautomerism, the data obtained up until about 1974 have been reviewed by Kwiatkowski and Pullman (Kwiatkowski & Pullman, 1975). In the solid phase, Cyt exists in a single keto-amino tautomeric state. However, experiments performed in the gas phase and in low-temperature inert matrices clearly demonstrated that Cyt exists as a mixture of several tautomeric forms (Bazsó et al., 2011; Brown et al., 1989b; Choi et al., 2005; Dong & Miller, 2002; Feyer et al., 2009; Govorun et al., 1992; Kostko et al., 2010; Lapinski et al., 2010; Min et al., 2009; Nir et al., 2001a, 2002a, 2002b; Nowak et al., 1989a; Radchenko et al., 1984; Szczesniak et al., 1988). Still, there are three matrix-isolation infrared spectroscopic studies on Cyt tautomerisation (Nowak et al., 1989a, 1989b; Radchenko et al., 1984; Szczesniak et al., 1988). Three tautomers of Cyt have been identified by molecular beam microwave spectroscopy with an estimated abundance of 1:1:0.25 for keto:amino-enol:keto-imino tautomers (Brown et al., 1989b). Several groups have explored the ultrafast excited-state dynamics of Cyt in molecular beams (Canuel et al., 2005; Kang et al., 2002; Kosma et al., 2009; Ullrich et al., 2004). As have been pointed out elsewhere (Fogarasi, 2002; Kosma et al., 2009; Ullrich, 2004), one ambiguity in these experiments is the coexistence of two or more tautomeric forms of Cyt in the gas phase.

The first experimental observation of amino-keto and amino-enol tautomeric forms of Gua has been performed on isolated species in cold inert gas matrix by ground state infrared spectroscopy (Sheina et al., 1987; Szczepaniak & Szczesniak, 1987). By using UV-UV, IR-UV hole burning (Nir et al., 2001b, 2002b) and resonance-enhanced multiphoton ionization (REMPI) (Nir et al., 1999, 2002b) spectroscopy, de Vries and co-workers found spectral features that they assigned to the N9H keto, N7H keto, and N9H enol (*cis*- or *trans*-) forms. However, the most intense band assigned to the N9H enol was later attributed by Mons and co-workers (Chin et al., 2004; Mons et al., 2002) to a higher-energy form of the N7H enol tautomer. Furthermore, they observed the fourth band, which they assigned to the N9H *cis*-enol form. Choi and Miller studied Gua molecules embedded in He droplets (Choi & Miller, 2006) and assigned the IR spectroscopic data to a mixture of the four more stable tautomeric forms: N7H keto, N9H keto, and N9H *cis*- and *trans*-enol. Mons et al. (Mons et al., 2006) later reported a new interpretation of the resonant two-photon ionization (R2PI) spectra. The authors suggested the occurrence of a fast nonradiative relaxation of the excited states of the N7H keto, N9H keto, and N9H *trans*-enol tautomeric forms that prevents the observation of these species in the R2PI spectra. The consistency between the experimental data obtained by molecular-beam Fourier-transform microwave (MB-FTMW) spectroscopy and theoretical calculations enabled Alonso and his collaborators to unequivocally identify the four most stable tautomers of guanine in the gas phase (Alonso et al., 2009). Recently also different tautomers of Gua were detected using vacuum ultraviolet (VUV) photoionization (Zhou et al., 2009). Theoretical calculations (Chen & Li, 2006; Elshakre, 2005; Hanus et al., 2003; Marian, 2007; Trygubenko et al., 2002) predict the existence of four low-energy tautomers with stabilities in the range 0–400 cm^{-1}, whereby the keto tautomers with a hydrogen atom at the N7 or N9 atoms are the most stable.

Besides its role as a nucleic acid building block, Ade and its derivatives are of interest in various other biochemical processes. For example, it is the main component of the energy-storing molecule adenosine triphosphate. Its high photostability under UV irradiation is an intriguing property that has been suggested to be essential for the preservation of genetic information (Crespo-Hernández et al., 2004).

Furthermore, the various tautomeric forms of Ade have been under substantial scrutiny (Hanus et al., 2004; Kwiatkowski & Leszczynski, 1992; Laxer et al., 2001; Mishra et al., 2000; Nowak et al., 1989b, 1991, 1994a, 1994b, 1996; Plützer et al., 2001; Plützer & Kleinermanns, 2002; Salter & Chaban, 2002), because of their proposed role in mutagenic and carcinogenic processes (Danilov et al., 2005; Harris et al., 2003; Topal & Fresco, 1976). Some of the first IR spectra of Ade recorded in low-temperature inert gas matrices in the 400 to 4000 cm^{-1} range date back to 1985 (Stepanian et al., 1985). This study was extended by comparing the experimental spectra with calculated IR frequencies at different levels of theory (Brovarets' & Hovorun, 2011b; Nowak et al., 1989b, 1991, 1994a, 1994b, 1996). It was concluded that the absorption of Ade was due to its 9H tautomer. Ade in the gas phase has been studied by UV photoelectron (Lin et al., 1980), microwave (Brown et al., 1989a), IR (Colarusso et al., 1997), jet-cooled REMPI (N.J. Kim et al., 2000; Lührs et al., 2001; Nir et al., 2001a, 2002b), IR–UV ion-dip (Nir et al., 2001a, 2002b; Plützer & Kleinermanns, 2002; Plützer et al., 2001; Van Zundert et al., 2011) and IR multiple-photon dissociation (IRMPD) spectroscopic investigations (Van Zundert et al., 2011). In all of these studies, it was suggested that the 9H amino tautomer of Ade is the dominant contributor to the spectra. The experimental results agree closely with calculations at different levels of theory and consistently show the 9H amino tautomer to be the most stable one (Brovarets' & Hovorun, 2011b; Hanus et al., 2004; Fonseca Guerra et al., 2006; Kwiatkowski & Leszczynski, 1992; Norinder, 1987; Nowak et al., 1989b, 1991, 1994a, 1994b, 1996; Sabio et al., 1990; Saha et al., 2006; Sygula & Buda, 1983; Wiorkiewicz-Kuczera & Karplus, 1990).

The existence of other Ade tautomers was evidenced by experimental studies (García-Terán et al., 2006; Gu & Leszczynski, 1999; Lührs et al., 2001; Stepanyugin et al., 2002a; Sukhanov et al., 2003), often in the presence of a metal (Samijlenko et al., 2004; Vrkic et al., 2004). It was also found that the Ade imino tautomer is more stabilized under the influence of charged platinum (Burda et al., 2000) or mercury (Zamora et al., 1997) cations.

It is generally believed that Thy exists in the canonical diketo form in the gas phase as well as in the aqueous solution (Kwiatkowski & Pullman, 1975), but there is experimental evidence of small amounts of its rare tautomeric forms in the gas phase (Fujii et al., 1986; Tsuchiya et al., 1988) and in the solution (Hauswirth & Daniels, 1971; Katritzky & Waring, 1962; Morsy et al., 1999; Samijlenko et al., 2010; Suwaiyan et al., 1995). Also laser ablation in combination with MB-FTMW spectroscopy spectroscopy has been used to establish unambiguously the presence of the diketo form of thymine in the gas phase and to obtain its structure (López et al., 2007). In some theoretical reports, there is also a substantial emphasis on the energetic and structural characteristics of the stable isolated tautomers of Thy (Basu et al., 2005; Fan et al., 2010; Mejía-Mazariegos & Hernández-Trujillo, 2009), indicating that the diketo is the most stable isomer both in the gas phase and in solution.

4.2 Intramolecular tautomerisation of the DNA bases

In this section the intramolecular tautomerisation of nucleotide bases as a factor in spontaneous mutagenesis is considered using quantum-chemical calculation methods. In particular, the forward and reverse barrier heights for proton transfer reactions in isolated DNA bases have been estimated and analysed.

The mutagenic tautomers of all DNA bases are depicted in Figure 2, while Table 2 shows their relative Gibbs free energies and kinetic parameters of the tautomerisation. As seen from Table 2 the mutagenic tautomers both of Cyt and Gua are energetically close to their

Conversion	$\Delta\Delta G_{TS}$, kcal/mol	k, s^{-1}	τ, s	$\tau_{1/2}$, s	$\tau_{99.9\%}$, s	ΔG, kcal/mol	K
Ade→Ade*	45.58	$9.67 \cdot 10^{-21}$	$1.03 \cdot 10^{20}$	$7.17 \cdot 10^{19}$	$5.14 \cdot 10^{10}$	14.00	$5.4 \cdot 10^{-11}$
Ade*→Ade	31.58	$1.79 \cdot 10^{-10}$	$5.59 \cdot 10^{9}$	$3.87 \cdot 10^{9}$			
Thy→Thy*	39.09	$5.60 \cdot 10^{-16}$	$1.79 \cdot 10^{15}$	$1.24 \cdot 10^{15}$	$4.73 \cdot 10^{7}$	11.64	$2.88 \cdot 10^{-9}$
Thy*→Thy	27.45	$1.95 \cdot 10^{-7}$	$5.14 \cdot 10^{6}$	$3.56 \cdot 10^{6}$			
Cyt→Cyt*	38.47	$1.61 \cdot 10^{-15}$	$6.22 \cdot 10^{14}$	$4.31 \cdot 10^{14}$	$1.35 \cdot 10^{14}$	2.21	$2.88 \cdot 10^{-9}$
Cyt*→Cyt	36.27	$6.68 \cdot 10^{-14}$	$1.50 \cdot 10^{13}$	$1.04 \cdot 10^{13}$			
Gua→Gua*	32.17	$6.50 \cdot 10^{-11}$	$1.54 \cdot 10^{10}$	$1.07 \cdot 10^{10}$	$6.28 \cdot 10^{10}$	0.13	$7.98 \cdot 10^{-1}$
Gua*→Gua	32.04	$8.15 \cdot 10^{-11}$	$1.23 \cdot 10^{10}$	$8.50 \cdot 10^{9}$			

$\Delta\Delta G_{TS}$ – the Gibbs free energy of activation for tautomerisation (T=298.15 K); k - the rate constant; τ – the lifetime; $\tau_{1/2}$ – the half-lifetime;; $\tau_{99.9\%}$ – the time necessary to reach 99.9% of the equilibrium concentration of rare tautomer in the system; ΔG – the relative Gibbs free energy of the tautomerized base (T=298.15 K); K – the equilibrium constant of tautomerisation

Table 2. Basic thermodynamic and kinetic characteristics of intramolecular tautomerisation of DNA bases obtained at the MP2/6-311++G(2df,pd)//B3LYP/6-311++G(d,p) level of theory in vacuum #

canonical tautomers that is in a complete agreement with the experimental data on Cyt and Gua tautomers (Nir et al., 1999, 2001a, 2001b, 2002a, 2002b) and their Gibbs free energy differences are only 2.21 and 0.13 kcal/mol, relatively. The considerably greater differences in energy of 14 and 12 kcal/mol were found for the mutagenic tautomers of Ade and Thy, respectively. This finding can explain why the mutagenic tautomers of Ade and Thy can not be detected experimentally.

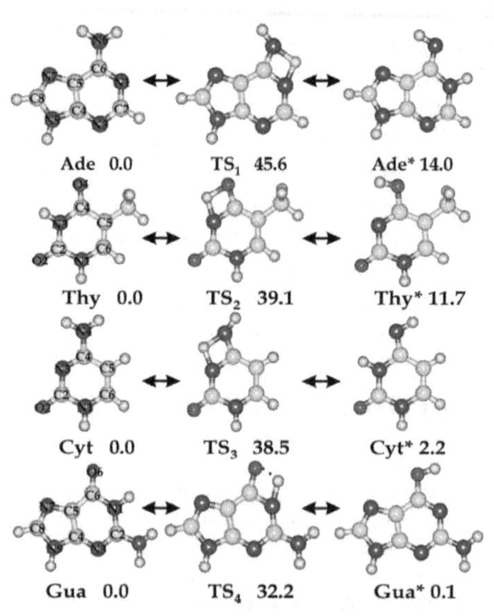

Fig. 2. Intramolecular tautomerisation of the DNA bases. The dotted line indicates intramolecular H-bond N1H...O6 in TS_4, while continuous lines show covalent bonds. Relative Gibbs free energy is presented near each structure in kcal/mol (T=298.15 K, in vacuum)

The intramolecular proton transfer schemes for isolated DNA bases are displayed in Figure 2. The MP2/6-311++G(2df,pd)//B3LYP/6-311++G(d,p) reaction barriers for the forward tautomerisation are 45.58 (Ade), 39.09 (Thy), 38.47 (Cyt) and 32.17 kcal/mol (Gua) and these values tightly correlate with the literature data (Basu et al., 2005; Brovarets' & Hovorun, 2010a, 2010f; Danilov et al., 2005; Fan et al., 2010; Fogarasi & Szalay, 2002; Fogarasi, 2008; Fonseca Guerra et al., 2006; Gorb & Leszczynski, 1998a, 1998b; Gorb et al., 2001; Gu & Leszczynski, 1999; Hanus et al., 2003, 2004; Kosenkov et al., 2009; Mejía-Mazariegos & Hernández-Trujillo, 2009; Saha et al., 2006). Very large kinetic barriers for intramolecular tautomerisation of all isolated DNA bases (above 32 kcal/mol) indicate that such tautomerisation will be very slow and this process may not occur readily in the isolated molecule. So, in such a way it is not possible to attain the equilibrium concentrations within biologically important period of time, namely during the replication of one base pair (*ca.* 10^{-4} s), as the value of $\tau_{99.9\%}$ amounts to more than 10^7 s. However, mutagenic tautomers, once formed, will be stable with a lifetime that by 3–10 orders exceeds the typical time of DNA replication in the cell (~10^3 s). This fact confirms that the postulate, on which the Watson-Crick tautomeric hypothesis of spontaneous transitions grounds, is adequate (Brovarets' & Hovorun, 2010a). It should be noted that equilibrium constants of Ade ($5.4 \cdot 10^{-11}$) and Thy ($2.88 \cdot 10^{-9}$) tautomerisation fall within the range of measured mutation frequency, but for the Cyt ($2.4 \cdot 10^{-2}$) and Gua ($8.0 \cdot 10^{-1}$) - remain above this value.

Of course, DNA bases are not isolated in living systems. In cellular DNA, the transition from canonical to mutagenic tautomers of nucleotide bases could be facilitated by the interactions with surrounding molecules. Also as suggested by Rodgers (Yang & Rodgers, 2004), bimolecular (intermolecular) tautomerisation may be much more feasible than monomolecular (intramolecular) tautomerisation.

4.3 The Löwdin's mechanism of the spontaneous point mutations

As seen from the literature survey, the possible tautomerisation of Gua·Cyt and Ade·Thy Watson-Crick base pairs occurs by Löwdin's mechanism (Fig. 3) through proton transfer along two neighbouring intermolecular H-bonds (Löwdin, 1963, 1965, 1966). However, the models exploring Löwdin's mechanism (Cerón-Carrasco et al., 2011; Cerón-Carrasco & Jacquemin, 2011; Florian et al., 1994, 1995; Florian & Leszczynski, 1996; Gorb et al., 2004; Villani, 2005, 2006, 2010) neglect the fact that electronic energy of reverse barriers of Gua·Cyt and Ade·Thy tautomerisation must exceed zero-point energy of vibrations causing this tautomerisation to provide dynamic stability (Gribov & Mushtakova, 1999) of the formed (Löwdin's) Gua* · Cyt* and Ade*·Thy* mispairs, accordingly. In addition, this barrier must exceed a dissociation energy of the formed mispair to allow such complex easily dissociate into mutagenic tautomers during DNA replication. The results of our calculations definitely demonstrated that the zero-point energy 1475.9 and 1674.6 cm^{-1} (Table 7) for Gua* · Cyt* and Ade* · Thy* base pairs, accordingly, of corresponding vibrational modes which frequencies become imaginary in the transition states of Gua·Cyt and Ade·Thy base pairs tautomerisation lies above (1800.8 cm^{-1}) and under (37.7 cm^{-1}) the value of the reverse barrier, accordingly (Table 3, 7). This means that Ade*·Thy* mispair is

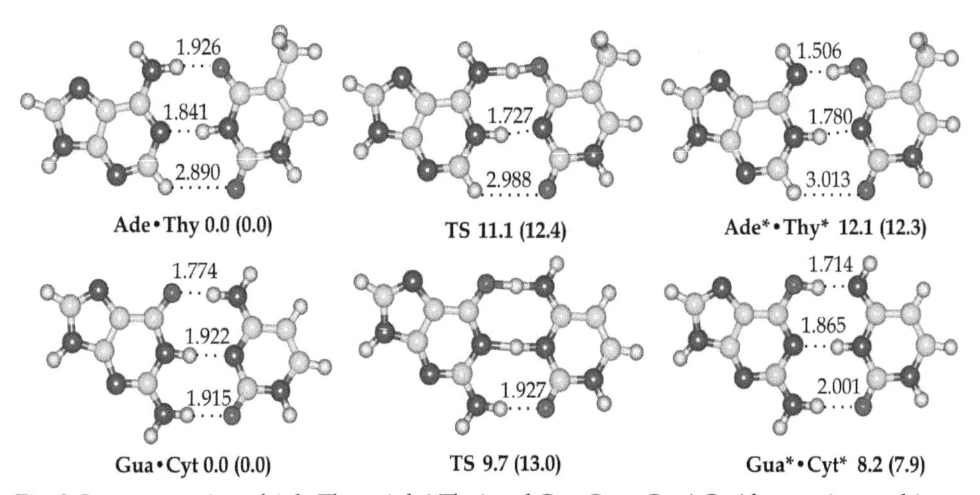

Fig. 3. Interconversion of Ade·Thy↔Ade*·Thy* and Gua·Cyt↔Gua*·Cyt* base pairs resulting from the mutagenic tautomerisation of DNA bases. Relative Gibbs free (T=298.15 K, in vacuum) and electronic (in brackets) energies are obtained at the MP2/6-311++G(2df,pd)//B3LYP/6-311++G(d,p) level of theory and reported near each structure in kcal/mol. The dotted lines indicate H-bonds AH…B (their lengths H…B are presented in angstroms), while continuous lines show covalent bonds

dynamically unstable, moreover, the value of its reverse barrier (in terms of Gibbs free energy) is negative (-1.01 kcal/mol) indicating that Ade*·Thy* minimum completely disappears from the Gibbs free energy surface. Therefore, Ade*·Thy* mispair really doesn't exist (Fig. 3). By comparing the values of zero-point energy (Table 3, 7) and the reverse barrier (Tables 3, 7) of the Gua·Cyt↔Gua*·Cyt* tautomerisation, we came to the conclusion that Gua*·Cyt* mispair is metastable.

Conversion	$\Delta\Delta G_{TS}$, kcal/mol	k, s^{-1}	τ, s	$\tau_{1/2}$, s	$\tau_{99.9\%}$, s	ΔG, kcal/mol	K
Ade·Thy→Ade*·Thy*	11.05	$5.49 \cdot 10^{4}$	$1.82 \cdot 10^{-5}$	$1.26 \cdot 10^{-5}$	$2.37 \cdot 10^{-13}$	12.07	$1.41 \cdot 10^{-9}$
Ade*·Thy*→Ade·Thy	-1.01	$3.88 \cdot 10^{13}$	$2.57 \cdot 10^{-14}$	$1.78 \cdot 10^{-14}$			
Gua·Cyt→Gua*·Cyt*	9.70	$9.02 \cdot 10^{5}$	$1.11 \cdot 10^{-6}$	$7.68 \cdot 10^{-7}$	$9.61 \cdot 10^{-12}$	8.22	$9.42 \cdot 10^{-7}$
Gua*·Cyt*→Gua·Cyt	1.49	$9.58 \cdot 10^{11}$	$1.04 \cdot 10^{-12}$	$7.24 \cdot 10^{-13}$			
Ade·Cyt*→Ade*·Cyt	7.76	$2.26 \cdot 10^{7}$	$4.42 \cdot 10^{-8}$	$3.06 \cdot 10^{-8}$	$4.63 \cdot 10^{-10}$	4.01	$1.14 \cdot 10^{-3}$
Ade*·Cyt→Ade·Cyt*	3.75	$1.99 \cdot 10^{10}$	$5.03 \cdot 10^{-11}$	$3.49 \cdot 10^{-11}$			
Gua·Thy*→Gua*·Thy	2.33	$2.74 \cdot 10^{11}$	$3.65 \cdot 10^{-12}$	$2.53 \cdot 10^{-12}$	$4.18 \cdot 10^{-12}$	1.16	$1.42 \cdot 10^{-1}$
Gua*·Thy→Gua·Thy*	1.17	$1.93 \cdot 10^{12}$	$5.18 \cdot 10^{-13}$	$3.59 \cdot 10^{-13}$			

see designations in Table 2

Table 3. Basic thermodynamic and kinetic characteristics of tautomerisation of Watson-Crick DNA base pairs obtained at the MP2/6-311++G(2df,pd)//B3LYP/6-311++G(d,p) level of theory in vacuum #

Comparatively with the reverse barriers heights of tautomerisation of the Gua*·Cyt* and Ade*·Thy* mispairs (5.15 and 0.11 kcal/mol, respectively) the values of their interaction energies (22.94 and 33.80 kcal/mol , respectively) are high enough for mispairs dissociation into mutagenic tautomers (Table 6).

Although the equilibrium constants of tautomerisation of the Gua*·Cyt* ($9.42 \cdot 10^{-7}$) and Ade* ·Thy* ($1.41 \cdot 10^{-9}$) (Table 3) mispairs involving mutagenic tautomers fall within the range of the mutation frequency (Drake, 1991), their lifetimes ($1.04 \cdot 10^{-12}$ s and $2.57 \cdot 10^{-14}$ s , accordingly, see Table 3) are negligible comparably with the time of one base pair dissociation during the enzymatic DNA replication (10^{-9} s) to cause spontaneous mutations. So, Löwdin's mispairs "escape from the hands" of replication apparatus.

These data indicate that Löwdin's mechanism is not sufficient to explain the mutagenic tautomers formation within Ade ·Thy and Gua ·Cyt base pairs of DNA.

4.4 Tautomerisation of the DNA bases facilitated by an isolated water molecule

It has been established quite some time ago that there is a shell of tightly bound water molecules at the surface of DNA with properties significantly different from those of bulk water and it seems that DNA interaction with water largely determines its conformation, stability, and ligand binding properties (J.H. Wang, 1955; Tunis & Hearst, 1968; Falk et al., 1970; Kubinec and Wemmer, 1992). The pure rotational spectra of the binary adducts of Ura and Thy with water were first observed by laser ablation molecular beam Fourier transform microwave spectroscopy (López et al., 2010). Investigation of the structure of the adducts from the rotational constants of the different isotopologues shows that the observed conformers of bases correspond to the most stable forms in which water closes a cycle with the nucleic acid bases through H-bonds (López et al., 2010).

In this work we for the first time present a complete study of the proton transfer kinetic of intramolecular water-assisted tautomerisation mechanism for all DNA bases (Fig. 4) by computing the rate constants with the conventional transition state theory (Atkins, 1998), including the Wigner's tunnelling correction (Wigner, 1932).

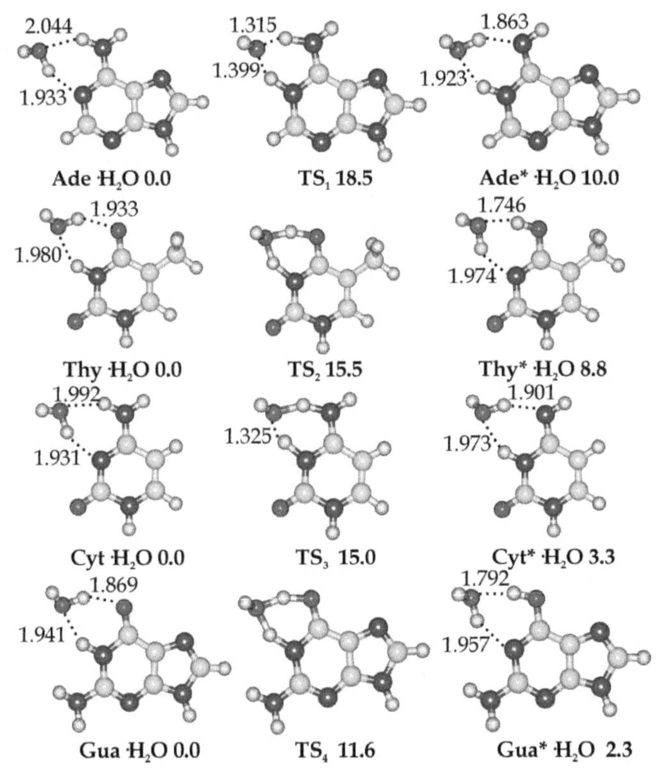

Fig. 4. Water-assisted tautomerisation of the DNA bases. The dotted lines indicate H-bonds AH...B (their lengths H...B are presented in angstroms), while continuous lines show covalent bonds. Relative Gibbs free energies (T=298.15 K, in vacuum) are obtained at the MP2/6-311++G(2df,pd)//B3LYP/6-311++G(d,p) level of theory and reported near each structure in kcal/mol

We found that the interaction of the canonical tautomers of the DNA bases with a water molecule at the Watson-Crick edge changes the gas-phase stability: the relative Gibbs free energies of the Ade and Thy decrease, while those of the Cyt and Gua - increase (Table 4). So, it means that in the case of complexes with water, the order of stability of Ade and Thy mutagenic tautomers remains the same as for isolated bases; moreover, they are stabilized in these complexes. On the contrary, the order of stability of Cyt and Gua mutagenic tautomers changes in their complexes with water. So, equilibrium constants of tautomerisation for the Ade·H_2O and Thy·H_2O complexes ($4.89 \cdot 10^{-8}$ and $3.39 \cdot 10^{-7}$, respectively) fall into the mutationally significant range, while for the Cyt·H_2O and Gua·H_2O complexes ($4.16 \cdot 10^{-3}$ and $2.16 \cdot 10^{-2}$, respectively)these values are considerably higher (Table 4).

For comparison, computation results reported by Gorb and Leszczynski (Gorb & Leszczynski, 1998a, 1998b) are of a special interest. As part of their comprehensive study of water-mediated proton transfer between canonical and mutagenic tautomers of Cyt and Gua, the authors have shown that the interaction with water changes the order of relative energies of cytosine tautomers.

Conversion	$\Delta\Delta G_{TS}$, kcal/mol	k, s^{-1}	τ, s	$\tau_{1/2}$, s	$\tau_{99.9\%}$, s	ΔG, kcal/mol	K
Ade·H_2O→Ade*·H_2O	18.53	$3.80 \cdot 10^{-1}$	2.63	1.82	$1.19 \cdot 10^{-6}$	9.97	$4.89 \cdot 10^{-8}$
Ade*·H_2O→Ade·H_2O	8.56	$7.77 \cdot 10^{6}$	$1.29 \cdot 10^{-7}$	$8.92 \cdot 10^{-8}$			
Thy·H_2O→Thy*·H_2O	15.51	$8.12 \cdot 10^{1}$	$1.23 \cdot 10^{-2}$	$8.54 \cdot 10^{-3}$	$3.84 \cdot 10^{-8}$	8.82	$3.39 \cdot 10^{-7}$
Thy*·H_2O→Thy·H_2O	6.69	$2.40 \cdot 10^{8}$	$4.17 \cdot 10^{-9}$	$2.89 \cdot 10^{-9}$			
Cyt·H_2O→Cyt*·H_2O	15.00	$1.80 \cdot 10^{2}$	$5.57 \cdot 10^{-3}$	$3.86 \cdot 10^{-3}$	$2.13 \cdot 10^{-4}$	3.25	$4.16 \cdot 10^{-3}$
Cyt*·H_2O→Cyt·H_2O	11.75	$4.32 \cdot 10^{4}$	$2.32 \cdot 10^{-5}$	$1.61 \cdot 10^{-5}$			
Gua·H_2O→Gua*·H_2O	11.63	$5.78 \cdot 10^{4}$	$1.73 \cdot 10^{-5}$	$1.20 \cdot 10^{-5}$	$3.37 \cdot 10^{-6}$	2.27	$2.16 \cdot 10^{-2}$
Gua*·H_2O→Gua·H_2O	9.36	$2.68 \cdot 10^{6}$	$3.74 \cdot 10^{-7}$	$2.59 \cdot 10^{-7}$			

see designations in Table 2

Table 4. Basic thermodynamic and kinetic characteristics of water-assisted tautomerisation of DNA bases obtained at the MP2/6-311++G(2df,pd)//B3LYP/6-311++G(d,p) level of theory in vacuum #

It should be noted that in the works devoted to the water-assisted tautomerisation (Fogarasi & Szalay, 2002; Furmanchuk et al., 2011; Gu & Leszczynski, 1999; H.-S. Kim et al., 2007; López et al., 2010; Michalkova et al., 2008; Sobolewski & Adamowicz, 1995) the authors did not justify their choice of the Watson-Crick edges of nucleotide bases (Watson & Crick, 1953a, 1953b) for interaction with a water molecule. This can be explained by the absence of the experimental or theoretical data on hydration of the isolated DNA bases. Up to date, the reported data include only the analysis of hydration of DNA bases in crystal structures of oligonucleotides of A- (Schneider et al., 1992), B- (Schneider et al., 1992, 1993; Schneider & Berman, 1995) and Z-forms of DNA (Schneider et al., 1992, 1993) and wide angle neutron scattering study of an A-DNA fiber (Langan et al., 1992). These studies revealed that sites of the preferred hydration of base pairs are localized in the major groove of DNA. Later on Fogarasi et al. (Fogarasi & Szalay, 2002) have demonstrated that the preferable position for water binding to Cyt is the O=C2-N1-H (H-O-C2=N1 in the enol form) moiety.

The energy barriers for water-assisted tautomerisation are greatly reduced (by 21-27 kcal/mol) as compared with the corresponding ones in the gas phase. Therefore, the explicit water molecules could accelerate by several orders the tautomerisation process from canonical to mutagenic tautomer. Such significant reduction in the internal tautomerisation barriers could be explained by the formation of the H-bonds between the water molecule and nucleic acid bases, which stabilize the transition state.

The time necessary to reach 99.9% of the equilibrium concentration of mutagenic tautomer in the system ($\tau_{99.9\%}$) for these barriers falls within the range $3.84 \cdot 10^{-8} \div 2.13 \cdot 10^{-4}$ s, which is by orders smaller, except Cyt, than the time of an elementary act of one base pair replication (*ca.* $4 \cdot 10^{-4}$ s). The barriers for the reverse reactions lead to a half-lifetime of about 10^{-9} s, and tunneling effects will further facilitate the reverse process. So, complexes "mutagenic tautomer-water" produced in the DPT process represent unstable intermediates, which quickly converted back into the complexes "canonical tautomer-water" in the time scale of the nucleotide-water interaction. However, if the dissociation of the water from the tautomerized complex occurs, the mutagenic tautomer would be a long-lived species, as the barrier for the reverse conversion to canonical tautomer is more than *ca.* 27 kcal/mol (see Table 2). It should be noted that electronic energy of the dissociation of the Ade*·H_2O and Thy*·H_2O complexes (Table 5) are lower than the corresponding reverse barriers. So, it can mean that these complexes more probably decay to the mutagenic tautomers and water molecule. To the contrary, in the case of Gua and Cyt – the Gua*·H_2O and Cyt*·H_2O transition to the complexes involving canonical tautomers will be more probable than the decay of the tautomerized complexes. Following the electronic energies of the interaction between bases and molecules of water, we could conclude that transition to the complexes containing mutagenic tautomers of Ade and Thy isn't preferential as they have larger electronic energies of the interaction that complicates their dissociation into mutagenic tautomers (Table 5). Interaction energy of the DNA bases with water is less than the energy of interaction with the complementary bases. So, the nucleotide bases competing with water for binding will displace water to the periphery of the interaction interface.

Complex	$-\Delta E_{int}$	$\Delta\Delta E$	$\Delta\Delta G$
Ade·H_2O	9.60	-	-
Ade*·H_2O	12.72	11.56	8.56
Thy·H_2O	8.74	-	-
Thy*·H_2O	12.48	9.55	6.69
Cyt·H_2O	11.26	-	-
Cyt*·H_2O	10.16	14.93	11.75
Gua·H_2O	11.52	-	-
Gua*·H_2O	9.72	12.40	9.36

ΔE_{int} – the counterpoise-corrected electronic energy of interaction; $\Delta\Delta E$ – the reverse barrier (difference in electronic energy) of tautomerisation; $\Delta\Delta G$ – the reverse barrier (difference in Gibbs free energy) of tautomerisation

Table 5. Electronic and Gibbs free energies (in kcal/mol) (T= 298.15 K) of complexes of DNA bases with water molecule obtained at the MP2/6-311++G(2df,pd)//B3LYP/6-311++G(d,p) level of theory in vacuum#

4.5 Tautomerisation of the DNA bases in dimers

Theoretical and experimental studies also explored agents other than water, which can enhance the stability of rare tautomers of DNA bases in the gas phase. Of particular interest were their interactions with amino acids (Fan et al., 2010; Samijlenko et al., 2001, 2004;

Stepanyugin et al., 2002a, 2002b) and protons or alkali metal cations (Lippert et al., 1986; Lippert & Gupta, 2009; Samijlenko et al., 2010; Šponer et al., 2001), as the extra positive charge could stabilize the structure of rare tautomers through an intramolecular salt bridge. Moreover, the coordination of metal ions to nucleobases is known to lead frequently to the stabilization of rare tautomeric forms (Burda et al., 2000; Lippert et al., 1986; Lippert & Gupta, 2009; Samijlenko et al., 2010), with numerous examples reported for various nucleobases (Lippert & Gupta, 2009; Lippert et al., 1986; Schoellhorn et al., 1989; Renn et al., 1991; Zamora et al., 1997). In these metal-stabilized rare tautomers, the metal is located at a position that is usually occupied by a proton, forcing the proton to move to another position and thereby generating the rare tautomer.

Yang and Rodgers (Yang & Rodgers, 2004) were probably the first to bring up the important question that a possible way of tautomerisation may be through dimerization.

In the literature, there are available papers devoted to the investigation of the tautomerisation of DNA bases by the different chemical compounds, e.g. glycine-assisted tautomerisation of Ura (Dąbkowska et al., 2005) and tautomerisation of Thy by methanol (Fan et al., 2010). However, it was established that such interactions result in the reducing of the internal barrier of tautomerisation and thermodynamic equilibrium could be easily attained at room temperature, the dynamical stability of the tautomerized in such a way complexes remained out of authors' eyeshot.

Providing *ab initio* quantum-chemical study of hydrogen-bonded complexes of acetic acid with canonical and mutagenic tautomers of DNA bases methylated at the glycosidic nitrogen atoms *in vacuo* and continuum with a low dielectric constant we established that all tautomerized complexes are dynamically unstable because their electronic energy barriers for the reverse tautomerisation reaction do not exceed zero-point energy of corresponding vibrational modes, frequencies of which become imaginary in the transition states of tautomerisation (Brovarets' et al., 2010c; Brovarets' et al., 2012) (Fig. 5).

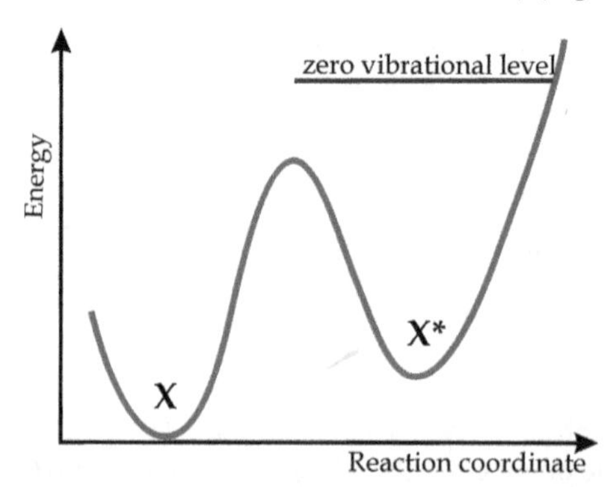

Fig. 5. Qualitative representation of potential energy profile of the X↔X* conversion. X and X* - complexes containing DNA base in canonical and mutagenic tautomeric forms, respectively

A potential pathway for the generation of the mutagenic amino-enol form of guanine is reported by Padermshoke et al. (Padermshoke et al., 2008), who investigated DPT reactions in three guanine-guanine dimers, a guanine-thymine wobble base pair, and a model compound 4(3H)-pyrimidinone dimer using *ab initio* MO calculations and liquid-phase IR spectroscopy. The calculations suggest that the DPT processes in these dimers are energetically accessible and temperature-dependent IR measurements of the model compound reveal that slight thermal energy can induce the DPT reaction, and hence the enol tautomer can appear.

5. Mispairs involving mutagenic tautomers of DNA bases

The mutagenic tautomers of DNA bases can form six possible purine-pyrimidine base pairs - Ade·Cyt*, Ade* ·Cyt, Gua* ·Thy, Gua ·Thy*, Ade* ·Thy* and Gua* ·Cyt* - thereby demonstrating the electronic and geometrical complementarity.

In a DNA double helix, Gua forms an H-bonded pair with Cyt. Meanwhile, the mutagenic enol form of Gua (Gua*) can pair with Thy (Brovarets' & Hovorun, 2010d; Danilov et al., 2005; Mejía-Mazariegos & Hernández-Trujillo, 2009) instead of Cyt. Similarly, the mutagenic imino form of Cyt (Cyt*) pairs with Ade (Danilov et al., 2005; Fonseca Guerra et al., 2006) instead of Gua. Then, during replication, when the two strands separate the Thy and Ade bases of the anomalous Gua*·Thy and Ade·Cyt* base pairs would combine with Ade and Thy instead of Cyt and Gua, respectively. Thus, the scheme postulated in (Watson & Crick, 1953a, 1953b) leads to a spontaneous transition Gua·Cyt→Ade·Thy in the subsequent rounds of replication if not repaired appropriately (Kunz et al., 2009). In DNA, the canonical form of Ade combines with the canonical form of Thy; however, the Ade* mutagenic imino tautomer combines with Cyt rather than with Thy, while the mutagenic enol form of Thy* forms a pair with Gua instead of Ade. After the strand separation, the counter-base pairs Gua·Cyt and Cyt·Gua instead of Ade·Thy and Thy·Ade are formed, respectively. As a result this leads to a spontaneous Ade·Thy→Gua·Cyt transition.

To gain more insight into the nature of the formed tautomeric base pairs, we have analysed their hydrogen-bonding mechanism and geometrical features to compare them with the same characteristics obtained for the natural Watson-Crick base pairs.

As shown by Kool et al. in the experiments on DNA replication (Guckian et al., 2000; Kool et al., 2000; Morales & Kool, 2000; Kool, 2002), an incoming nucleotide must be able to form, with its partner in the template, a base pair which sterically resembles the natural Watson-Crick base pair(Ade·Thy or Gua·Cyt). In addition, it was recently shown that the ability of the incoming base to form H-bonds with the template base is also of great importance (Bebenek et al., 2011; W. Wang et al., 2011). Bebenek et al. (Bebenek et al., 2011) have shown that a human DNA polymerase λ poised to misinsert dGTP opposite a template Thy can form a mismatch with Watson–Crick-like geometry and Wang et al. (W. Wang et al., 2011) observed that the Ade·Cyt mismatch can mimic the shape of cognate base pairs at the site of incorporation.

According to the geometric selection mechanism of bases as a principal determinant of DNA replication fidelity (Echols & Goodman, 1991; Goodman, 1997; Sloane et al., 1988), the geometrical and electrostatic properties of the polymerase active site are likely to have a profound influence on nucleotide-insertion specificities. This influence would strongly favor the insertion of the base pairs having an optimal geometry, in which the distance between C1 atoms of paired nucleotides and the N9–C1(Pur)–C1(Pyr) and N1–C1(Pyr)–C1(Pur) angles characterizing the nucleotide pair in double helix are most closely approximated to those of the Watson–Crick base pairs. These values for the irregular base pair as distinguished from the Watson–Crick base pairs reflect the distortion of double helix conformation and can be factor taking into account the recognition of the structural invariants of the sugar-phosphate backbone by the polymerase.

Detailed study of the geometric characteristics for the optimized mutagenic and Watson–Crick base pairs leads to the following results. The distance between the bonds joining the bases to the deoxyribose groups in the Gua*·Thy and Gua·Thy* mutagenic base pairs is close to the corresponding canonical distance in the Gua·Cyt base pair, and the corresponding distance in the Ade*·Cyt and Ade·Cyt* base pairs is close to that in the Ade·Thy base pair. Moreover, in each pair of stereoisomers (Gua*·Thy, Gua·Thy* and Ade*·Cyt, Ade·Cyt*), the N9–C1-C1 and N1–C1–C1 glycosidic angles are close to the corresponding value in one of the Watson–Crick canonical base pairs. Analogous conclusions were made earlier by Topal and Fresco (Topal & Fresco, 1976) and Danilov et al. (Danilov et al., 2005), who studied each of the above-mentioned mutagenic base pairs by model building and by ab initio methods, respectively, and showed that these pairs are sterically compatible with the Watson–Crick base pairs.

Finally, according to the molecular mechanism of recognition of the complementary base pairs of nucleic acids by DNA polymerase (Li & Waksman, 2001), the key role in the selection of the correct substrate is the interactions of the certain amino acid residues in the recognition site of DNA polymerase with the invariant arrangement of the N3 purine and O2 pyrimidine atoms (Beard & Wilson, 1998, 2003; Poltev et al., 1998). These hydrogen-bonding interactions may provide a means of detecting misincorporation at this position. Our data show that the structural invariants of the mutagenic nucleotide pairs are very close to those of the correct nucleotide pairs. In other words, the mutual position of the atoms and atomic groups is practically the same both for the correct and the irregular pairs, so that the DNA polymerase (more exactly its recognizing site) can play the role of additional matrix under the inclusion of the nucleotides. Therefore, we conclude that the formation of the DNA mutagenic base pairs satisfies the geometric constraints of the standard double helical DNA. If these mutagenic base pairs would be incorporated into a standard Watson–Crick double helix, the helix would not likely experience significant distortion and its stability would not be greatly deteriorated.

The comparison of the formation energies of the canonical and mutagenic base pairs (Table 6) shows that the Löwdin's Ade*·Thy* base pair, which electronic formation energy is -33.80 kcal/mol, is the most stable among all the studied base pairs. At the same time, the formation of the Gua*·Thy and Ade*·Cyt mispairs is more favorable than that of the Ade·Thy canonical base pair, Gua·Thy* and Ade·Cyt* mispairs which have -14.92; -33.39 and -23.50 kcal/mol formation energy, respectively (Table 6). From the other point of view, it may evidence that dissociation of the Gua*·Thy and Ade*·Cyt mispairs will be complicated during the strand separation. These data therefore confirm that Ade·Cyt* and Gua*·Thy mispairs are suitable candidates for the spontaneous point mutations arising in DNA (Fig. 6). The Ade*·Cyt and Gua·Thy* lifetimes ($3.49 \cdot 10^{-11}$ s and $3.59 \cdot 10^{-13}$ s, accordingly) are too short comparably with the time of one base pair dissociation during the enzymatic DNA replication (10^{-9} s). This means that these mispairs will "slip away" from replication machinery: they transfer to Ade·Cyt* and Gua*·Thy accordingly (Fig. 6). In this way Ade*·Cyt and Gua·Thy* mispairs act as intermediates in this reaction.

Base pair	$-\Delta E_{int}$	$\Delta\Delta E$	$\Delta\Delta G$
Ade·Thy	14.92	-	-
Ade*·Thy*	33.80	0.11	-1.01
Gua·Cyt	29.28	-	-
Gua*·Cyt*	22.94	5.15	1.49
Ade·Cyt*	15.73	-	-
Ade*·Cyt	23.50	6.44	3.75
Gua·Thy*	33.39	-	-
Gua*·Thy	19.82	4.16	1.17

see designations in Table 5

Table 6. Electronic and Gibbs free energies (in kcal/mol) (T=298.15 K) of base pairs obtained at the MP2/6-311++G(2df,pd)//B3LYP/6-311++G(d,p) level of theory in vacuum#

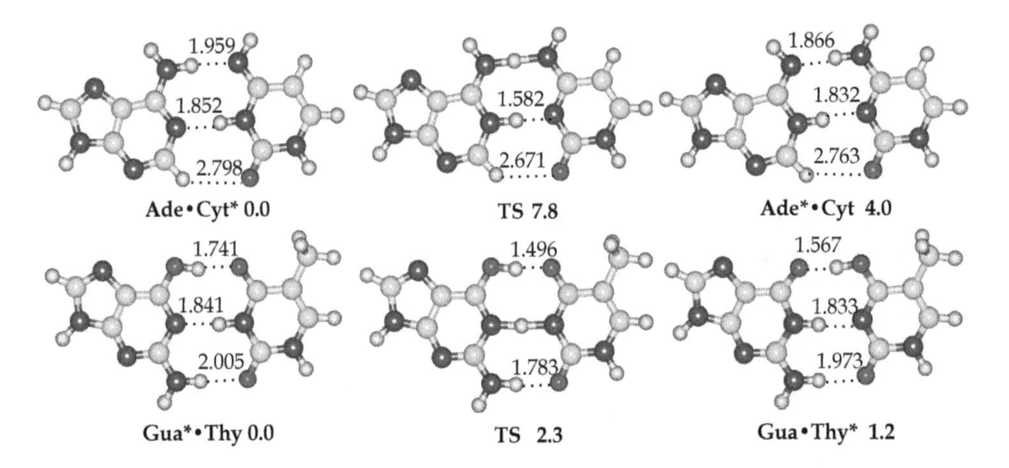

Fig. 6. Interconversion of Ade·Cyt*↔Ade*·Cyt and Gua*·Thy↔Gua·Thy* mispairs involving mutagenic tautomers of DNA bases. Relative Gibbs free energies (T=298.15 K, in vacuum) are obtained at the MP2/6-311++G(2df,pd)//B3LYP/6-311++G(d,p) level of theory and reported near each structure in kcal/mol. The dotted lines indicate H-bonds AH...B (their lengths H...B are presented in angstroms), while continuous lines show covalent bonds

The obtained Gibbs free energies of interaction indicate that Gua*·Thy and Ade·Cyt* are more favorable than Gua·Thy* and Ade*·Cyt. It was established that the Ade*· Cyt and Gua*·Cyt* base pairs are metastable and easily (i.e., without facing significant barrier) „slip" into the energetically more favorable Ade · Cyt* and Gua·Cyt base pairs, respectively. The comparison of reverse electronic barriers of interconversion with the zero-point energies of competent vibrational modes (Table 7) of the tautomerized complexes allows concluding that Ade* ·Thy* and Gua ·Thy* complexes are dynamically unstabletheir electronic barriers of the reverse transition are noticeably lower than zero-point energy of corresponding vibrational modes.

Tautomerisation reaction	ΔE, kcal/mol	$\Delta\Delta E_{TS}$, kcal/mol	$\Delta\Delta E$		ν, cm^{-1}
			kcal/mol	cm^{-1}	
Gua·Cyt↔Gua*·Cyt*	7.87	13.02	5.15	1800.8	2951.8
Ade·Thy↔Ade*·Thy*	12.26	12.37	0.11	37.7	3349.2
Ade·Cyt*↔Ade*·Cyt	3.67	10.11	6.44	2253.5	3024.9
Gua* • Thy↔Gua • Thy*	1.13	5.29	4.16	1455.4	3155.7

ΔE – the relative electronic energy of the tautomerized complex; $\Delta\Delta E_{TS}$ – the activation barrier of tautomerisation in terms of electronic energy; $\Delta\Delta E=\Delta\Delta E_{TS}-\Delta E$ – the reverse barrier of tautomerisation in terms of electronic energy; ν – the frequency of the vibrational mode of the tautomerized complex which becomes imaginary in the transition state of tautomerisation

Table 7. Energetic characteristics of DNA bases tautomerisation in studied base pairs obtained at the MP2/6-311++G(2df,pd)//B3LYP/6-311++G(d,p) level of theory in vacuum#

6. Conclusions

In this study, we made an attempt to answer some actual questions related to physico-chemical nature of spontaneous point mutations in DNA induced by prototropic tautomerism of its bases.

It was shown that the lifetime of mutagenic tautomers of all four canonical DNA bases exceeds by many orders not only the time required for replication machinery to enzymatically incorporate one incoming nucleotide into structure of DNA double helix ($\sim 4 \cdot 10^{-4}$ s), and even a typical time of DNA replication in cell ($\sim 10^{3}$ s). The high stability of mutagenic tautomers of DNA bases is mainly determined by the absence of intramolecular H-bonds in their canonical and mutagenic forms.

This finding substantially supports the tautomeric hypothesis of the origin of spontaneous point mutations, for instance replication errors, removing all doubts on instability of mutagenic tautomers of isolated DNA bases, which are sometimes expressed by biologists.

Notwithstanding a tremendous heuristic and methodological role of the classical Löwdin's mechanism of the origin of spontaneous point mutations during DNA replication, it was demonstrated that this mechanism probably has substantial limitations. From the physico-chemical point of view, the advantage of Löwdin's mechanism lies in the fact that the tautomerisation of base pairs does not disturb standard Watson-Crick base-pairing geometry. Its main disadvantage is the instability of Ade*·Thy* base pair and metastability of Gua* • Cyt* base pair. The lifetime of tautomerized (Löwdin's) Ade* ·Thy* and Gua* ·Cyt* base pairs is less by orders than a characteristic time required for replication machinery to separate any Watson-Crick base pair ($\sim 10^{-9}$ s). Figuratively speaking, the Löwdin's base pairs "slip away" from replication apparatus: they transform to canonical base pairs and then dissociate without losing their canonical coding properties, as they haven't enough time to dissociate to mutagenic tautomers. These facts put the possibility of such mispairs involving mutagenic tautomers formation under a doubt, not to mention their complicated dissociation into mutagenic tautomers.

In this context, a topic of current importance is the search of novel physico-chemical mechanisms of tautomerisation of DNA bases in Watson-Crick base pairs: the pioneering, but encouraging steps have been already made in this direction (Brovarets', 2010; Cerón-Carrasco et al., 2009a, 2009b, 2011; Cerón-Carrasco & Jacquemin, 2011; Kryachko & Sabin, 2003).

It was found that a specific interaction of a single water molecule with the site of mutagenic tautomerisation in each of four canonical DNA bases could transform to into mutagenic tautomeric form in a definite time notably less than $\sim 4 \cdot 10^{-4}$ s. The most vulnerable point of this model of origin of replication error in DNA is a complete lack of experimental and especially theoretical support for a probability of the penetration of water molecules at a replication fork per one Watson-Crick base pair. Most likely such a probability is very low, since a compact, essentially hydrophobic organization of replisome (Marians, 2008; Pomerantz & O'Donnell, 2007) is supposed to minimize this probability.

In this work it was found that among all purine-pyrimidine base pairs with Watson-Crick geometry involving one base in mutagenic tautomeric form - Ade·Cyt*, Gua*·Thy, Ade*·Cyt and Gua·Thy*, Gua·Thy* mispair is dynamically unstable and Ade*·Cyt mispair has very small lifetime ($<<10^{-9}$ s) and therefore plays an intermediate role in DNA replication cycle, "sliding down" to the Ade·Cyt* mispair. This fact substantially alters the Löwdin's scheme (Löwdin, 1963, 1965, 1966) of replication point errors fixation arising due to the prototropic tautomerism of DNA bases, which treats all four base pairs Ade·Cyt*, Ade*·Cyt, Gua*·Thy and Gua·Thy* as stable structures.

In our opinion, the results reported here not only provide more evidence in support of Watson and Crick classical tautomeric hypothesis of point mutations, but also fill it with concrete physico-chemical content.

By combining the data from the literature with our findings, we concluded that the tautomeric mechanism of the origin of mutations in DNA should satisfy the following thermodynamic and kinetic criteria:

- the time needed to reach tautomerisation equilibrium in the complex $\tau_{99.9\%}$ should be considerably less than a specific time of one elementary DNA replication event (several ms);
- the tautomerized complex should be dynamically stable and moreover should have the lifetime significantly exceeding a specific time required for a replication machinery to forcibly dissociate a Watson-Crick base pair into monomers (several ns);
- a dissociation energy of the tautomerized complex should not exceed a dissociation energy of the complex with canonical tautomer participation;
- a thermodynamic population (equilibrium constant of tautomerisation) of the pair with a mutagenic tautomer participation relative to the basic tautomeric state should be within the range of 10^{-8}-10^{-11}, that agrees fully with biological experimental data.

Finishing our conclusions, we hope that this theoretical study gives valuable and thorough information on the chemically intriguing and biologically relevant questions of the DNA bases tautomerism. Our results presented here are believed to provide a new insight into the molecular nature of spontaneous point mutations in DNA and also be a promising and perspective tool for experimentalists working in the field of DNA mutagenesis.

7. Acknowledgments

This work was partly supported by the State Fund for Fundamental Research of Ukraine within the Ukrainian-Russian (0111U006629) and Ukrainian-Slovenian (0111U007526) research bilateral projects. Authors thank Bogolyubov Institute for Theoretical Physics of the National Academy of Sciences of Ukraine and Ukrainian-American Laboratory of Computational Chemistry (President, Prof. Dr. Jerzy Leszczynski) for providing calculation resources and software allocation.

8. References

Aamouche, A., Ghomi, M., Grajcar, L., Baron, M.H., Romain, F., Baumruk, V., Stepanek, J., Coulombeau, C., Jobic, H. & Berthier, G. (1997). Neutron inelastic scattering, optical spectroscopies and scaled quantum mechanical force fields for analyzing the vibrational dynamics of pyrimidine nucleic acid bases: 3. Cytosine. *J. Phys. Chem. A*, Vol. 101, No. 51, (December 1997), pp. 10063-10074, ISSN: 1089-5639 (Print), 1520-5215 (Electronic).

AIMAll (Version 10.07.01), Keith, T.A., 2010 (aim.tkgristmill.com).

Alonso, J.L., Peña, I., López, J.C. & Vaquero, V. (2009). Rotational spectral signatures of four tautomers of guanine. *Angew. Chem. Int. Ed.*, Vol. 48, No. 33, (August 2009), pp. 6141–6143, ISSN: 1521-3773 (Electronic).

Atkins, P.W. (January 1998). *Physical Chemistry* (6th edition), Oxford University Press, ISBN-10: 0198501013, ISBN-13: 978-0198501015, Oxford, UK.

Basu, S., Majumdar, R., Das, G.K. & Bhattacharyya, D. (2005). Energy barrier and rates of tautomeric transitions in DNA bases: *ab initio* quantum chemical study. *Indian J. Biochem. Biophys.*, Vol. 42, No. 6, (December 2005), pp. 378-385, ISSN: 0301-1208 (Print), 0975-0959 (Electronic).

Bazsó, G., Tarczay, G., Fogarasi, G. & Szalay, P.G. (2011). Tautomers of cytosine and their excited electronic states: a matrix isolation spectroscopic and quantum chemical study. *Phys. Chem. Chem. Phys.*, Vol. 13, No. 15, (April 2011), pp. 6799-6807, ISSN: 1463-9076 (Print), 1463-9084 (Electronic).

Beard, W.A. & Wilson, S.H. (1998). Structural insights into DNA polymerase β fidelity: hold tight if you want it right. *Chem. Biol.*, Vol. 5, No. 1, (January 1998), pp. R7-R13, ISSN: 1074-5521 (Print), 1879-1301 (Electronic).

Beard, W.A. & Wilson, S.H. (2003). Structural insights into the origins of DNA polymerase fidelity. *Structure*, Vol. 11, No. 5, (May 2003), pp. 489–496, ISSN: 0969-2126 (Print), 1878-4186 (Electronic).

Bebenek, K., Pedersen, L.C. & Kunkel, T.A. (2011). Replication infidelity *via* a mismatch with Watson-Crick geometry. *Proc. Natl. Acad. Sci. U.S.A.*, Vol. 108, No. 5, (February 2011), pp. 1862-1867, ISSN: 0027-8424 (Print), 1091-6490 (Electronic).

Becke, A.D. (1993). Density-functional thermochemistry. III. The role of exact exchange. *J. Chem. Phys.*, Vol. 98, No. 7, (April 1993), pp. 5648-5652, ISSN: 0021-9606 (Print), 1089-7690 (Electronic).

Bludský, O., Šponer, J., Leszczynski, J., Špirko, V. & Hobza, P. (1996). Amino groups in nucleic acid bases, aniline, aminopyridines, and aminotriazine are nonplanar: results of correlated *ab initio* quantum chemical calculations and anharmonic analysis of the aniline inversion motion. *J. Chem. Phys.*, Vol. 105, No. 24, (December 1996), pp. 11042-11050, ISSN: 0021-9606 (Print), 1089-7690 (Electronic).

Boys, S.F. & Bernardi, F. (1970). The calculation of small molecular interactions by the differences of separate total energies. Some procedures with reduced errors. *Mol. Phys.*, Vol. 19, No. 4, (1970), pp. 553-566, ISSN: 0026-8976 (Print), 1362-3028 (Electronic).

Brauer, N.B., Smolarek, S., Zhang, X., Buma, W.J. & Drabbels M. (2011). Electronic spectroscopy of aniline ions embedded in helium nanodroplets. *J. Phys. Chem. Lett.*, Vol. 2, No. 13, (June 2011), pp. 1563–1566, ISSN: 1948-7185 (Electronic).

Brovarets', O.O. & Hovorun, D.M. (2010a). How stable are the mutagenic tautomers of DNA bases? *Biopolym. Cell*, Vol. 26, No.1, (January-February 2010), pp. 72-76, ISSN: 0233–7657 (Print), 1993-6842 (Electronic).

Brovarets', O.O. & Hovorun, D.M. (2010b). Intramolecular tautomerisation and the conformational variability of some classical mutagens - DNA purine bases derivatives: quantum chemical study. *Physics of the Alive (Fizyka zhyvoho)*, Vol. 18, No. 1, (January-February 2010), pp. 5-17, ISSN: 1023-2427.

Brovarets', O.O., Bulavin, L.A. & Hovorun, D.M. (2010c). Can the proteins tautomerize the DNA base pairs: the physical answer to the biologically important question. *Reports of the National Academy of Sciences of Ukraine*, No. 2, (February 2010), pp. 76-82, ISSN: 1025-6415.

Brovarets', O.O. & Hovorun, D.M. (2010d). By how many characters is the genetic information written in DNA? *Reports of the National Academy of Sciences of Ukraine*, No. 6, (June 2010), pp. 175-179, ISSN: 1025-6415.

Brovarets', O.O., Zhurakivsky, R.O. & Hovorun, D.M. (2010e). Is there adequate ionization mechanism of the spontaneous transitions? Quantum-chemical investigation. *Biopolym. Cell*, Vol. 26, No. 5, (September-October 2010), pp. 398-405, ISSN: 0233–7657 (Print), 1993-6842 (Electronic).

Brovarets', O.O. & Hovorun, D.M. (2010f). Stability of mutagenic tautomers of uracil and its halogen derivatives: the results of quantum-mechanical investigation. *Biopolym. Cell*, Vol. 26, No. 4, (July-August 2010), pp. 295-298, ISSN: 0233–7657 (Print), 1993-6842 (Electronic).

Brovarets', O.O. PhD Thesis: Physico-chemical nature of the spontaneous and induced by the mutagens transitions and transversions, Kyiv, 2010.

Brovarets', O.O. & Hovorun, D.M. (2011a). Intramolecular tautomerization and the conformational variability of some classical mutagens – cytosine derivatives: quantum chemical study. *Biopolym. Cell*, Vol. 27, No. 3, (May-June 2011), pp. 221–230, ISSN: 0233–7657 (Print), 1993-6842 (Electronic).

Brovarets', O.O. & Hovorun, D.M. (2011b). IR vibrational spectra of H-bonded complexes of adenine, 2-aminopurine and 2-aminopurine$^+$ with cytosine and thymine: quantum-chemical study. *Opt. Spectrosc.*, Vol. 111, No. 5, (November 2011), pp. 750–757, ISSN: 0030-400X (Print), 1562-6911 (Electronic).

Brovarets', O.O., Yurenko, Y.P., Dubey, I.Ya. & Hovorun, D.M. (2012). Can DNA-binding proteins of replisome tautomerize nucleotide bases? *Ab initio* model study. *J. Biol. Struct. Dynam.*, ISSN: 0739-1102, (in press).

Brown, T., Kennard, O., Kneale, G. & Rabinovich, D. (1985). High-resolution structure of a DNA helix containing mismatched base pairs. *Nature,* Vol. 315, No. 6020, (June 1985), pp.604-606, ISSN: 0028-0836 (Print), 1476-4687 (Electronic).

Brown, R.D., Godfrey, P.D., McNaughton, D. & Pierlot, A.P. (1989a). A study of the major gas-phase tautomer of adenine by microwave spectroscopy. *Chem. Phys. Lett.,* Vol. 156, No. 1, (March 1989), pp. 61-63, ISSN: 0009-2614 (Print).

Brown, R.D., Godfrey, P.D., McNaughton, D. & Pierlot, A.P. (1989b). Tautomers of cytosine by microwave spectroscopy. *J. Am. Chem. Soc.,* Vol. 111, No. 6, (March 1989), pp. 2308–2310, ISSN: 0002-7863 (Print), 1520-5126 (Electronic).

Burda, J.V., Šponer, J. & Leszczynski, J. (2000). The interactions of square platinum(II) complexes with guanine and adenine: a quantum-chemical *ab initio* study of metalated tautomeric forms. *J. Biol. Inorg. Chem.,* Vol. 5, No. 2, (April 2000), pp. 178-188, ISSN: 0949-8257 (Print), 1432-1327 (Electronic).

Canuel, C., Mons, M., Piuzzi, F., Tardivel, B., Dimicoli, I. & Elhanine, M. (2005). Excited states dynamics of DNA and RNA bases: characterization of a stepwise deactivation pathway in the gas phase. *J. Chem. Phys.,* Vol. 122, No. 7, (February 2005), pp. 074316-074321, ISSN: 0021-9606 (Print), 1089-7690 (Electronic).

Cerón-Carrasco, J.P., Requena, A., Michaux, C., Perpète, E.A. & Jacquemin, D. (2009a). Effects of hydration on the proton transfer mechanism in the adenine-thymine base pair. *J. Phys. Chem. A,* Vol. 113, No. 127, (June 2009), pp. 7892-7898, ISSN: 1089-5639 (Print), 1520-5215 (Electronic).

Cerón-Carrasco, J.P., Requena, A., Zúñiga, J., Michaux, C., Perpète, E. A. & Jacquemin, D. (2009b). Intermolecular proton transfer in microhydrated guanine-cytosine base pair: a new mechanism for spontaneous mutation in DNA. *J. Phys. Chem. A,* Vol. 113, No. 39, (September 2009), pp. 10549-10556, ISSN: 1089-5639 (Print), 1520-5215 (Electronic).

Cerón-Carrasco, J.P., Zúñiga, J., Requena, A., Perpète, E. A., Michaux, C. & Jacquemin, D. (2011a). Combined effect of stacking and solvation on the spontaneous mutation in DNA. *Phys. Chem. Chem. Phys.,* Vol. 13, No. 32, (August 2011), pp. 14584-14589, ISSN: 1463-9076 (Print), 1463-9084 (Electronic).

Cerón-Carrasco, J.P. & Jacquemin, D. (2011b). Influence of Mg^{2+} on the guanine-cytosine tautomeric equilibrium: simulations of the induced intermolecular proton transfer. *Chem. Phys. Chem.,* Vol. 12, No. 14, (October 2011), pp. 2615-2623, ISSN: 1439-4235 (Print), 1439-7641 (Electronic).

Chatake, T., Hikima, T., Ono, A., Ueno, Y., Matsuda, A. & Takenaka, A. (1999). Crystallographic studies on damaged DNAs. II. N-6-methoxyadenine can present two alternate faces for Watson–Crick base-pairing, leading to pyrimidine transition mutagenesis. *J. Mol. Biol.,* Vol. 294, No. 5, (December 1999), pp. 1223-1230, ISSN: 0022-2836 (Print), 1089-8638 (Electronic).

Chen, H. & Li, S. (2006). Theoretical study on the excitation energies of six tautomers of guanine: evidence for the assignment of the rare tautomers. *J. Phys. Chem. A,* Vol. 110, No. 45, (November 2006), pp. 12360–12362, ISSN: 1089-5639 (Print), 1520-5215 (Electronic).

Chin, W., Mons, M., Piuzzi, F., Tardivel, B., Dimicoli, I., Gorb, L. & Leszczynski, J. (2004). Gas phase rotamers of the nucleobase 9-methylguanine enol and its monohydrate: optical spectroscopy and quantum mechanical calculations. *J. Phys. Chem. A,* Vol.

108, No. 40, (October 2004), pp. 8237–8243, ISSN: 1089-5639 (Print), 1520-5215 (Electronic).

Choi, M.Y., Dong, F. & Miller, R.E. (2005). Multiple tautomers of cytosine identified and characterized by infrared laser spectroscopy in helium nanodroplets: probing structure using vibrational transition moment angles. *Phil. Trans. R. Soc. A*, Vol. 363, No. 1827, (February 2005), pp. 393–413, ISSN: 1471-2962 (Electronic).

Choi, M.Y. & Miller, R.E. (2006). Four tautomers of isolated guanine from infrared laser spectroscopy in helium nanodroplets. *J. Am. Chem. Soc.*, Vol. 128, No. 22, (June 2006), pp. 7320–7328, ISSN: 0002-7863 (Print), 1520-5126 (Electronic).

Choi, M.Y., Dong, F., Han, S.W. & Miller, R.E. (2008). Nonplanarity of adenine: vibrational transition moment angle studies in helium nanodroplets. *J. Phys. Chem. A*, Vol. 112, No. 31, (August 2008), pp. 7185–7190, ISSN: 1089-5639 (Print), 1520-5215 (Electronic).

Colarusso, P., Zhang, K., Guo, B. & Bernath, P.F. (1997). The infrared spectra of uracil, thymine, and adenine in the gas phase. *Chem. Phys. Lett.*, Vol. 269, No. 1-2, (April 1997), pp. 39-48, ISSN: 0009-2614 (Print).

Crespo-Hernández, C.E., Cohen, B., Hare, P.M. & Kohler, B. (2004). Ultrafast excited-state dynamics in nucleic acids. *Chem. Rev.*, Vol. 104, No. 4, (April 2004), pp. 1977-2020, ISSN: 0009-2665 (Print), 1520-6890 (Electronic).

Crick, F.H. (1966). Codon-anticodon pairing: the wobble hypothesis. *J. Mol. Biol.*, Vol. 19, No. 2, (August 1966), pp. 548–555, ISSN: 0022-2836 (Print), 1089-8638 (Electronic).

Dąbkowska, I., Gutowski, M. & Rak, J. (2005). Interaction with glycine increases stability of a mutagenic tautomer of uracil. A density functional theory study. *J. Am. Chem. Soc.*, Vol. 127, No. 7, (February 2005), pp. 2238-2248, ISSN: 0002-7863 (Print), 1520-5126 (Electronic).

Danilov, V.I., Anisimov, V.M., Kurita, N. & Hovorun, D. (2005). MP2 and DFT studies of the DNA rare base pairs: the molecular mechanism of the spontaneous substitution mutations conditioned by tautomerism of bases. *Chem. Phys. Lett.*, Vol. 412, No. 4-6, (September 2005), pp. 285-293, ISSN: 0009-2614 (Print).

Dolinnaya, N.G. & Gromova, E.S. (1983). Complementation interactions of oligonucleotides. *RUSS CHEM REV*, Vol. 52, No. 1, (1983), pp. 79-95, ISSN: 0036-021X.

Dolinnaya, N.G. & Gryaznova, O.I. (1989). Complexes of oligo(poly)nucleotides with structural anomalies. *RUSS CHEM REV*, Vol. 58, No. 8, (1989), pp. 758-777, ISSN: 0036-021X.

Dong, F. & Miller, R.E. (2002). Vibrational transition moment angles in isolated biomolecules: a structural tool. *Science*, Vol. 298, No. 5596, (November 2002), pp. 1227–1230, ISSN: 0036-8075 (Print), 1095-9203 (Electronic).

Drabløs, F., Feyzi, E., Aas, P.A., Vaagbø, C.B., Kavli, B., Bratlie, M.S., Peña-Diaz, J., Otterlei, M., Slupphaug, G. & Krokan, H.E. (2004). Alkylation damage in DNA and RNA—repair mechanisms and medical significance. *DNA Repair*, Vol. 3, No. 11, (November 2004), pp. 1389–1407, ISSN: 1568-7864 (Print), 1568-7856 (Electronic).

Drake, J.W. (1991). A constant rate of spontaneous mutation in DNA-based microbes. *Proc. Natl. Acad. Sci. U.S.A.*, Vol. 88, No. 16, (August 1991), pp. 7160–7164, ISSN: 0027-8424 (Print), 1091-6490 (Electronic).

Dreyfus, M., Bensaude, O., Dodin, G. & Dubois, J.E. (1976). Tautomerism in cytosine and 3-methylcytosine. A thermodynamic and kinetic study. *J. Am. Chem. Soc.*, Vol. 98, No. 20, (September 1976), pp. 6338-6349, ISSN: 0002-7863 (Print), 1520-5126 (Electronic).

Echols, H. & Goodman, M.F. (1991). Fidelity mechanisms in DNA replication. *Annu. Rev. Biochem.*, Vol. 60, (July 1991), pp. 477-511, ISSN: 0066-4154.

Ehrlich, M., Norris, K.F., Wang, R.Y., Kuo, K.C. & Gehrke, C.W. (1986). DNA cytosine methylation and heat-induced deamination. *Biosci. Rep.*, Vol. 6, No. 4, (April 1986), pp. 387–393, ISSN: 0144-8463 (Print), 1573-4935 (Electronic).

Elshakre, M. (2005). *Ab initio* study of guanine tautomers in the S_0 and D_0 states. *Int. J. Quantum Chem.*, Vol. 104, No. 1, (2005), pp. 1-15, ISSN: 0020-7608 (Print), 1097-461X (Electronic).

Falk, M., Poole, A.G. & Goymour, C.G. (1970). Infrared study of the state of water in the hydration shell of DNA. *Can. J. Chem.*, Vol. 48, No. 10, (May 1970), pp. 1536-1542, ISSN: 0008-4042 (Print), 1480-3291 (Electronic).

Fan, J.C., Shang, Z.C., Liang, J., Liu, X.H. & Jin, H. (2010). Systematic theoretical investigations on the tautomers of thymine in gas phase and solution. *J. Mol. Struct.: THEOCHEM*, Vol. 939, No. 1-3, (January 2010), pp. 106-111, ISSN: 0166-1280 (Print).

Fersht, A.R. & Knill-Jones, J.W. (1983). Fidelity of replication of bacteriophage X174 DNA *in vitro* and *in vivo*. *J. Mol. Biol.*, Vol. 165, No. 4, (April 1983), pp. 633-654, ISSN: 0022-2836 (Print), 1089-8638 (Electronic).

Feyer, V., Plekan, O., Richter, R., Coreno, M., Vall-llosera, G., Prince, K.C., Trofimov, A.B., Zaytseva, I.L., Moskovskaya T.E., Gromov, E.V. & Schirmer, J. (2009). Tautomerism in cytosine and uracil: an experimental and theoretical core level spectroscopic study. *J. Phys. Chem. A*, Vol. 113, No. 19, (May 2009), pp. 5736–5742, ISSN: 1089-5639 (Print), 1520-5215 (Electronic).

Feyer, V., Plekan, O., Richter, R., Coreno, M., de Simone, M., Prince, K.C., Trofimov, A.B., Zaytseva, I.L. & Schirmer, J. (2010). Tautomerism in cytosine and uracil: a theoretical and experimental X-ray absorption and resonant auger study. *J. Phys. Chem. A*, Vol. 114, No. 37, (September 2010), pp. 10270–10276, ISSN: 1089-5639 (Print), 1520-5215 (Electronic).

Florian, J., Hrouda, V. & Hobza, P. (1994). Proton transfer in the adenine-thymine base pair. *J. Am. Chem. Soc.*, Vol. 116, No. 4, (February 1994), pp. 1457-1460, ISSN: 0002-7863 (Print), 1520-5126 (Electronic).

Florian, J., Leszczynski, J. & Scheiner, S. (1995). *Ab initio* study of the structure of guanine-cytosine base pair conformers in gas phase and polar solvents. *Mol. Phys.*, Vol. 84, No. 3, (1995), pp. 469–480, ISSN: 0026-8976 (Print), 1362-3028 (Electronic).

Florian, J. & Leszczynski, J. (1996). Spontaneous DNA mutations induced by proton transfer in the guanine cytosine base pairs: an energetic perspective. *J. Am. Chem. Soc.*, Vol. 118, No. 12, (March 1996), pp. 3010–3017, ISSN: 0002-7863 (Print), 1520-5126 (Electronic).

Fogarasi, G. & Szalay, P.G. (2002). The interaction between cytosine tautomers and water: an MP2 and coupled cluster electron correlation study. *Chem. Phys. Lett.*, Vol. 356, No. 3, (April 2002), pp. 383–390, ISSN: 0009-2614 (Print).

Fogarasi, G. (2002). Relative stabilities of three low-energy tautomers of cytosine: a coupled cluster electron correlation study. *J. Phys. Chem. A*, Vol. 106, No. 7, (February 2002), pp. 1381–1390, ISSN: 1089-5639 (Print), 1520-5215 (Electronic).

Fogarasi, G. (2008). Water-mediated tautomerization of cytosine to the rare imino form: an *ab initio* dynamics study. *Chem. Phys.*, Vol. 349, No. 1-3, (June 2008), pp. 204–209, ISSN: 0301-0104 (Print).

Fonseca Guerra, C., Bickelhaupt, F.M., Saha, S. & Wang, F. (2006). Adenine tautomers: relative stabilities, ionization energies, and mismatch with cytosine. *J. Phys. Chem. A*, Vol. 110, No. 11, (March 2006), pp. 4012-4020, ISSN: 1089-5639 (Print), 1520-5215 (Electronic).

Friedberg, E.C., Walker, G.C., Siede, W., Wood, R.D., Schultz, R.A. & Ellenberger, T. (May 2006). *DNA Repair and Mutagenesis* (2nd edition), ASM Press, ISBN: 1-55581-319-4, Washington, D.C.,USA.

Frisch, M.J.; Trucks, G.W.; Schlegel, H.B.; Scuseria, G.E.; Robb, M.A.; Cheeseman, J.R.; Montgomery, J.A., Jr.; Vreven, T.; Kudin, K.N.; Burant, J.C.; Millam, J.M.; Iyengar, S.S.; Tomasi, J.; Barone, V.; Mennucci, B.; Cossi, M.; Scalmani, G.; Rega, N.; Petersson, G.A.; Nakatsuji, H.; Hada, M.; Ehara, M.; Toyota, K.; Fukuda, R.; Hasegawa, J.; Ishida, M.; Nakajima, T.; Honda, Y.; Kitao, O.; Nakai, H.; Klene, M.; Li, X.; Knox, J.E.; Hratchian, H.P.; Cross, J.B.; Adamo, C.; Jaramillo, J.; Gomperts, R.; Stratmann, R.E.; Yazyev, O.; Austin, A.J.; Cammi, R.; Pomelli, C.; Ochterski, J.W.; Ayala, P.Y.; Morokuma, K.; Voth, G.A.; Salvador, P.; Dannenberg, J.J.; Zakrzewski, V.G.; Dapprich, S.; Daniels, A.D.; Strain, M.C.; Farkas, O.; Malick, D.K.; Rabuck, A.D.; Raghavachari, K.; Foresman, J.B.; Ortiz, J.V.; Cui, Q.; Baboul, A.G.; Clifford, S.; Cioslowski, J.; Stefanov, B.B.; Liu, G.; Liashenko, A.; Piskorz, P.; Komaromi, I.; Martin, R.L.; Fox, D.J.; Keith, T.; Al-Laham, M.A.; Peng, C.Y.; Nanayakkara, A.; Challacombe, M.; Gill, P.M.W.; Johnson, B.; Chen, W.; Wong, M.W.; Gonzalez, C.; Pople, J.A. *Gaussian 03, Revision C.02*, Gaussian, Inc.: 2003.

Fujii, M., Tamura, T., Mikami, N. & Ito, M. (1986). Electronic spectra of uracil in a supersonic jet. *Chem. Phys. Lett.*, Vol. 126, No. 6, (May 1986), pp. 583–587, ISSN: 0009-2614 (Print).

Furmanchuk, A., Isayev, O., Gorb, L., Shishkin, O.V., Hovorun, D.M. & Leszczynski, J. (2011). Novel view on the mechanism of water-assisted proton transfer in the DNA bases: bulk water hydration. *Phys. Chem. Chem. Phys.*, Vol. 13, No. 10, (2011), pp. 4311–4317, ISSN: 1463-9076 (Print), 1463-9084 (Electronic).

García-Terán, J.P., Castillo, O., Luque, A., García-Couceiro, U., Beobide, G. & Román, P. (2006). Supramolecular architectures assembled by the interaction of purine nucleobases with metal-oxalato frameworks. Non-covalent stabilization of the 7H-adenine tautomer in the solid-state. *Dalton Trans.*, No. 7, (February 2006), pp. 902-911, ISSN: 1477-9226.

Goodman, M.F. (1997). Hydrogen bonding revisited: geometric selection as a principal determinant of DNA replication fidelity. *Proc. Natl. Acad. Sci. U.S.A.*, Vol. 94, No. 20, (September 1997), pp. 10493-10495, ISSN: 0027-8424 (Print), 1091-6490 (Electronic).

Govorun, D.N., Danchuk, V.D., Mishchuk, Ya.R., Kondratyuk, I.V., Radomsky, N.F. & Zheltovsky, N.V. (1992). AM1 calculation of the nucleic acid bases structure and vibrational spectra. *J. Mol. Struct.*, Vol. 267, (March 1992), pp. 99-103, ISSN: 0022-2860.

Gonzalez, C. & Schlegel, H.B. (1989). An improved algorithm for reaction path following. *J. Chem. Phys.*, Vol. 90, No. 4, (February 1989), pp. 2154-2161, ISSN: 0021-9606 (Print), 1089-7690 (Electronic).

Gorb, L. & Leszczynski, J. (1998a). Intramolecular proton transfer in monohydrated tautomers of cytosine: an *ab initio* post-Hartree–Fock study. *Int. J. Quant. Chem.*, Vol. 70, No. 4-5, (1998), pp. 855-862, ISSN: 1097-461X.

Gorb, L. & Leszczynski, J. (1998b). Intramolecular proton transfer in mono- and dihydrated tautomers of guanine: an *ab initio* post Hartree-Fock study. *J. Am. Chem. Soc.*, Vol. 120, No. 20, (May 1998), pp. 5024-5032, ISSN: 0002-7863 (Print), 1520-5126 (Electronic).

Gorb, L., Podolyan, Y., Leszczynski, J., Siebrand, W., Fernandez-Ramos, A. & Smedarchina, Z. (2001). A quantum-dynamics study of the prototropic tautomerism of guanine and its contribution to spontaneous point mutations in *Escherichia coli*. *Biopolymers*, Vol. 61, No. 1, (2001/2002), p. 77-83, ISSN: 0006-3525 (Print), 1097-0282 (Electronic).

Gorb, L., Podolyan, Y., Dziekonski, P., Sokalski, W.A. & Leszczynsky, J. (2004). Double-proton transfer in adenine-thymine and guanine-cytosine base pairs. A post-Hartree-Fock *ab initio* study. *J. Am. Chem. Soc.*, Vol. 126, No. 32, (August 2004), pp. 10119-10129, ISSN: 0002-7863 (Print), 1520-5126 (Electronic).

Gorb, L., Kaczmarek, A., Gorb, A., Sadlej, A.J. & Leszczynski, J. (2005). Thermodynamics and kinetics of intramolecular proton transfer in guanine. Post Hartree-Fock study. *J. Phys. Chem. B*, Vol. 109, No. 28, (July 2005), pp. 13770–13776, ISSN: 1520-6106 (Print), 1520-5207 (Electronic).

Gribov, L.A. & Mushtakova, S.P. (1999). *Quantum Chemistry: Textbook (Kvantovaya Khimiya: Uchebnik)*, Gardariki, ISBN: 5-8297-0017-4, Moscow, Russian Federation, pp 317-319.

Gu, J. & Leszczynski, J. (1999). A DFT study of the water-assisted intramolecular proton transfer in the tautomers of adenine. *J. Phys. Chem. A.*, Vol. 103, No. 15, (March 1999), pp. 2744-2750, ISSN: 1089-5639 (Print), 1520-5215 (Electronic).

Guckian, K.M., Krugh, T.R. & Kool, E.T. (2000). Solution structure of a nonpolar, non-hydrogen-bonded base pair surrogate in DNA. *J. Am. Chem. Soc.*, Vol. 122, No. 29, (July 2000), pp. 6841-6847, ISSN: 0002-7863 (Print), 1520-5126 (Electronic).

Gutowski, M., Van Lenthe, J.H., Verbeek, J., Van Duijneveldt, F.B. & Chalasinski, G. (1986). The basis set superposition error in correlated electronic structure calculations. *Chem. Phys. Lett.*, Vol. 124, No. 4, (1986), pp. 370-375, ISSN: 0009-2614 (Print).

Hanus, M., Ryjáček, F., Kabeláč, M., Kubař, T., Bogdan, T.V., Trygubenko, S.A. & Hobza, P. (2003). Correlated *ab initio* study of nucleic acid bases and their tautomers in the gas phase, in a microhydrated environment and in aqueous solution. Guanine: surprising stabilization of rare tautomers in aqueous solution. *J. Am. Chem. Soc.*, Vol. 125, No. 25, (June 2003), pp. 7678-7688, ISSN: 0002-7863 (Print), 1520-5126 (Electronic).

Hanus, M., Kabeláč, M., Rejnek, J., Ryjáček, F. & Hobza, P. (2004). Correlated *ab initio* study of nucleic acid bases and their tautomers in the gas phase, in a microhydrated environment, and in aqueous solution. Part 3. Adenine. *J. Phys. Chem. B*, Vol. 108, No. 6, (February 2004), pp. 2087-2097, ISSN: 1520-6106 (Print), 1520-5207 (Electronic).

Harris, V.H., Smith, C.L., Jonathan Cummins, W., Hamilton, A.L., Adams, H., Dickman, M., Hornby, D.P. & Williams, D.M. (2003). The effect of tautomeric constant on the specificity of nucleotide incorporation during DNA replication: support for the rare tautomer hypothesis of substitution mutagenesis. *J. Mol. Biol.*, Vol. 326, No. 5, (March 2003), pp. 1389-1401, ISSN: 0022-2836 (Print), 1089-8638 (Electronic).

Hauswirth, W. & Daniels, M. (1971). Fluorescence of thymine in aqueous solution at 300° K. *Photochem. Photobiol.*, Vol. 13, No. 2, (February 1971), pp. 157-163, ISSN: 1751-1097 (Electronic).

Hobza, P. & Šponer, J. (1999). Structure, energetics, and dynamics of the nucleic acid base pairs: nonempirical *ab initio* calculations. *Chem. Rev.*, Vol. 99, No. 11, (November 1999), pp. 3247-3276, ISSN: 0009-2665 (Print), 1520-6890 (Electronic).

Hovorun, D.M., Danchuk, V.D., Mishchuk, Ya.R., Kondratyuk, I.V. & Zheltovsky, M.V. (1995a). About the non-planarity and dipole non-stability of the canonical nucleotide bases methylated at the glycosidic nitrogen atom. *Reports of the National Academy of Sciences of Ukraine,* No. 6, (June 1995), pp. 117-119, ISSN: 1025-6415.

Hovorun, D.M., Kondratyuk, I.V., Mishchuk, Ya.R. & Zheltovsky, M.V. (1995b). Non-equivalence of the amine hydrogen atoms in the canonical nucelotide bases. *Reports of the National Academy of Sciences of Ukraine,* No. 8, (August 1995), pp. 130-132, ISSN: 1025-6415.

Hovorun, D.M. & Kondratyuk, I.V. (1996). Anisotropy of the rotational mobility of the amino group in the canonical nucleotide bases. *Reports of the National Academy of Sciences of Ukraine*, No. 10, (October 1996), pp. 152-155, ISSN: 1025-6415.

Hovorun, D.M., Gorb, L. & Leszczynski, J. (1999). From the nonplanarity of the amino group to the structural nonrigidity of the molecule: a post-Hartree-Fock *ab initio* study of 2-aminoimidazole. *Int. J. Quant. Chem.*, Vol. 75, No. 3, (1999), pp. 245-253, ISSN: 1097-461X.

Hunter, W.N., Brown, T., Anand, N.N. & Kennard, O. (1986). Structure of an adenine-cytosine base pair in DNA and its implications for mismatch repair. *Nature,* Vol. 320, No. 6062, (April 1986), pp. 552-555, ISSN: 0028-0836 (Print), 1476-4687 (Electronic).

Joyce, C.M. & Benkovic, S.J. (2004). DNA polymerase fidelity: kinetics, structure, and checkpoints. *Biochemistry*, Vol. 43, No. 45, (November 2004), pp. 14317–14324, ISSN: 0006-2960 (Print), 1520-4995 (Electronic).

Kang, H., Lee, K.T., Jung, B., Ko, Y.J. & Kim, S.K. (2002). Intrinsic lifetimes of the excited state of DNA and RNA bases. *J. Am. Chem. Soc.*, Vol. 124, No., 44, (November 2002), pp. 12958-12959, ISSN: 0002-7863 (Print), 1520-5126 (Electronic).

Katritzky, A.R. & Waring, A.J. (1962). 299. Tautomeric azines. Part I. The tautomerism of 1-methyluracil and 5-bromo-1-methyluracil. *J. Chem. Soc.*, No. 0, (1962), pp. 1540–1544, ISSN: 0368-1769.

Kennard, O. (1985). Structural studies of DNA fragments: the G·T wobble base pair in A, B and Z DNA; the G·A base pair in B-DNA. *J. Biomol. Struct. Dyn.*, Vol. 3, No. 2, (October 1985), pp. 205–226, ISSN: 0739-1102 (Print), 1538-0254 (Electronic).

Kim, H.-S., Ahn, D.-S., Chung, S.-Y., Kim, S.K. & Lee, S. (2007). Tautomerization of adenine facilitated by water: computational study of microsolvation. *J. Phys. Chem. A*, Vol. 111, No. 32, (August 2007), pp. 8007–8012, ISSN: 1089-5639 (Print), 1520-5215 (Electronic).

Kim, N.J., Jeong, G., Kim, Y.S., Sung, J., Kim, S.K. & Park, Y.D. (2000). Resonant two-photon ionization and laser induced fluorescence spectroscopy of jet-cooled adenine. *J. Chem. Phys.*, Vol. 113, No. 22, (December 2000), pp. 10051-10055, ISSN: 0021-9606 (Print), 1089-7690 (Electronic).

Komarov, V.M. & Polozov, R.V. (1990). Nonplanar structure of aminosubstituted nitrogenous bases. *Biofizika,* Vol. 35, No. 2, (1990), pp. 367-368, ISSN: 0006-3509 (Print), 1555-6654 (Electronic).

Komarov, V.M., Polozov, R.V. & Konoplev, G.G. (1992). Non-planar structure of nitrous bases and non-coplanarity of Watson-Crick pairs. *J. Theor. Biol.*, Vol. 155, No. 3, (April 1992), pp. 281-294, ISSN: 0022-5193.

Kondratyuk, I.V., Samijlenko, S.P., Kolomiets, I.M. & Hovorun, D.M. (2000). Prototropic molecular-zwitterionic tautomerism of xanthine and hypoxanthine. *J. Mol. Struct.*, Vol. 523, No. 1-3, (May 2000), pp. 109-118, ISSN: 0022-2860.

Kool, E.T., Morales, J.C. & Guckian, K.M. (2000). Mimicking the structure and function of DNA: insights into DNA stability and replication. *Angew. Chem. Int. Ed. Engl.*, Vol. 39, No. 6, (March 2000), pp. 990-1009, ISSN: 1521-3773 (Electronic).

Kool, E.T. (2002). Active site tightness and substrate fit in DNA replication. *Annu. Rev. Biochem.*, Vol. 71, (July 2002), pp. 191–219, ISSN: 0066-4154.

Kornberg, A. & Baker, T.A. (January 1992). *DNA Replication,* W. H. Freeman, ISBN-10: 0716720035, ISBN-13: 978-0716720034, New York, USA.

Kosenkov, D., Kholod, Y., Gorb, L., Shishkin, O., Hovorun, D.M., Mons, M. & Leszczynski, J. (2009). *Ab initio* kinetic simulation of gas-phase experiments: tautomerization of cytosine and guanine. *J. Phys. Chem. B,* Vol. 113, No. 17, (April 2009), pp. 6140-6150, ISSN: 1520-6106 (Print), 1520-5207 (Electronic).

Kosma, K., Schroter, C., Samoylova, E., Hertel, I.V. & Schultz, T. (2009). Excited-state dynamics of cytosine tautomers. *J. Am. Chem. Soc.*, Vol. 131, No. 46, (November 2009), pp. 16939-16943, ISSN: 0002-7863 (Print), 1520-5126 (Electronic).

Kostko, O., Bravaya, K., Krylov, A. & Ahmed, M. (2010). Ionization of cytosine monomer and dimer studied by VUV photoionization and electronic structure calculations. *Phys. Chem. Chem. Phys.*, Vol. 12, No. 12, (March 2010), pp. 2860–2872, ISSN: 1463-9076 (Print), 1463-9084 (Electronic).

Kow, Y.W. (2002). Repair of deaminated bases in DNA. *Free Radic. Biol. Med.*, Vol. 33, No. 7, (October 2002), pp. 886–893, ISSN: 0891-5849 (Print), 1873-4596 (Electronic).

Kryachko, E.S. & Sabin, J.R. (2003). Quantum chemical study of the hydrogen-bonded patterns in A•T base pair of DNA: origins of tautomeric mispairs, base flipping, and Watson-Crick → Hoogsteen conversion. *Int. J. Quant. Chem.*, Vol. 91, No. 6, (2003), pp. 695-710, ISSN: 1097-461X.

Kubinec, M.G., & Wemmer, D.E. (1992). NMR evidence for DNA bound water in solution. *J. Am. Chem. Soc.*, Vol. 114, No. 22, (October 1992), pp. 8739-8740, ISSN: 0002-7863 (Print), 1520-5126 (Electronic).

Kunz, C., Saito, Y. & Schär, P. (2009). DNA repair in mammalian cells: mismatched repair: variations on a theme. *Cell. Mol. Life Sci.*, Vol. 66, No. 6, (March 2009), pp. 1021-1038, ISSN: 1420-682X (Print), 1420-9071 (Electronic).

Kwiatkowski, J.S. & Pullman, B. (1975). Tautomerism and electronic structure of biological pyrimidines, In: *Advances in Heterocyclic Chemistry,* Katritzky, A.R., Boulton, A.J., Vol. 18, pp. 199-335, Academic Press, ISBN: 0-12-020618-8, New York, USA.

Kwiatkowski, J.S. & Leszczynski, J. (1992). An *ab initio* quantum-mechanical study of tautomerism of purine, adenine and guanine. *J. Mol. Struct.: THEOCHEM,* Vol. 208, No. 1-2, (August 1992), pp. 35-44, ISSN: 0166-1280 (Print).

Kydd, R.A. & Krueger, P.J. (1977). The far-infrared vapour phase spectra of aniline-ND_2 and aniline-NHD. *Chem. Phys. Lett.,* Vol. 49, No. 3, (August 1977), pp. 539-543, ISSN: 0009-2614 (Print).

Kydd, R.A. & Krueger, P.J. (1978). The far-infrared vapor phase spectra of some halosubstituted anilines. *J. Chem. Phys.,* Vol. 69, No. 2, (July 1978), pp. 827-832, ISSN: 0021-9606 (Print), 1089-7690 (Electronic).

Labet, V., Grand, A., Morell, C., Cadet, J. & Eriksson, L.A. (2008). Proton catalyzed hydrolytic deamination of cytosine: a computational study. *Theor. Chem. Acc.,* Vol. 120, No. 4-6, (July 2008), pp. 429–435, ISSN: 1432-881X (Print), 1432-2234 (Electronic).

Langan, P., Forsyth, V.T., Mahendrasingam, A., Pigram, W.J., Mason, S.A. & Fuller, W. (1992). A high angle neutron fiber diffraction study of the hydration of the A conformation of the DNA double helix. *J. Biomol. Struct. Dyn.,* Vol. 10, No. 3, (December 1992), pp. 489-503, ISSN: 0739-1102 (Print), 1538-0254 (Electronic).

Lapinski, L., Nowak, M.J., Reva, I., Rostkowska, H. & Fausto, R. (2010). NIR-laser-induced selective rotamerization of hydroxy conformers of cytosine. *Phys. Chem. Chem. Phys.,* Vol. 12, No. 33, (September 2010), pp. 9615–9618, ISSN: 1463-9076 (Print), 1463-9084 (Electronic).

Larsen, N.W., Hansen, E.L. & Nicolaisen, F.M. (1976). Far infrared investigation of aniline and 4-fluoroaniline in the vapour phase. Inversion and torsion of the amino group. *Chem. Phys. Lett.,* Vol. 43, No. 3, (November 1976), pp. 584-586, ISSN: 0009-2614 (Print).

Laxer, A., Major, D.T., Gottlieb, H.E. & Fischer, B. (2001). ($^{15}N_5$)-labeled adenine derivatives: synthesis and studies of tautomerism by ^{15}N NMR spectroscopy and theoretical calculations. *J. Org. Chem.,* Vol. 66, No. 16, (August 2001), pp. 5463-5481, ISSN: 0022-3263 (Print), 1520-6904 (Electronic).

Lee, C., Yang, W. & Parr, R.G. (1988). Development of the Colle-Salvetti correlation-energy formula into a functional of the electron density. *Phys. Rev. Condens. Matter,* Vol. 37, No. 2, (January 1988), pp. 785-789, ISSN: 0163-1829 (Print).

Li, Y. & Waksman, G. (2001). Crystal structures of a ddATP-, ddTTP-, ddCTP-, and ddGTP-trapped ternary complex of Klentaq1: insights into nucleotide incorporation and selectivity. *Protein Science,* Vol. 10, No. 6, (June 2001), pp. 1225–1233, ISSN: 0961-8368 (Print), 1469-896X (Electronic).

Lin, J., Yu, C., Peng, S., Akiyama, I., Li, K., Lee, L.K. & LeBreton, P.R. (1980). Ultraviolet photoelectron studies of the ground-state electronic structure and gas-phase tautomerism of purine and adenine. *J. Am. Chem. Soc.,* Vol. 102, No. 14, (July 1980), pp. 4627-4631, ISSN: 0002-7863 (Print), 1520-5126 (Electronic).

Lippert, B., Schoellhorn, H. & Thewalt, U. (1986). Metal-stabilized rare tautomers of nucleobases. 1. Iminooxo form of cytosine: formation through metal migration and estimation of the geometry of the free tautomer. *J. Am. Chem. Soc.,* Vol. 108, No. 21, (October 1986), pp. 6616–6621, ISSN: 0002-7863 (Print), 1520-5126 (Electronic).

Lippert, B. & Gupta, D. (2009). Promotion of rare nucleobase tautomers by metal binding. *Dalton Trans.,* No. 24, (2009), pp. 4619–4634, ISSN: 1477-9226 (Print), 1477-9234 (Electronic).

Lister, D.G., Tyler, J.K., Hog, J.H. & Larsen, N.W. (1974). The microwave spectrum, structure and dipole moment of aniline. *J. Mol. Struct.*, Vol. 23, No. 2, (November 1974), pp. 253-264, ISSN: 0022-2860.

Loeb, L.A. (2001). A mutator phenotype in cancer. *Cancer Res.*, Vol. 61, No. 8, (April 2001), pp. 3230–3239, ISSN: 0008-5472 (Print), 1538-7445 (Electronic).

López, J.C., Peña, M.I., Sanz, M.E. & Alonso, J.L. (2007). Probing thymine with laser ablation molecular beam Fourier transform microwave spectroscopy. *J. Chem. Phys.*, Vol. 126, No. 19, (May 2007), pp. 191103-191106, ISSN: 0021-9606 (Print), 1089-7690 (Electronic).

López, J.C., Alonso, J.L., Peña, I. & Vaquero, V. (2010). Hydrogen bonding and structure of uracil–water and thymine–water complexes. *Phys. Chem. Chem. Phys.*, Vol. 12, No. 42, (2010), pp. 14128–14134, ISSN: 1463-9076 (Print), 1463-9084 (Electronic).

Löwdin, P.-O. (1963). Proton tunneling in DNA and its biological implications. *Rev. Mod. Phys.*, Vol. 35, No. 3, (July-September 1963), pp. 724–732, ISSN: 0034-6861 (Print), 1539-0756 (Electronic).

Löwdin, P.-O. (1965). Isotope effect in tunneling and its influence on mutation rates. *Mutat. Res.*, Vol. 2, No. 3, (June 1965), pp. 18-221, ISSN: 0027-5107 (Print).

Löwdin, P.-O. (1966). Quantum genetics and the aperiodic solid: some aspects on the biological problems of heredity, mutations, aging, and tumors in view of the quantum theory of the DNA molecule, In: *Advances in Quantum Chemistry*, Löwdin, P.-O., Vol. 2, pp. 213-360, Academic Press, ISBN: 978-0-12-386477-2, New York, USA, London, UK.

Lührs, D.C., Viallon, J. & Fischer, I. (2001). Excited state spectroscopy and dynamics of isolated adenine and 9-methyladenine. *Phys. Chem. Chem. Phys.*, Vol. 3, No. 10, (2001), pp. 1827-1831, ISSN: 1463-9076 (Print), 1463-9084 (Electronic).

Marian, C.M. (2007). The guanine tautomer puzzle: quantum chemical investigation of ground and excited states. *J. Phys. Chem. A*, Vol. 111, No. 8, (March 2007), pp. 1545–1553, ISSN: 1089-5639 (Print), 1520-5215 (Electronic).

Marians, K.J. (2008). Understanding how the replisome works. *Nat. Struct. Mol. Biol.*, Vol. 15, No. 2, (February 2008), pp. 125-127, ISSN: 1545-9993 (Print), 1545-9985 (Electronic).

Mejía-Mazariegos, L. & Hernández-Trujillo, J. (2009). Electron density analysis of tautomeric mechanisms of adenine, thymine and guanine and the pairs of thymine with adenine or guanine. *Chem. Phys. Lett.*, Vol. 482, No. 1-3, (November 2009), pp. 24–29, ISSN: 0009-2614 (Print).

Michalkova, A., Kosenkov, D., Gorb, L. & Leszczynski, J. (2008). Thermodynamics and kinetics of intramolecular water assisted proton transfer in Na^+-1-methylcytosine water complexes. *J. Phys. Chem. B*, Vol. 112, No. 29, (July 2008), pp. 8624–8633, ISSN: 1520-6106 (Print), 1520-5207 (Electronic).

Min, A., Lee, S.J., Choi, M.Y. & Miller, R.E. (2009). Electric field dependence experiments and *ab initio* calculations of three cytosine tautomers in superfluid helium nanodroplets. *Bull. Korean Chem. Soc.*, Vol. 30, No. 12, (December 2009), pp. 3039–3044, ISSN: 0253-2964.

Mishra, S.K., Shukla, M.K. & Mishra, P.C. (2000). Electronic spectra of adenine and 2-aminopurine: an *ab initio* study of energy level diagrams of different tautomers in gas phase and aqueous solution. *Spectrochim. Acta A: Mol. Biomol. Spectrosc.*, Vol. 56, No. 7, (June 2000), pp. 1355-1384, ISSN: 1386-1425 (Print), 1873-3557 (Electronic).

Mons, M., Dimicoli, I., Piuzzi, F., Tardivel, B. & Elhanine, M. (2002). Tautomerism of the DNA base guanine and its methylated derivatives as studied by gas-phase Infrared and Ultraviolet Spectroscopy. *J. Phys. Chem. A*, Vol. 106, No. 20, (May 2002), pp. 5088–5094, ISSN: 1089-5639 (Print), 1520-5215 (Electronic).

Mons, M., Piuzzi, F., Dimicoli, I., Gorb, L. & Leszczynski, J. (2006). Near-UV resonant two-photon ionization spectroscopy of gas phase guanine: evidence for the observation of three rare tautomers. *J. Phys. Chem. A*, Vol. 110, No. 38, (September 2006), pp. 10921–10924, ISSN: 1089-5639 (Print), 1520-5215 (Electronic).

Morales, J.C. & Kool, E.T. (2000). Varied molecular interactions at the active sites of several DNA polymerases: nonpolar nucleoside isosteres as probes. *J. Am. Chem. Soc.*, Vol. 122, No. 6, (February 2000), pp. 1001-1007, ISSN: 0002-7863 (Print), 1520-5126 (Electronic).

Morsy, M.A., Al-Somali, A.M. & Suwaiyan, A. (1999). Fluorescence of thymine tautomers at room temperature in aqueous solutions. *J. Phys. Chem. B*, Vol. 103, No. 50, (December 1999), pp. 11205–11210, ISSN: 1520-6106 (Print), 1520-5207 (Electronic).

Nakabeppu, Y., Tsuchimoto, D., Yamaguchi, H. & Sakumi, K. (2007). Oxidative damage in nucleic acids and Parkinson's disease. *J. Neurosci. Res.*, Vol. 85, No. 5, (April 2007), pp. 919–934, ISSN: 1097-4547 (Electronic).

Nikolaienko, T.Yu., Bulavin, L.A. & Hovorun, D.M. (2011a). Conformational capacity of 5'-deoxyguanylic acid molecule investigated by quantum-mechanical methods. *Biopolym. Cell*, Vol. 27, No. 4, (July-August 2011), pp. 291-299, ISSN: 0233-7657 (Print), 1993-6842 (Electronic).

Nikolaienko, T.Yu., Bulavin, L.A. & Hovorun, D.M. (2011b). The 5`-deoxyadenylic acid molecule conformational capacity: quantum-mechanical investigation using density functional theory (DFT). *Ukr. Biochem. J. (Ukr. Biokhim. Zh.)*, Vol. 83, No. 4, (July-August 2011), pp. 16-28, ISSN: 0201-8470.

Nikolaienko, T.Yu., Bulavin, L.A. & Hovorun, D.M. (2011c). Structural flexibility of canonical 2`-deoxyribonucleotides in DNA-like conformers. *Ukr. Biochem. J. (Ukr. Biokhim. Zh.)*, Vol. 83, No. 5, (September-October 2011), pp. 22-32, ISSN: 0201-8470.

Nir, E., Grace, L., Brauer, B. & de Vries, M.S. (1999). REMPI spectroscopy of jet-cooled guanine. *J. Am. Chem. Soc.*, Vol. 121, No. 20, (May 1999), pp. 4896–4897, ISSN: 0002-7863 (Print), 1520-5126 (Electronic).

Nir, E., Kleinermanns, K., Grace, L. & de Vries, M.S. (2001a). On the photochemistry of purine nucleobases. *J. Phys. Chem. A*, Vol. 105, No. 21, (May 2001), pp. 5106-5110, ISSN: 1089-5639 (Print), 1520-5215 (Electronic).

Nir, E., Janzen, Ch., Imhof, P., Kleinermanns, K. & de Vries, M.S. (2001b). Guanine tautomerism revealed by UV-UV and IR-UV hole burning spectroscopy. *J. Chem. Phys.*, Vol. 115, No. 10, (September 2001), pp. 4604-4611, ISSN: 0021-9606 (Print), 1089-7690 (Electronic).

Nir, E., Muller, M., Grace, L.I. & de Vries, M.S. (2002a). REMPI spectroscopy of cytosine. *Chem. Phys. Lett.*, Vol. 355, No. 1-2, (March 2002), pp. 59–64, ISSN: 0009-2614 (Print).

Nir, E., Plützer, Chr., Kleinermanns, K. & de Vries, M. (2002b). Properties of isolated DNA bases, base pairs and nucleosides examined by laser spectroscopy. *Eur. Phys. J. D*, Vol. 20, No. 3, (September 2002), pp. 317–329, ISSN: 1434-6060 (Print), 1434-6079 (Electronic).

Norinder, U. (1987). A theoretical reinvestigation of the nucleic bases adenine, guanine, cytosine, thymine and uracil using AM1. *J. Mol. Struct.: THEOCHEM,* Vol. 151, (May 1987), pp. 259-269, ISSN: 0166-1280 (Print).

Nowak, M.J., Lapinski, L. & Fulara, J. (1989a). Matrix isolation studies of cytosine: the separation of the infrared spectra of cytosine tautomers. *Spectrochim. Acta A: Mol. Spectrosc.*, Vol. 45, No. 2, (February 1989), pp. 229-242, ISSN: 1386-1425.

Nowak, M.J., Lapinski, L. & Kwiatkowski, J.S. (1989b). An infrared matrix isolation study of tautomerism in purine and adenine. *Chem. Phys. Lett.*, Vol. 157, No. 1-2, (April 1989), pp. 14-18, ISSN: 0009-2614 (Print).

Nowak, M.J., Lapinski, L., Kwiatkowski, J.S. & Leszczynski, J. (1991). Infrared matrix isolation and *ab initio* quantum mechanical studies of purine and adenine. *Spectrochim. Acta A: Mol. Spectrosc.*, Vol. 47, No. 1, (1991), pp. 87-103, ISSN: 0584-8539.

Nowak, M.J., Rostkowska, H., Lapinski, L., Kwiatkowski, J.S. & Leszczynski, J. (1994a). Tautomerism N(9)H↔N(7)H of purine, adenine, and 2-chloroadenine: combined experimental IR matrix isolation and *ab initio* quantum mechanical studies. *J. Phys. Chem.*, Vol. 98, No. 11, (March 1994), pp. 2813-2816, ISSN: 0022-3654 (Print).

Nowak, M.J., Rostkowska, H., Lapinski, L., Kwiatkowski, J.S. & Leszczynski, J. (1994b). Experimental matrix isolation and theoretical *ab initio* HF/6-31G(d, p) studies of infrared spectra of purine, adenine and 2-chloroadenine. *Spectrochim. Acta A: Mol. Spectrosc.*, Vol. 50, No. 6, (June 1994), pp. 1081-1094, ISSN: 1386-1425.

Nowak, M.J., Lapinski, L., Kwiatkowski, J.S. & Leszczynski, J. (1996). Molecular structure and infrared spectra of adenine. Experimental matrix isolation and density functional theory study of adenine ^{15}N isotopomers. *J. Phys. Chem.*, Vol. 100, No. 9, (February 1996), pp. 3527-3534, ISSN: 0022-3654 (Print).

Padermshoke, A., Katsumoto, Y., Masaki, R. & Aida, M. (2008). Thermally induced double proton transfer in GG and wobble GT base pairs: a possible origin of the mutagenic guanine. *Chem. Phys. Lett.*, Vol. 457, No. 1-3, (May 2008), pp. 232–236, ISSN: 0009-2614 (Print).

Patel, D.J., Kozlowski, S.A., Marky, L.A., Rice, J.A., Broka, C., Dallas, J., Itakura, K. & Breslauer, K.J. (1982a). Structure, dynamics, and energetics of deoxyguanosine-thymidine wobble base pair formation in the self-complementary d(CGTGAATTCGCG) duplex in solution. *Biochemistry,* Vol. 21, No. 3, (February 1982), pp. 437-444, ISSN: 0006-2960 (Print), 1520-4995 (Electronic).

Patel, D.J., Pardi, A. & Itakura, K. (1982b). DNA conformation, dynamics, and interactions in solution. *Science,* Vol. 216, No. 4546, (May 1982), pp. 581-590, ISSN: 0036-8075 (Print), 1095-9203 (Electronic).

Patel, D.J., Kozlowski, S.A., Ikuta, S. & Itakura, K. (1984a). Deoxyadenosine-deoxycytidine pairing in the d(C-G-C-G-A-A-T-T-C-A-C-G) duplex: conformation and dynamics at and adjacent to the dA.dC mismatch site. *Biochemistry,* Vol. 23, No. 14, (July 1984), pp. 3218–3226, ISSN: 0006-2960 (Print), 1520-4995 (Electronic).

Patel, D.J., Kozlowski, S.A., Ikuta, S. & Itakura, K. (1984b). Dynamics of DNA duplexes containing internal G·T, G·A, A·C, and T·C pairs: hydrogen exchange at and adjacent to mismatch sites. *Fed. Proc.*, Vol. 43, No. 11, (August 1984), pp. 2663-2670, ISSN: 0014-9446 (Print).

Peng, C. & Schlegel, H.B. (1993). Combining synchronous transit and quasi-Newton methods to find transition states. *Isr. J. Chem.*, Vol. 33, No. 4, (1993), pp. 449-454, ISSN: 0021-2148 (Print), 1869-5868 (Electronic).

Peng, C., Ayala, P.Y., Schlegel, H.B. & Frisch, M.J. (1996). Using redundant internal coordinates to optimize equilibrium geometries and transition states. *J. Comput. Chem.*, Vol. 17, No. 1, (January 1996), pp. 49-56, ISSN: 0192-8651 (Print), 1096-987X (Electronic).

Plekan, O., Feyer, V., Richter, R., Coreno, M., Vall-llosera, G., Prince, K.C., Trofimov, A.B., Zaytseva, I.L., Moskovskaya, T.E., Gromov, E.V. & Schirmer, J. (2009). An experimental and theoretical core-level study of tautomerism in guanine. *J. Phys. Chem. A*, Vol. 113, No. 33, (August 2009), pp. 9376-9385, ISSN: 1089-5639 (Print), 1520-5215 (Electronic).

Plützer, Chr., Nir, E., de Vries, M.S. & Kleinermanns, K. (2001). IR–UV double-resonance spectroscopy of the nucleobase adenine. *Phys. Chem. Chem. Phys.*, Vol. 3, No. 24, (2001), pp. 5466-5469, ISSN: 1463-9076 (Print), 1463-9084 (Electronic).

Plützer, Chr. & Kleinermanns, K. (2002). Tautomers and electronic states of jet-cooled adenine investigated by double resonance spectroscopy. *Phys. Chem. Chem. Phys.*, Vol. 4, No. 20, (2002), pp. 4877-4882, ISSN: 1463-9076 (Print), 1463-9084 (Electronic).

Poltev, V.I., Shulyupina, N.V. & Bruskov, V.I. (1998). Fidelity of nucleic acid biosynthesis. Comparison of computer modeling results with experimental data. *Molecular Biology (Molekuliarnaia biologiia)*, Vol. 32, No. 2, (1998), pp. 233-240, ISSN: 0026-8933 (Print), 1608-3245 (Electronic).

Pomerantz, R.T. & O'Donnell, M. (2007). Replisome mechanics: insights into a twin DNA polymerase machine. *Trends in Microbiology*, Vol. 15, No. 4, (April 2007), pp. 156–164, ISSN: 0966-842X (Print), 1878-4380 (Electronic).

Privé, G.G., Heinemann, U., Chandrasegaran, S., Kan, L.-S., Kopka, M.L. & Dickerson, R. E. (1987). Helix geometry, hydration, and G·A mismatch in a B-DNA decamer. *Science*, Vol. 238, No. 4826, (October 1987), pp. 498-504, ISSN: 0036-8075 (Print), 1095-9203 (Electronic).

Quack, M. & Stockburger, M. (1972). Resonance fluorescence of aniline vapour. *J. Mol. Spectrosc.*, Vol. 43, No. 1, (July 1972), pp. 87-116, ISSN: 0022-2852 (Print).

Radchenko, E.D., Sheina, G.G., Smorygo, N.A. & Blagoi, Yu.P. (1984). Experimental and theoretical studies of molecular structure features of cytosine. *J. Mol. Struct.*, Vol. 116, No. 3-4, (May 1984), pp. 387–396, ISSN: 0022-2860.

Renn, O., Lippert, B. & Albinati, A. (1991). Metal-stabilized rare tautomers of nucleobases 3. (1-methylthyminato-*N3*) (1-methylthymine-*N3*)-*cis*-diammineplatinum(II) hemihexachloroplatinate(IV) dihydrate. *Inorganica Chim. Acta*, Vol. 190, No. 2, (December 1991), pp. 285–289, ISSN: 0020-1693 (Print).

Robinson, H., Gao, Y.G., Bauer, C., Roberts, C., Switzer, C. & Wang, A.H.J. (1998). 2'-Deoxyisoguanosine adopts more than one tautomer to form base pairs with thymidine observed by high-resolution crystal structure analysis. *Biochemistry*, Vol. 37, No. 31, (August 1998), pp. 10897-10905, ISSN: 0006-2960 (Print), 1520-4995 (Electronic).

Sabio, M., Topiol, S. & Lumma, W.C. (1990). An investigation of tautomerism in adenine and guanine. *J. Phys. Chem.*, Vol. 94, No. 4, (February 1990), pp. 1366-1372, ISSN: 0022-3654 (Print).

Saha, S., Wang, F. & Brunger, M.J. (2006). Intramolecular proton transfer in adenine imino tautomers. *Molecular Simulation*, Vol. 32, No. 15, (December 2006), pp. 1261–1270, ISSN: 0892-7022 (Print), 1029-0435 (Electronic).

Salter, L.M. & Chaban, G.M. (2002). Theoretical study of gas phase tautomerization reactions for the ground and first excited electronic states of adenine. *J. Phys. Chem. A*, Vol. 106, No. 16, (April 2002), pp. 4251-4256, ISSN: 1089-5639 (Print), 1520-5215 (Electronic).

Samijlenko, S.P., Bogdan, T.V., Trygubenko, S.A., Potyahaylo, A.L. & Hovorun, D.M. (2000). Deprotonated carboxylic group of amino acids transforms adenine into its rare prototropic tautomers. *Ukr. Biochem. J.* (*Ukr. Biokhim. Zh.*), Vol. 72, No. 6, (November-December 2000), pp. 92-95, ISSN: 0201-8470.

Samijlenko, S.P., Potyahaylo, A.L., Stepanyugin, A.V., Kolomiets, I.M. & Hovorun, D.M. (2001). Recognition modes of hypoxanthine, xanthine and their derivatives by amino acid carboxylic group: UV spectroscopic and quantum chemical data. *Ukr. Biochem. J.* (*Ukr. Biokhim. Zh.*), Vol. 73, No. 6, (November 2001), pp. 61-72, ISSN: 0201-8470.

Samijlenko, S.P., Krechkivs'ka, O.M., Kosach, D.A. & Hovorun, D.M. (2004). Transition to high tautomeric states can be induced in adenine by interactions with carboxylate and sodium ions: DFT calculation data. *J. Mol. Struct.*, Vol. 708, No. 1-3, (December 2004), pp. 97-104, ISSN: 0022-2860.

Samijlenko, S.P., Yurenko, Y.P., Stepanyugin, A.V. & Hovorun, D.M. (2010). Tautomeric equilibrium of uracil and thymine in model protein - nucleic acid contacts. Spectroscopic and quantum chemical approach. *J. Phys. Chem. B*, Vol. 114, No. 3, (January 2010), pp. 1454-1461, ISSN: 1520-6106 (Print), 1520-5207 (Electronic).

Schneider, B., Cohen, D. & Berman, H.M. (1992). Hydration of DNA bases: analysis of crystallographic data. *Biopolymers*, Vol. 32, No. 7, (July 1992), pp. 725-750, ISSN: 0006-3525 (Print), 1097-0282 (Electronic).

Schneider, B., Cohen, D.M., Schleifer, L., Srinivasan, A.R., Olson, W.K. & Berman, H.M. (1993). A systematic method for studying the spatial distribution of water molecules around nucleic acid bases. *Biophys. J.*, Vol. 65, No. 6, (December 1993), pp. 2291-2303, ISSN: 0006-3495 (Print), 1542-0086 (Electronic).

Schneider, B. & Berman, H.M. (1995). Hydration of the DNA bases is local. *Biophys. J.*, Vol. 69, No. 6, (December 1995), pp. 2661-2669, ISSN: 0006-3495 (Print), 1542-0086 (Electronic).

Schoellhorn, H., Thewalt, U. & Lippert, B. (1989). Metal-stabilized rare tautomers of nucleobases. 2. 2-Oxo-4-hydroxo form of uracil: crystal structures and solution behavior of two platinum(II) complexes containing iminol tautomers of 1-methyluracil. *J. Am. Chem. Soc.*, Vol. 111, No. 18, (August 1989), pp. 7213–7221, ISSN: 0002-7863 (Print), 1520-5126 (Electronic).

Sheina, G.G., Stepanian, S.G., Radchenko, E.D. & Blagoi, Yu.P. (1987). IR spectra of guanine and hypoxanthine isolated molecules. *J. Mol. Struct.*, Vol. 158, No. 3, (May 1987), pp. 275-292, ISSN: 0022-2860.

Sinclair, W.E. & Pratt, D.W. (1996). Structure and vibrational dynamics of aniline and aniline-Ar from high resolution electronic spectroscopy in the gas phase. *J. Chem. Phys.*, Vol. 105, No. 18, (November 1996), pp. 7942-7956, ISSN: 0021-9606 (Print), 1089-7690 (Electronic).

Sloane, D.L., Goodman, M.F. & Echols, H. (1988). The fidelity of base selection by the polymerase subunit of DNA polymerase III holoenzyme. *Nucleic Acids Res.*, Vol. 16, No. 14A, (July 1988), pp. 6465-6475, ISSN: 0305-1048 (Print), 1362-4962 (Electronic).

Sobolewski, A.L. & Adamowicz, L. (1995). Theoretical investigations of proton transfer reactions in a hydrogen bonded complex of cytosine with water. *J. Chem. Phys.*, Vol. 102, No. 14, (April 1995), pp. 5708-5718, ISSN: 0021-9606 (Print), 1089-7690 (Electronic).

Sordo, J.A., Chin, S. & Sordo, T.L. (1988). On the counterpoise correction for the basis set superposition error in large systems. *Theor. Chim. Acta.*, Vol. 74, No. 2, (August 1988), pp. 101-110, ISSN: 0040-5744 (Print).

Sordo, J.A. (2001). On the use of the Boys–Bernardi function counterpoise procedure to correct barrier heights for basis set superposition error. *J. Mol. Struct.: THEOCHEM*, Vol. 537, No. 1-3, (March 2001), pp. 245-251, ISSN: 0166-1280 (Print).

Sowers, L.C., Fazakerley, G.V., Kim, H., Dalton, L. & Goodman, M.F. (1986). Variation of nonexchangable proton resonance chemical shifts as a probe of aberrant base pair formation in DNA. *Biochemistry,* Vol. 25, No. 14, (July 1986), pp. 3983–3988, ISSN: 0006-2960 (Print), 1520-4995 (Electronic).

Sowers, L.C., Shaw, B.R., Veigl, M.L. & Sedwick, W.D. (1987). DNA base modification: ionized base pairs and mutagenesis. *Mutat. Res.*, Vol. 177, No. 2, (April 1987), pp. 201- 218, ISSN: 0027-5107 (Print), 1873-135X (Electronic).

Šponer, J. & Hobza, P. (1994). Nonplanar geometries of DNA bases. *Ab initio* second-order Moeller-Plesset study. *J. Phys. Chem.*, Vol. 98, No. 12, (March 1994), pp. 3161-3164, ISSN: 0022-3654 (Print).

Šponer, J., Leszczynski, J. & Hobza, P. (2001). Hydrogen bonding, stacking and cation binding of DNA bases. *J. Mol. Struct.: THEOCHEM,* Vol. 573, No. 1-3, (October 2001), pp. 43-53, ISSN: 0166-1280 (Print).

Stepanian, S.G., Sheina, G.G., Radchenko, E.D. & Blagoi, Yu.P. (1985). Theoretical and experimental studies of adenine, purine and pyrimidine isolated molecule structure. *J. Mol. Struct.*, Vol. 131, No. 3-4, (November 1985), pp. 333-346, ISSN: 0022-2860.

Stepanyugin, A.V., Kolomiets', I.M., Potyahaylo, A.L., Samijlenko, S.P. & Hovorun, D.M. (2002a). UV spectra of adenine methyl and glycosyl derivatives and their transformation induced by amino acid carboxylic groups. *Ukr. Biochem. J.* (*Ukr. Biokhim. Zh.*), Vol. 74, No. 3, (May 2002), pp. 73-81, ISSN: 0201-8470.

Stepanyugin, A.V., Potyahaylo, A.L., Kolomiets, I.M., Samijlenko, S.P. & Hovorun, D.M. (2002b). UV spectra of guanine methyl and glycosyl derivatives and their transformations induced by interactions with amino acids *via* carboxylic group in dimethylsulfoxide. *Ukr. Biochem. J.* (*Ukr. Biokhim. Zh.*), Vol. 74, No. 2, (March 2002), pp. 73-85, ISSN: 0201-8470.

Sukhanov, O.S., Shishkin, O.V., Gorb, L., Podolyan, Y. & Leszczynski, J. (2003). Molecular structure and hydrogen bonding in polyhydrated complexes of adenine: a DFT study. *J. Phys. Chem. B*, Vol. 107, No. 12, (March 2003), pp. 2846-2852, ISSN: 1520-6106 (Print), 1520-5207 (Electronic).

Suwaiyan, A., Morsy, M.A. & Odah, K.A. (1995). Room temperature fluorescence of 5-chlorouracil tautomers. *Chem. Phys. Lett.*, Vol. 237, No. 3-4, (May 1995), pp. 349–355, ISSN: 0009-2614 (Print).

Sygula, A. & Buda, A. (1983). MNDO study of the tautomers of nucleic bases: Part II. Adenine and guanine. *J. Mol. Struct.: THEOCHEM,* Vol. 92, No., 3-4, (April 1983), pp. 267-277, ISSN: 0166-1280 (Print).

Szczepaniak, K. & Szczesniak, M. (1987). Matrix isolation infrared studies of nucleic acid constituents: Part 4. Guanine and 9-methylguanine monomers and their keto-enol tautomerism. *J. Mol. Struct.*, Vol. 156, No. 1-2, (January 1987), pp. 29-42, ISSN: 0022-2860.

Szczesniak, M., Szczepaniak, K., Kwiatkowski, J.S., KuBulat, K. & Person, W.B. (1988). Matrix isolation infrared studies of nucleic acid constituents. 5. Experimental matrix-isolation and theoretical *ab initio* SCF molecular orbital studies of the infrared spectra of cytosine monomers. *J. Am. Chem. Soc.,* Vol. 110, No. 25, (December 1988), pp. 8319–8330, ISSN: 0002-7863 (Print), 1520-5126 (Electronic).

Topal, M.D. & Fresco, J.R. (1976). Complementary base pairing and the origin of substitution mutations. *Nature,* Vol. 263, No. 5575, (September 1976), pp. 285-289, ISSN: 0028-0836 (Print), 1476-4687 (Electronic).

Trygubenko, S.A., Bogdan, T.V., Rueda, M., Orozco, M., Luque, F.J., Šponer, J., Slavíček, P. & Hobza, P. (2002). Correlated *ab initio* study of nucleic acid bases and their tautomers in the gas phase, in a microhydrated environment and in aqueous solution. Part 1. Cytosine. *Phys. Chem. Chem. Phys.,* Vol. 4, No. 17, (2002), pp. 4192–4203, ISSN: 1463-9076 (Print), 1463-9084 (Electronic).

Tsuchiya, Y., Tamura, T., Fujii, M. & Ito, M. (1988). Keto-enol tautomer of uracil and thymine. *J. Phys. Chem.*, Vol. 92, No. 7, (April 1988), pp. 1760–1765, ISSN: 0022-3654 (Print).

Tunis, M.J.B., & Hearst, J.E. (1968). On the hydration of DNA. II. Base composition dependence of the net hydration of DNA. *Biopolymers,* Vol. 6, No. 9, (September 1968), pp. 1345-1353, ISSN: 0006-3525 (Print), 1097-0282 (Electronic).

Ullrich, S., Schultz, T., Zgierski, M. Z. & Stolow, A. (2004). Electronic relaxation dynamics in DNA and RNA bases studied by time-resolved photoelectron spectroscopy. *Phys. Chem. Chem. Phys.,* Vol. 6, No. 10, (2004), pp. 2796-2801, ISSN: 1463-9076 (Print), 1463-9084 (Electronic).

Villani, G. (2005). Theoretical investigation of hydrogen transfer mechanism in the adenine–thymine base pair. *Chem. Phys.,* Vol. 316, No. 1-3, (September 2005), pp. 1–8, ISSN: 0301-0104 (Print).

Villani, G. (2006). Theoretical investigation of hydrogen transfer mechanism in the guanine–cytosine base pair. *Chem. Phys.,* Vol. 324, No. 2-3, (May 2006), pp. 438–446, ISSN: 0301-0104 (Print).

Villani, G. (2010). Theoretical investigation of hydrogen atom transfer in the cytosine-guanine base pair and its coupling with electronic rearrangement. Concerted *vs* stepwise mechanism. *J. Phys. Chem. B,* Vol. 114, No. 29, (July 2010), pp. 9653–9662, ISSN: 1520-6106 (Print), 1520-5207 (Electronic).

Vrkic, A.K., Taverner, T., James, P.F. & O'Hair, R.A.J. (2004). Gas phase ion chemistry of biomolecules, part 38. Gas phase ion chemistry of charged silver (I) adenine ions *via* multistage mass spectrometry experiments and DFT calculations. *Dalton Trans.,* No. 2, (2004), pp. 197-208, ISSN: 1477-9226 (Print), 1477-9234 (Electronic).

Wang, J.H. (1955). The hydration of desoxyribonucleic acid. *J. Am. Chem. Soc.*, Vol. 77, No. 2, (January 1955), pp. 258-260, ISSN: 0002-7863 (Print), 1520-5126 (Electronic).

Wang, S. & Schaefer III, H.F. (2006). The small planarization barriers for the amino group in the nucleic acid bases. *J. Chem. Phys.*, Vol. 124, No. 4, (January 2006), pp. 044303-044310, ISSN: 0021-9606 (Print), 1089-7690 (Electronic).

Wang, W., Hellinga, H.W., Beese, L.S. (2011). Structural evidence for the rare tautomer hypothesis of spontaneous mutagenesis. *Proc. Natl. Acad. Sci. U.S.A.*, Vol. 108, No. 43, (October 2011), pp. 17644-17648, ISSN: 0027-8424 (Print), 1091-6490 (Electronic).

Wang, Y., Saebo, S. & Pittman, C.U. Jr. (1993). The structure of aniline by *ab initio* studies. *J. Mol. Struct.: THEOCHEM*, Vol. 281, No. 2-3, (April 1993), pp. 91-98, ISSN: 0166-1280 (Print).

Watson, J.D. & Crick, F.H.C. (1953a). The structure of DNA. *Cold Spring Harbor Symp. Quant. Biol.*, Vol. 18, 1953, pp. 123-131. ISSN: 0091-7451 (Print), 1943-4456 (Electronic).

Watson, J.D. & Crick, F.H.C. (1953b). Molecular structure of nucleic acids: a structure for deoxyribose nucleic acid. *Nature*, Vol. 171, No. 4356, (April 1953), pp. 737-738, ISSN: 0028-0836 (Print), 1476-4687 (Electronic).

Wigner, E. (1932). Über das Überschreiten von Potentialschwellen bei chemischen Reaktionen. *Z. Phys. Chem.*, Vol. B19, (1932), pp. 203–216, ISSN: 0044-3336.

Wiorkiewicz-Kuczera, J. & Karplus, M. (1990). *Ab initio* study of the vibrational spectra of N9-H and N7-H adenine and 9-methyladenine. *J. Am. Chem. Soc.*, Vol. 112, No. 13, (June 1990), pp. 5324-5340, ISSN: 0002-7863 (Print), 1520-5126 (Electronic).

Yang, Z. & Rodgers, M.T. (2004). Theoretical studies of the unimolecular and bimolecular tautomerization of cytosine. *Phys. Chem. Chem. Phys.*, Vol. 6, No. 10, (2004), pp. 2749–2757, ISSN: 1463-9076 (Print), 1463-9084 (Electronic).

Yu, H., Eritja, R., Bloom, L.B. & Goodman, M.F. (1993). Ionization of bromouracil and fluorouracil stimulates base mispairing frequencies with guanine. *J. Biol. Chem.*, Vol. 268, No. 21, (July 1993), pp. 15935–15943, ISSN: 0021-9258 (Print), 1083-351X (Electronic).

Yurenko, Y.P., Zhurakivsky, R.O., Ghomi, M., Samijlenko, S.P. & Hovorun, D.M. (2007a). Comprehensive conformational analysis of the nucleoside analogue 2′-β-deoxy-6-azacytidine by DFT and MP2 calculations. *J. Phys. Chem. B*, Vol. 111, No. 22, (June 2007), pp. 6263-6271, ISSN: 1520-6106 (Print), 1520-5207 (Electronic).

Yurenko, Y.P., Zhurakivsky, R.O., Ghomi, M., Samijlenko, S.P. & Hovorun, D.M. (2007b). How many conformers determine the thymidine low-temperature matrix infrared spectrum? DFT and MP2 quantum chemical study. *J. Phys. Chem. B*, Vol. 111, No. 32, (August 2007), pp. 9655-9663, ISSN: 1520-6106 (Print), 1520-5207 (Electronic).

Yurenko, Y.P., Zhurakivsky, R.O., Samijlenko, S.P., Ghomi, M. & Hovorun, D.M. (2007c). The whole of intramolecular H-bonding in the isolated DNA nucleoside thymidine. AIM electron density topological study. *Chem. Phys. Lett.*, Vol. 447, No. 1-3, (October 2007), pp. 140-146, ISSN: 0009-2614 (Print).

Yurenko, Y.P., Zhurakivsky, R.O., Ghomi, M., Samijlenko, S.P. & Hovorun, D.M. (2008). *Ab initio* comprehensive conformational analysis of 2′-deoxyuridine, the biologically significant DNA minor nucleoside, and reconstruction of its low-temperature matrix infrared spectrum. *J. Phys. Chem. B*, Vol. 112, No. 4, (January 2008), pp. 1240-1250, ISSN: 1520-6106 (Print), 1520-5207 (Electronic).

Yurenko, Y.P., Zhurakivsky, R.O. & Hovorun, D.M. 37. (2009). Intramolecular hydrogen bonds CH...O in biologically significant conformers of canonical 2'-deoxyribonucleosides: *ab initio* topological analysis of the electron density. *Physics of the Alive (Fizyka zhyvoho),* Vol. 17, No. 1, (January-February 2009), pp. 44-53, ISSN: 1023-2427.

Zamora, F., Kunsman, M., Sabat, M. & Lippert, B. (1997). Metal-stabilized rare tautomers of nucleobases. 6-Imino tautomer of adenine in a mixed-nucleobase complex of mercury(II). *Inorg. Chem.*, Vol. 36, No. 8, (April 1997), pp. 1583–1587, ISSN: 0020-1669 (Print), 1520-510X (Electronic).

Zhao, Z.-M., Zhang, Q.R., Gao, C.Y. & Zhuo, Y.Z. (2006). Motion of the hydrogen bond proton in cytosine and the transition between its normal and imino states. *Phys. Lett. A,* Vol. 359, No. 1, (November 2006), pp. 10–13, ISSN: 0375-9601 (Print).

Zhou, J., Kostko, O., Nicolas, C., Tang, X., Belau, L., de Vries, M.S. & Ahmed, M. (2009). Experimental observation of guanine tautomers with VUV photoionization. *J. Phys. Chem. A,* Vol. 113, No. 17, (April 2009), pp. 4829–4832, ISSN: 1089-5639 (Print), 1520-5215 (Electronic).

Zhurakivsky, R.O. & Hovorun, D.M. (2006). Conformational properties of cytidine: the DFT quantum mechanical investigation. *Physics of the Alive (Fizyka zhyvoho),* Vol. 14, No. 3, (May-June 2006), pp. 33-46, ISSN: 1023-2427.

Zhurakivsky, R.O. & Hovorun, D.M. (2007a). Complete conformational analysis of deoxyadenosine by density functional theory. *Biopolym. Cell,* Vol. 23, No. 1, (January-February 2007), pp. 45-53, ISSN: 0233–7657 (Print), 1993-6842 (Electronic).

Zhurakivsky, R.O. & Hovorun, D.M. (2007b). The comprehensive conformational analysis of 2'-deoxyguanosine molecule by the quantum-chemical density functional method. *Reports of the National Academy of Sciences of Ukraine,* No. 4, (April 2007), pp. 187-195, ISSN: 1025-6415.

Zierkiewicz, W., Komorowski, L., Michalska, D., Cerny, J. & Hobza, P. (2008). The amino group in adenine: MP2 and CCSD(T) complete basis set limit calculations of the planarization barrier and DFT/B3LYP study of the anharmonic frequencies of adenine. *J. Phys. Chem. B,* Vol. 112, No. 51, (December 2008), pp. 16734-16740, ISSN: 1520-6106 (Print), 1520-5207 (Electronic).

Van Zundert, G.C.P., Jaeqx, S., Berden, G., Bakker, J.M., Kleinermanns, K., Oomens, J. & Rijs, A.M. (2011). IR spectroscopy of isolated neutral and protonated adenine and 9-methyladenine. *ChemPhysChem.,* Vol. 12, No. 10, (July 2011), pp. 1921-1927, ISSN: 1439-4235 (Print), 1439-7641 (Electronic).

Part 3

Molecules to Nanodevices

6

Quantum Transport and Quantum Information Processing in Single Molecular Junctions

Tomofumi Tada
Department of Materials Engineering, Global COE for Mechanical System Innovation, The University of Tokyo
Japan

1. Introduction

Superposition states and entanglement in quantum bits (qubits) are inherently required in quantum computations (Benenti et al., 2004; Miyano & Furusawa, 2008; Nielsen & Chuang, 2004; Sagawa & Yoshida, 2003). Electron and nuclear spins have been identified as attractive candidates for qubits (Ladd et al., 2010), and the prominent properties involved in quantum spins have been observed in liquid state molecules (Vandersypen et al., 2001) and solid state materials such as doped silicon (Kane, 1998) and nitrogen-vacancy (NV) center in diamond (Childress et al., 2006). An impressive demonstration of quantum computations on Shor's algorithm was carried out by Vandersypen and co-workers by using a liquid state system, in which each molecule includes seven nuclear spin qubits (Vandersypen et al., 2001). The operations for single and double qubits were implemented through bulk nuclear magnetic resonance (NMR) technique, in which radio-frequency (RF) pulse sequences were constructed so as to manipulate nuclear spin states along the design of quantum gates for the factorization. The RF pulse applications were succeeded in the precise control of nuclear spin states, and in turn in the factorization of a small number (N=15) in Shor's algorithm. However, the liquid NMR signals are inherently averaged signals from a huge number of molecules, and therefore problems on initialization of qubits and pseudo-entanglement appear in the liquid NMR system, which make the liquid system difficult for the quantum computations using a larger number of qubits, although the operations in NMR are in principle robust.

The difficulties lying on the spin ensemble study are resolved when operations and readout are implemented on a single spin. Optical excitation and fluorescence of a single electron spin of NV center in diamond are powerful for the observations of coherent dynamics and readout of the single electron spin states (Cappellaro et al., 2009; Childress et al., 2006; Jacques et al., 2009; Jelezko et al., 2004a;b; Neumann et al., 2008; Smeltzer et al., 2009). The microwave (MW) and RF pulses designed by taking the hyperfine structures into account also lead to the robust control and readout of single/few nuclear spin states. Besides, the multiparticle entanglement among single electron/nuclear spins was confirmed in the NV center (Neumann et al., 2008).

The electron-nuclear spins for qubits have been investigated extensively also in P-doped silicon, which has a field effect transistor (FET) structure (Kane, 1998; Lo et al., 2007; McCamey et al., 2006; Mccamey et al., 2009; Morton et al., 2008; Stegner et al., 2006). In Kane's silicon-based quantum computer, the two kinds of gates, named as A- and J-gates, were introduced to control the resonance frequency of a ^{31}P nuclear spin and electron-mediated couplings between adjacent nuclear spins, respectively (Kane, 1998). In the silicon systems, the electrical detection of ^{31}P nuclear spin states is also possible. The hyperfine interactions between electron and ^{31}P nuclear spins are again key interactions for the control and readout of spin states. These pioneering works show that the electron-nuclear spin pairs in solid state materials have prerequisite properties for quantum computations, and that we can observe and control the quantum properties in practical ways.

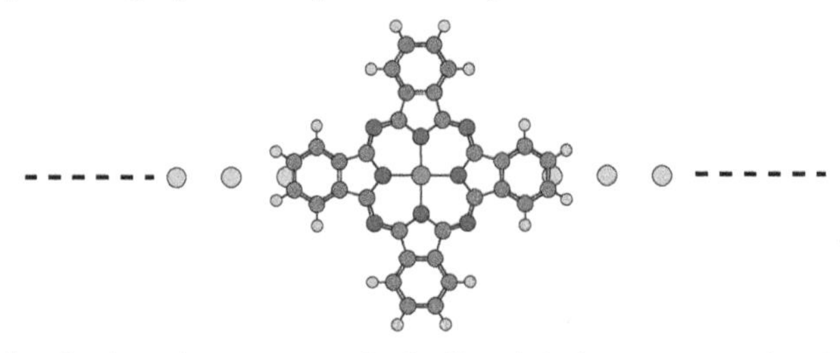

Fig. 1. Single molecular junction composed of a Cu-phthalocyanine molecule and one-dimensional gold chain. The yellow, dark gray, light gray, blue, and purple atoms are gold, carbon, hydrogen, nitrogen, and copper, respectively.

Another key issue for the design of quantum computers is high scalability. The NV center in diamond and doped-silicon systems of course have scalable designs in its original concept because the operation and readout of single/few spin(s) in these systems are available and highly robust. However, we have to *perfectly* control the doping of P in Si and NV center in diamond in terms of the concentration and doping positions to realize the solid state quantum computers including many qubits. This is not a trivial task if we use a top-down fashion from bulk materials. On the other hand, bottom-up approaches have been succeeded in the fabrication of the one-dimensional atomic chains and molecular chains between electrodes or on substrates (Carroll & Gorman, 2002; Joachim, 2000). For example, a well-defined one-dimensional structure composed of gold atomic chains and a metal-phthalocyanine molecule (Fig. 1) on NiAl(110) surface (Nazin et al., 2003) has a large feasibility for high-scalable and well-defined spin arrays by repeating the same junction structure on a substrate through self-assembling processes. Since a single organic molecule can be the source of electron and nuclear spins and one-dimensional atomic chains connected to the single molecules will be useful for electrical detection of spin states, the spin array structure composed of single molecular junctions (see Fig. 2) is a new candidate for the device structure designed for quantum computations with high scalability. Here we use the term *single molecular junctions* to express the nano-scale junctions composed of a single molecule and electrodes. However, despite the potential ability in this type of device structures, the investigations of molecular junctions intended for quantum computations seldom have been reported so far (Tada, 2008). Therefore this chapter is devoted to investigate the possibility of operation and readout of single spins in molecular junctions in the framework of the first principles electronic structure calculations.

This chapter is organized as follows; Section2: the introduction for the classical and quantum computers, Section 3: the control of spin states using the rotating magnetic field, Section 4: the electronic conduction in single molecular junctions, and Section 5: the operation-readout robust switching of a single nuclear spin in the single molecular junctions for the quantum information processing.

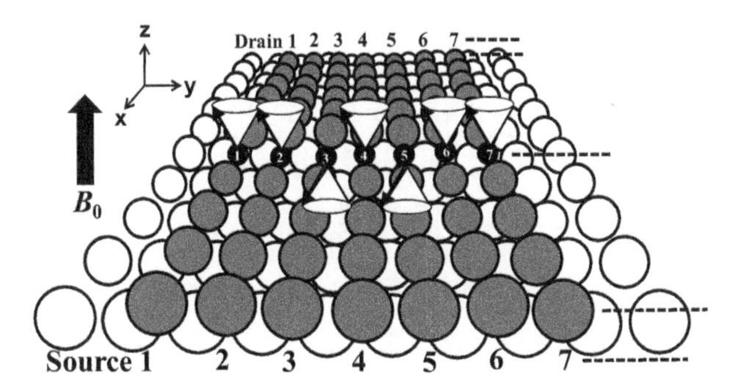

Fig. 2. Schematic of the solid-state nuclear-spin (black) array on a substrate (white). The gray symbols are atoms in electrodes connected to the nuclear spins.

2. Classical computers and quantum computers

2.1 Classical and quantum bits

To briefly understand the difference between the classical and quantum computers, let us consider three classical/quantum bits in an electronic device. The information of "0" or "1" can be stored in a classical bit, whereas a superposition state (e.g., "0+1") can be allowed in a quantum bit (qubit). In the three classical bits, the eight combinations (i.e., "0,0,0" , "0,0,1" , "0,1,0" , "0,1,1" , "1,0,0" , "1,0,1" , "1,1,0" , "1,1,1") can be realized, but the bit information we can store concurrently is just a single combination (e.g, "0,0,1"). When a operation A is operated to the eight combinations, the eight operations of A in a sequential manner are required in the classical three bits device. On the other hand, in the three qubits device, we can prepare the eight combinations concurrently by using the superposition states (i.e., "0+1,0+1,0+1" $\rightarrow$ the eight combinations). Thus the eight operations of A in the classical bits can be reduced to a single operation A in the three qubits device. When the number of bits increases much more, the advantage of quantum computers is absolutely obvious. We thus sometimes call the quantum computer the *super-parallel* computer.

2.2 Classical and quantum logic gates

The electronic circuits in electronic devices currently used are designed on the basis of the classical logic gates. The "AND" and "NOT" gates shown in Fig. 3(a) are the typical classical gates, in which the input of two/one bit information is converted to a single bit as an output in the "AND"/"NOT" gate. It is well-known that any logic gates in classical computers can be constructed from the "AND" and "NOT" gates. In quantum computers, on the other hand, the "unitary (U)" and "controlled-NOT (c-NOT)" gates are the key logic gates in the quantum computations (Benenti et al., 2004; Miyano & Furusawa, 2008; Nielsen & Chuang, 2004; Sagawa & Yoshida, 2003). Fig. 3(b) shows the rules in the bit transformation of the U and c-NOT gates. The characteristic properties in these quantum gates are (i) a superposition state is generated in the U gate, and (ii) the entanglement of the arbitrary two qubits are necessary for the c-NOT gate. Thus the next concern is how we should realize the two quantum gates in physical systems.

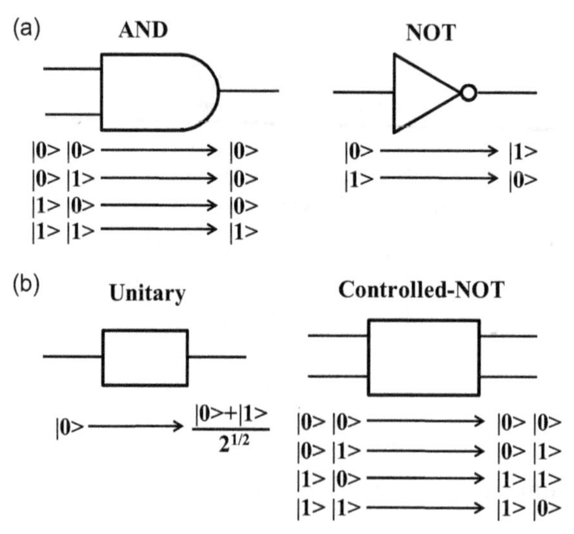

Fig. 3. Schematic of the typical logic gates and the rules of the bit transformation in (a) classical computers and in (b) quantum computers

The trapped ion/atom, nuclear/electron spins, and superconducting charge/phase bit have been extensively investigated as the potential candidates for the qubit, and the superposition and entanglement properties have been confirmed in these systems. In addition to the two properties (superposition and entanglement), the ability for the retention of the qubit information is also an important property for robust systems. According to the review by Ladd and co-workers (Ladd et al., 2010), the nuclear spin of ^{29}Si in ^{28}Si shows the longest T_2 time among the trapped ion/atom, nuclear/electron spins, and superconducting charge/phase bit; T_2 is the phase coherence time and hence T_2 is the measure of the ability of the retention time. The long coherence (25 s of T_2 in ^{29}Si nuclear spin) is the result of the weak interactions between the nuclear spin and its environment. Therefore, the nuclear spins are quite attractive target for the qubit in terms of the long retention time.

3. Control of the nuclear spin states using a rotating magnetic field

3.1 Unitary and controlled NOT gates

Since the advantage of nuclear spins for the retention of qubit information is described in the previous section, let us next consider how we can operate the nuclear spin states. Figure 4 shows a single nuclear spin of 1/2 in the static magnetic field of B_0 directed to the z-axis. In the condition, the nuclear spin states split into the two states, $s_z = +1/2$ and $-1/2$, and the states can be recognized as "0" and "1" for the bit information. To use the nuclear spin as a qubit, a superposition state of "0" and "1" is required through the unitary transformation (U-gate). The application of a rotating magnetic field B_1 perpendicular to B_0 enables us to operate the nuclear spin. In the presence of the rotating magnetic field B_1 with the angular velocity of ω, the Hamiltonian is written as

$$H = -\gamma[B_0 s_z + B_1\{(\cos\omega t)s_x + (\sin\omega t)s_y\}], \tag{1}$$

where γ is an effective g-factor for the nucleus. Using the time-dependent Schrödinger equation $i\hbar[d|\psi(t)\rangle/dt] = H|\psi(t)\rangle$ and a rotating reference frame $|\psi(t)\rangle = \exp(i\omega t s_z)|\phi(t)\rangle$, the wave function of the nuclear spin can be written as

$$|\phi(t)\rangle = \exp\{it/\hbar[(\gamma B_0 - \hbar\omega)s_z + \gamma B_1 s_x]\}||\phi(0)\rangle. \tag{2}$$

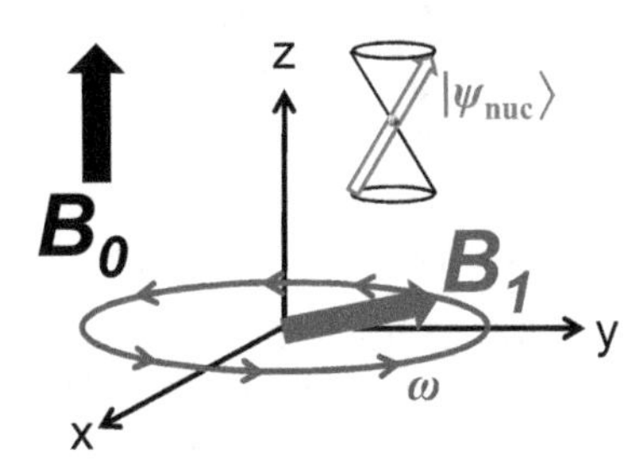

Fig. 4. Single nuclear spin of 1/2 in the static magnetic field B_0 and rotating magnetic field B_1 with the angular velocity of ω.

This equation reveals that when we apply a magnetic field B_1 with $\omega = \gamma B_0/\hbar$, the time dependence of the wave function is completely controlled by the magnetic field B_1, and thus any superposition state of "0" and "1" can be generated by tunning the application period of the rotating magnetic field (Sagawa & Yoshida, 2003). Since the γ depends on the kinds of the nuclei and its environment, the unitary transformation can be carried out in a selective manner.

The c-NOT operation is feasible by using the spin-spin interaction J between two nuclear spins, leading to the interaction term $Js_z^1 s_z^2$ in the Hamiltonian, where s_z^1 and s_z^2 are the spin components along the z-axis for the first and second nuclear spins, respectively. Since Js_z^1 can be regarded as an effective magnetic field for the second nuclear spin, a rotating magnetic field being in resonance to the spin-spin interaction is useful to rotate the second nuclear spin with respect to the first nuclear spin. The magnitude of the spin-spin interaction J depends on the selected two spins, and thus the selective c-NOT is in principle possible.

3.2 Liquid state NMR for quantum computations

Vandersypen and co-workers reported that the nuclear magnetic resonance (NMR) technique is useful for the unitary transformation and c-NOT operations, and that the factorization of a small number (N=15) in Shor's algorithm was succeeded in the application of the radio-frequency (RF) pulse sequence designed for factorization (Vandersypen et al., 2001). Figure 5 shows the seven qubits molecule used in the NMR quantum computations. According to Vandersypen's study, the resonance frequencies ($\omega/2\pi$) for F(2), F(3), C(6), and C(7) are 0.4895, 25.0883, −4.5191, and 4.2443 kHz, respectively, and those for J-couplings between F(2)-F(3), F(2)-C(7), F(1)-C(7) are 0.0039, 0.0186, and −0.2210 kHz, respectively. The selective operations on each qubit and on the coupled two qubits are supported in the

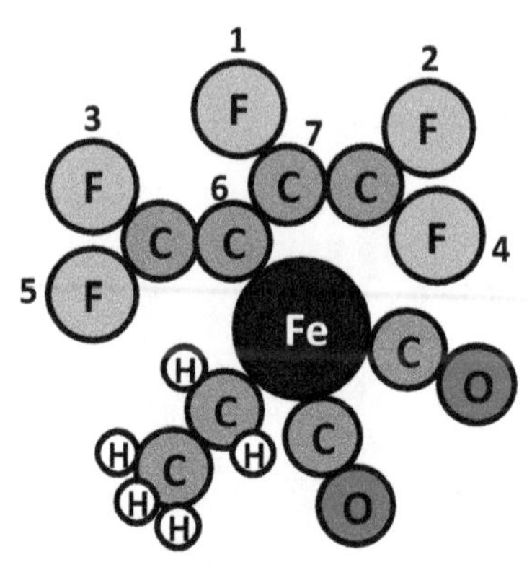

Fig. 5. Seven-bits organic molecule used in the liquid-NMR quantum computations. The numbered atoms are the qubits used in the quantum computations.

magnetic resonance conditions. If we can use a larger molecule designed for a large number of qubits, we may thank that such a large molecule is a promising candidate for a practical quantum computer. However, the large molecule including a large number of qubits has a problem that the resonance frequencies for qubits become closer with each other, resulting in the difficulty in the selective operations. In addition, we have to pay attention to the situation that the NMR experiments are carried out in the liquid phase, and therefore the NMR system includes the huge number of the qubit molecules. At the initial stage of the quantum computations, we have to prepare a suitable initialized state for the qubits. When the number of qubits in a molecule becomes larger, the procedure for the initialization becomes more complicated and time consuming. Thus, it is now believed that a molecule including 10-qubits is the largest molecule available for the liquid NMR quantum computations.

3.3 Solid state NMR for quantum computations

The difficulties involved in the liquid NMR systems will be resolved when operations and readout are implemented on a single nuclear spin. The ^{13}C-NMR studies of the NV center in diamond show the powerfulness of the solid state NMR for the applications in quantum computations. However, as described in Introduction, to construct the practical solid state quantum computers, we have to perfectly control the positions of the nuclear spins in diamond. The probability of the precise control in the atomistic level is quite doubtful if we use a top-down fashion from bulk materials. The alternative is the bottom-up approach in which the nano-scale contact between a single molecule (i.e., a spin) and probes is controlled.

Figure 2 shows the schematic of the solid-state nuclear-spin array on a substrate fabricated in the bottom-up approach. The each nuclear spin is connected to the source and drain electrodes for the selective detection of the nuclear spin states from the current measurements. The operations on nuclear spins are carried out using the rotating magnetic field, like in the liquid state NMR. The selectivity in the solid-state nuclear-spin array is achieved by the selective modulation of the resonance frequencies using the probes. The details for the detection and operation using the probes, which is the main topic in this chapter, will be described in the later section. Before the detail explanation of the solid-state nuclear-spin system, the brief introduction of a single molecular junction is given in the next section.

4. Single molecular junctions

4.1 Measurements of the conductance in single molecular junctions

The single molecular junction is composed of electrodes and a single molecule sandwiched in between the electrodes. The single molecular junction designed for an atomic-scale diode was originally proposed by Aviram and Ratner in 1974 (Aviram & Ratner, 1974). After 23 years from the pioneering theoretical proposal, an experimental work on a single molecular junction was firstly reported by Reed and co-workers in 1997 (Reed et al., 1997), in which a single benzen-1,4-dithiolate molecule was sandwiched between gold electrodes. They used mechanically controllable break junction (MCBJ) technique to fabricate the single molecular junction, and measured the electronic current through the junction. In the MCBJ technique (Ruitenbeek et al., 2005), a notch cut on the bulk electrode will be the breaking position by the push-and-pull breaking motion of the electrodes, and thereby a single molecular junction will be obtained at this breaking position (see Figure 6(a)). However, the length of the notch on the electrode may spread over the macro scale, and thus it is sometimes quite difficult to confirm that the number of the molecule sandwiched between the electrodes is exactly one. The break junction technique thus has been improved to be more precise method, and one of the experimental technique frequently used now is the break junction using the scanning tunneling microscope (STM) (Cui et al., 2001; Xu & Tao, 2003). We call it STM-BJ method. Since the length scale of the apex of the STM probe carefully prepared is in the nano-scale, the STM-BJ technique is more suitable for the fabrication and current measurements of the single molecular junctions (see Figure 6(b)). Like in MCBJ, the STM-tip is pushed or pulled to fabricate the single molecular junction.

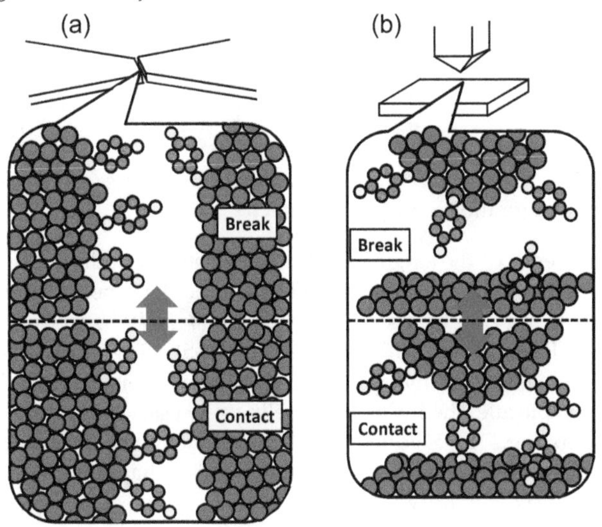

Fig. 6. Schematics of single molecular junctions fabricated during the push-and-pull motion of electrodes in (a) MCBJ and (b) STM-BJ techniques. The blown atoms represent electrodes, and a tiny structure composed of the gray and yellow atoms represents a single molecule.

Even for the STM-BJ technique, we still have the question that the number of the molecule sandwiched between the STM probe and substrate is exactly one or not. The standard strategy to distinguish the number of sandwiched molecules is the tracing of the measured current during the pulling motion of the STM-tip. Figure 7(a) shows the typical trace of the measured conductance, in which we can confirm several plateaus followed by a drop. If there is no bridge (i.e., a molecule) between the STM-tip and substrate, we can expect the current of zero. Thus the last plateau indicated with an arrow in Fig. 7(a) is the signature of the single molecular chain or atomic chain sandwiched between the electrodes. Since the conductance values of the single molecular junctions are typically less than 0.1 G_0, and those of the atomic chain of metal larger than 1.0 G_0 in general, we can speculate the presence of the single molecule between electrodes from the value of the conductance; G_0 is the quantum unit of conductance, $2e^2/h$.

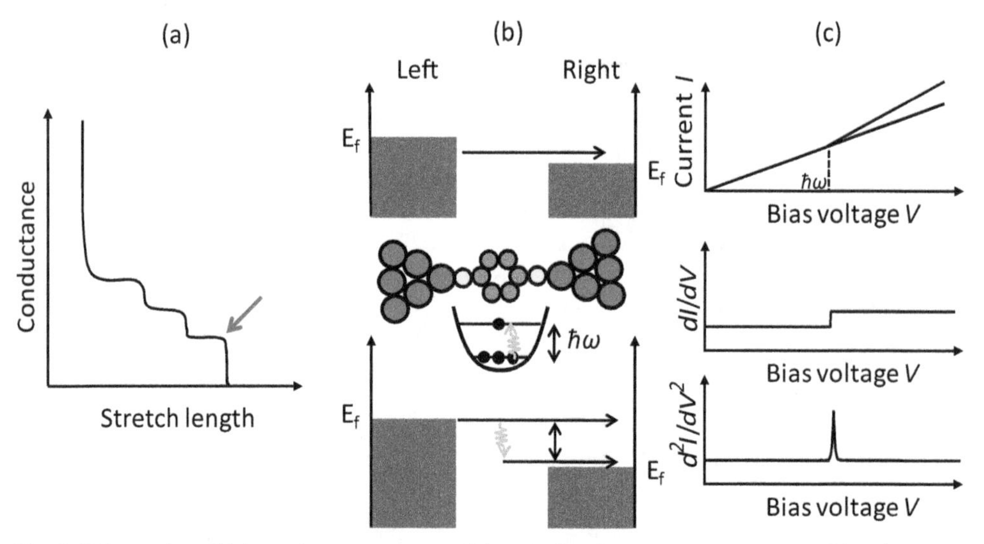

Fig. 7. Schematics of (a) conductance traces, (b) an inelastic process accompanied by the excitation of molecular vibration, and (c) IETS spectra.

To determine the presence of the single molecule between electrodes more precisely, we can use inelastic electron tunneling spectroscopy (IETS). In the IETS measurements, the signature of the single molecule appears as a sharp peak in the d^2I/dV^2 spectrum at the energy of molecular vibration (Ruitenbeek, 2010; Wang et al., 2005). The appearance of the peak is comprehensible as follows: when the applied bias voltage becomes larger than the vibration energy of the single molecule, the inelastic tunneling process accompanied by the excitation of molecular vibration (Fig. 7(b)) is added to the total tunneling current, leading to the abrupt enhancement of the tunneling current at the voltage. This enhancement results in the step in dI/dV and the sharp peak in d^2I/dV^2 (Fig. 7(c)).

Using these experimental techniques, STM-BJ and IETS, the measurements of single molecular junctions have been reported for many kinds of molecules (e.g., hydrogen molecule, saturated hydrocarbon chains, π conjugated organic molecules, DNA molecules, and so on.) (Metzger, 2005; Porath et al., 2005; Ruitenbeek et al., 2005; Wang et al., 2005). In addition to the sophisticated experimental method, the theoretical computations for the conductance of the single molecular junctions have been developed and applied in order to understand the experimental results precisely. As for the theoretical methods for conductance calculation, there are many types of theoretical frameworks; scattering wave function method (Choi & Ihm, 1999; Gohda et al., 2000; Smogunov et al., 2004), Lippmann-Schwinger equation (Emberly & Kirczenow, 1999; Lang, 1995), the recursion-transfer-matrix method (Hirose & Tsukada, 1994; 1995), and non-equilibrium Green's function (NEGF) method (Brandbyge et al., 2002; Taylor et al., 2001) are the frequently adopted methods. In the next section, the NEGF and wave-packet scattering methods for conductance calculations in the coherent regime are described.

4.2 Conductance calculations in single molecular junctions

4.2.1 Wave-packet scattering approach

In this section, we consider how we can calculate the electronic conductance in molecular nano-scale junctions. We focus our attention on a one-dimensional system in which the Hamiltonian is written in the tight-binding framework in order to represent the scattering events caused by the sandwiched molecule.

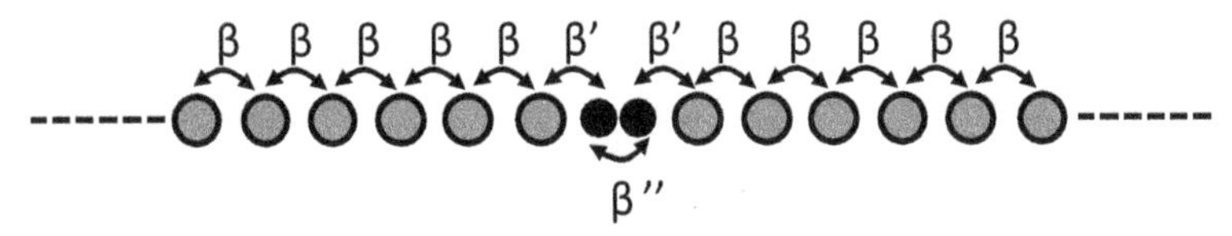

Fig. 8. One-dimensional diatomic molecular junction. The black dots are the single diatomic molecule sandwiched between the one-dimensional electrodes.

Figure 8 shows the one-dimensional molecular junction in which the diatomic molecule is sandwiched in between the one-dimensional electrodes. In the tight-binding model with the hopping integrals β, β', β'' and on-site energy α, α', the Hamiltonian matrix can be written as

$$\mathbf{H} = \left(\begin{array}{c|c|c} \mathbf{H_e} & \mathbf{t}^\dagger & \mathbf{0} \\ \hline \mathbf{t} & \mathbf{H_c} & \mathbf{t}^\dagger \\ \hline \mathbf{0} & \mathbf{t} & \mathbf{H_e}, \end{array}\right) \tag{3}$$

where

$$\mathbf{H_e} = \begin{array}{c} \\ \vdots \\ i{-}1 \\ i \\ i{+}1 \\ i{+}2 \\ \vdots \end{array} \begin{array}{c} \cdots \quad i{-}1 \quad\; i \quad\; i{+}1 \quad i{+}2 \quad \cdots \\ \left(\begin{array}{cccccc} \ddots & \ddots & \ddots & & & \\ \ddots & \alpha & -\beta & & & \\ \ddots & -\beta & \alpha & -\beta & & \\ & & -\beta & \alpha & -\beta & \ddots \\ & & & -\beta & \alpha & \ddots \\ & & & \ddots & \ddots & \ddots \end{array}\right) \end{array}, \tag{4}$$

$$
\mathbf{H_c} = \begin{array}{c} \vdots \\ m{-}2 \\ m{-}1 \\ m \\ m{+}1 \\ m{+}2 \\ m{+}3 \\ \vdots \end{array}
\overset{\begin{array}{cccccccc} \cdots & m{-}2 & m{-}1 & m & m{+}1 & m{+}2 & m{+}3 & \cdots \end{array}}{\begin{pmatrix}
\ddots & \ddots & & & & & & \\
\ddots & \alpha & -\beta & & & & & \\
& -\beta & \alpha & -\beta' & & & & \\
& & -\beta' & \alpha' & -\beta'' & & & \\
& & & -\beta'' & \alpha' & -\beta' & & \\
& & & & -\beta' & \alpha & -\beta & \\
& & & & & -\beta & \alpha & \ddots \\
& & & & & & \ddots & \ddots
\end{pmatrix}}, \tag{5}
$$

and

$$
\mathbf{t} = \begin{pmatrix} \cdots & 0 & -\beta \\ & & 0 \\ & & \vdots \end{pmatrix}. \tag{6}
$$

In these matrix elements, i is the site number in the one-dimensional electrode, and m and m+1 are the sites of the sandwiched molecule. The on-site energy α and hopping integral in electrodes β are respectively set to be 0 and 1. Let us firstly explain the dynamics of a wave-packet scattered by the sandwiched molecule using the wave-packet propagation. Since we can straightforwardly understand the scattering process in the wave-packet dynamics, the comparison between the results from the wave-packet and Green's function is useful to understand what is expressed in Green's function method.

Using the Hamiltonian matrix, we can propagate the wave-packets in the Crank-Nicholson scheme (Press et al., 1992), in which the norm of the wave-packet is completely conserved.

$$
\psi(x_i, t + \Delta t) = \frac{1 + \frac{1}{i\hbar}\frac{\Delta t}{2} H}{1 - \frac{1}{i\hbar}\frac{\Delta t}{2} H} \psi(x_i, t). \tag{7}
$$

This expression is straightforwardly derived from the time-dependent schrödinger equation. The number of the sites in the whole system is 2,000 in total. As for the initial wave-packet, we constructed the Gaussian packets from the eigenvectors of the one-dimensional electrode.

Figure 9 shows a typical wave-packet dynamics in the one-dimensional molecular junction, in which the energy of the wave-packet is set to be the Fermi level E_f of the electrode. The velocity of the initial wave-packet is oriented to the right direction. When the packet reaches the sites of the diatomic molecule, m and $m+1$ (the dots in Fig. 9), a portion of the packet is reflected or transmitted by the presence of the hopping integral β'. Counting the amplitudes of transmitted packet in the right electrode, we can calculate transmission probabilities as a function of the energy as

$$
T(E) = \frac{\sum_{i \in R} |\psi^E(x_i, t_1)|^2}{\sum_{i \in L} |\psi^E(x_i, t_0)|^2}, \tag{8}
$$

where L/R is respectively the left/right electrode and E is the energy of the propagating wave-packet. t_0 is the initial time, and t_1 is an arbitrary time after the transmission and reflection events (e.g., 50 fs in Fig. 9).

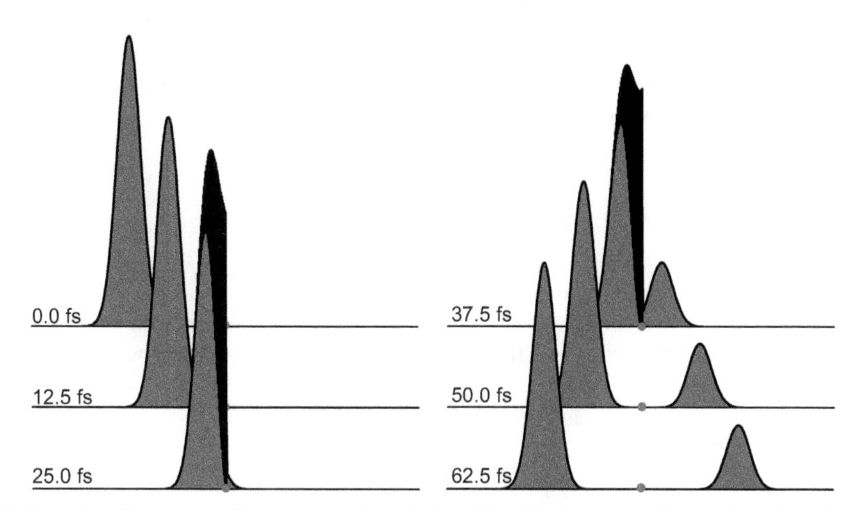

Fig. 9. Simulated wave-packet propagation in the tight binding one-dimensional molecular junction. The tight-binding parameters adopted in the propagation are 0.5 β for β', 1.0 β for β'', and 0 for α'. The red dots indicate the position of the diatomic molecule.

We calculate the electronic conductance using the transmission probability as

$$G = \frac{1}{V}\frac{2e}{h}\int_{E_f-eV/2}^{E_f+eV/2} dE\ T(E). \tag{9}$$

When the energy dependence on the transmission probability is negligibly small, the Ladauer's formula for conductance is obtained as

$$G = \frac{2e^2}{h}T(E_f). \tag{10}$$

The calculation of transmission probabilities is thus the key in the calculations of the conductance in molecular junctions.

4.2.2 Green's function approach

In this section we also adopted the same molecular junction used in the previous section, and thereby the matrix elements are also the same in Eqs. 3-6. Since the details of the relationship between the wave functions and Green's functions are described in many sophisticated text books (Bruus & Flensberg, 2004; Datta, 1997; 2005; Haug & Jauho, 1998; Stokbro et al., 2005; Ventra, 2008), we foucus only on the important equations in Green's function approach for conductance in this section.

In the matrix representation of Green's function approach for electronic conduction, the transmission probability is represented as

$$T(E) = \mathrm{Tr}[i\{\mathbf{\Sigma}_L^R(E) - \mathbf{\Sigma}_L^A(E)\}\mathbf{G}^R(E)i\{\mathbf{\Sigma}_R^R(E) - \mathbf{\Sigma}_R^A(E)\}\mathbf{G}^A(E)], \tag{11}$$

where $\mathbf{G}^{A/R}$ is the advanced/retarded Green's functions describing the scattering processes, and $\mathbf{\Sigma}^{A/R}$ is the advanced/retarded self-energies including the interactions between the molecular region and electrodes; the subscript L/R means the left/right electrodes and the advanced functions are the Hermitian conjugate of the corresponding retarded functions. The matrix expressions for these functions are

$$\mathbf{G}^R(E) = [E\mathbf{1} - \mathbf{H}_{mol} - \mathbf{\Sigma}_L^R(E) - \mathbf{\Sigma}_R^R(E)]^{-1} \tag{12}$$

and

$$\mathbf{\Sigma}^R(E) = \mathbf{t}\mathbf{g}^R(E)\mathbf{t}^\dagger, \tag{13}$$

where $\mathbf{g}$ is the surface Green's function of the electrode, which is obtained with a recursive method. In the simple one-dimensional junction shown in Fig. 8, the matrix $\mathbf{H}_{mol}$ is the 2×2 matrix in the basis of m and $m+1$ and the matrix elements in the 2×2 self energies are

$$(\mathbf{\Sigma}_L^R)_{i,j}(E) = \beta'^2 g^R(E)\delta_{ij}\delta_{i1} \tag{14}$$

and

$$(\mathbf{\Sigma}_R^R)_{i,j}(E) = \beta'^2 g^R(E)\delta_{ij}\delta_{i2}, \tag{15}$$

where the surface Green's function of the one-dimensional electrode g can be expressed in the analytical form (Emberly & Kirczenow, 1999) as

$$g^R(E) = \frac{i}{2\beta'} \frac{1 - \exp[i2y_0(E)]}{\sin y_0(E)} \tag{16}$$

and

$$y_0(E) = \arccos(E/2\beta'). \tag{17}$$

Using the Green's functions, we calculated the transmission probabilities of the one-dimensional molecular junction. Figure 10 shows the calculated transmission probabilities, together with those calculated from the wave-packet scattering approach. Both methods show the same transmission probabilities and we thus understand that the scattering event described in wave-packet approach is exactly represented in Green's function approach.

4.2.3 Nonequilibrium Green's function approach in the framework of the first-principles method

Let us briefly describe the conductance calculations in the nonequilibrium Green's function approach within the framework of the first-principles methods, especially for what is modified and what is added compared with the simple tight-binding framework. The major differences are listed as follows:

(i) We have to solve the scattering problems using non-orthogonal basis sets, which are generally used in the first-principles method.

(ii) Many non-diagonal elements show non-zero values because of the more spreading basis sets in the first-principles method.

(iii) The matrix elements depend on the electron densities, and we thus have to determine the electron densities in a self-consistent manner.

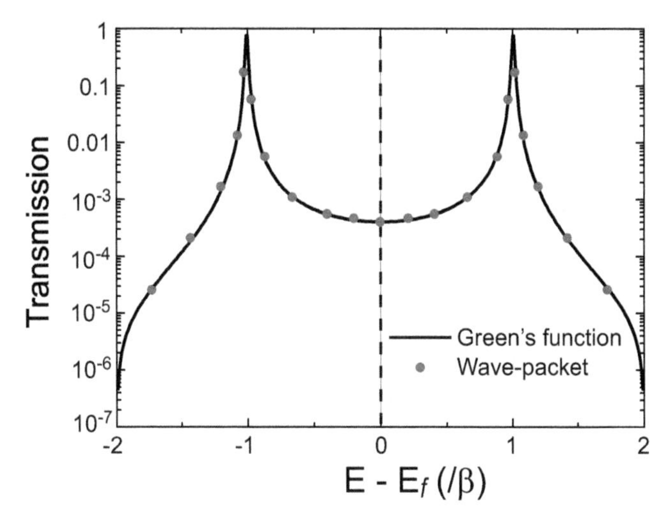

Fig. 10. Calculated transmission probabilities of the one-dimensional molecular junction. The solid line is obtained using Green' function method, and symbols (dots) are wave-packet scattering approach. The tight-binding parameters adopted in these calculations are the same to those in Fig. 9: 0.5 β for β', 1.0 β for β'', and 0 for α'.

(iv) The electrostatic potential has to be determined to hold the boundary conditions.

Since the Green's function in the non-orthogonal basis sets is expressed as $\mathbf{G}(E) = [E\mathbf{S} - \mathbf{H}_{mol} - \mathbf{\Sigma}_L(E) - \mathbf{\Sigma}_R(E)]^{-1}$ using the overlap matrix $\mathbf{S}$, the first point is trivial. This is also true for the Green's functions of electrodes. In Point (ii), we have to take care in the matrix division so as not to having non-zero elements in the non-diagonal positions between the left and right electrodes. Thereby the scattering (i.e., molecular) region sandwiched between the two electrodes must be larger than that in the simple tight-binding model where the matrix size of $\mathbf{H}_{mol}$ is 2×2. In general, the scattering region includes several metallic atoms/layers. The appropriate size of the scattering region is determined by confirming that the calculated result does not depend on the adopted size of the scattering region. As for Point (iii), we have to take care that the electron density is represented by the lesser Green's function $\mathbf{G}^{<}(= i\mathbf{G}^R[i(\mathbf{\Sigma}_L^R - \mathbf{\Sigma}_L^A)f_L + i(\mathbf{\Sigma}_R^R - \mathbf{\Sigma}_R^A)f_R]\mathbf{G}^A)$, where $f_{L/R}$ is the Fermi distribution function of the left/right electrode, respectively. The density matrix ρ of the scattering region is calculated using the lesser Green's function as

$$\rho = \int dE \frac{-i}{2\pi} \mathbf{G}^{<}(E). \tag{18}$$

When the two Fermi distribution functions have the same value, the lesser Green's function can be represented with the retarded Green's function as $-i\mathbf{G}^{<} = \mathrm{Im}[\mathbf{G}^R]$. Since the retarded Green's function has the analytic continuity property, the numerical integral with respect to energy in Eq. 18 can be extensively reduced by means of the contour integral on the complex plane (Taylor et al., 2001). The energy integral in Eq. 18 is thus divided into two parts: (i) the complex contour integral in the energy range out of the bias windows and (ii) the integral on the real axis in the bias window. The submatrix inversion method will be useful for the design of the contour path when the core/semi-core states are included in the calculations (Tada & Watanabe, 2006). In Point (iv), the Poisson equation with appropriate boundary conditions is solved, and the voltage drop over the sandwiched molecule is reasonably determined.

5. Operation-readout robust switching for the single nuclear spin qubit

5.1 Readout from current measurements

Now that we have introduced the computational tools for the calculations of the conductance of the single molecular junctions, let us consider the application for the quantum computers constituted by the single molecular junctions. Since we consider the nuclear spin qubit for quantum computations in this chapter, the single molecular junction must include a nuclear spin, and we thus regard each qubit in Fig. 2 as a single atom with the nuclear spin of $\frac{1}{2}$.

Figure 11 shows the concept of the readout of a nuclear spin state in tunneling current measurements. The keys in the readout are the hyperfine interactions between the nuclear and tunneling electron spins and the inelastic tunneling current caused by the hyperfine interactions. Assuming that the spin of the incoming electron is polarized as down-spin ($s_z = -\frac{1}{2}; -$) by spin valve α, we expect the following tunneling processes: (i) when the nuclear spin I_z is equal to $\frac{1}{2}$ (up-spin; +), a new conduction channel of the tunneling electron opens through the spin flip of $|s_z I_z\rangle = |-+\rangle \rightarrow |+-\rangle$ when applied bias voltage is larger than the Zeeman energy $\hbar\omega_0$ of the nuclear spin, resulting in the inelastic current with up-spin (Fig. 11(a)), and (ii) when the nuclear spin I_z is equal to $-\frac{1}{2}$ (down-spin), spin flip does not occur because of the spin conservation, leading to the elastic current with down-spin only (Fig. 11(b)). The inelastic process in the former case is comprehensible in the analogy of IETS described in Section 4. Note that we also assumed that there are no localized electron spins in the single nuclear spin-flip (SNuSF) region, and that nuclei in electrodes (i.e., source/drain electrodes) have no nuclear spins to avoid unexpected spin-flip processes. Thus the detection of tunneling current with up-spin in the drain electrode is the proof of the nuclear spin I_z of $\frac{1}{2}$. The detection of the tunneling electron with up-spin can be achieved by using another spin valve β in the drain electrode. Since the event we should detect is the single spin-flip process, the measured current accompanied by the spin-flip must be originated from the single electron tunneling at most. This is a clear difference from IETS for the detection of the molecular vibrations. Thus the spin valves connected to the single molecular junction is essentially important in the determination of nuclear spin states from tunneling current measurements.

5.2 Initialization through dynamic nuclear polarization

In order to confirm the plausibility of the electron-nuclear spin-flip process, we consider a situation where a bias voltage is applied to all the pairs of the source and drain electrodes in the single nuclear spin array (Fig. 2). Since the appropriate spin valves α and β are connected to the spin array and the spins of the incoming electrons passing through valve α are perfectly polarized, the spin-flip process shown in Fig. 11(a) occurs only for the nuclei with up-spin by waiting an enough time for the spin-flip process. Once the spin-flip occurs, the spin-flipped nucleus has the down-spin, and the spin-flip process for the nucleus is not expected to occur any more. That is, all the nuclear spins will be polarized to the down-spin by the bias

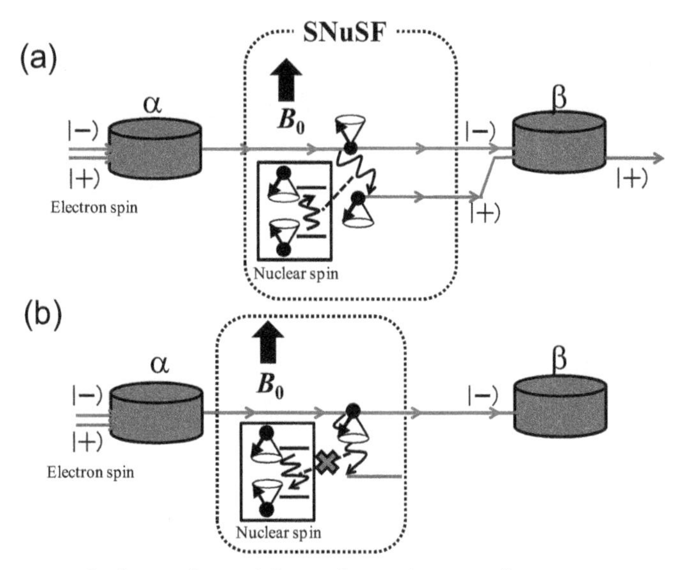

Fig. 11. Basis concept in the readout of the nuclear spin states from current measurements. The red and blue arrows represent the down and up spins of conduction electrons, respectively. The inelastic processes are allowed for (a) the nucleus with up-spin and forbidden for (b) the nucleus with down-spin.

application as shown in Fig. 12. This is the dynamic nuclear polarization by applying the bias voltage in the one-dimensional nuclear spin array.

The dynamic nuclear polarizations by the spin current or in spin-selective systems have already been confirmed in quantum dots, silicon substrates including phosphorus, and NV centers in diamond; the reported polarization rate is 38 − 52 % in quantum dots (Baugh et al., 2007; Petta et al., 2008), 68 % in phosphorus in silicon (Mccamey et al., 2009), and 98 % in NV centers in diamond (Jacques et al., 2009). These experimental observations thus guarantee the plausibility of the electron-nuclear spin-flip process discussed in this chapter. The dynamic nuclear polarization in the one-dimensional nuclear spin array is thus quite useful for the initialization process in a large number of qubits.

5.3 Operations in a selective manner

In the previous sections for the readout and initialization, we did not mention what the enough time for the single nuclear spin-flip is. In fact, this will be described using the first-principles NEGF calculations in the later section, but the point we should stress here is that we can create a special situation where the nuclear spin does not experience the spin-flip (i.e., the nuclear-spin conserved situation) even when we apply a bias voltage to the target nucleus. Note that the probability of the nuclear spin-flip is the issue of the spin-flip time: If the spin-flip time is long enough, the nucleus is lying on the situation of the spin conserved, and if it is short the nuclear spin states will be influenced by tunneling current (i.e., the readout situation). We will explain in the later section that these two situations can be exchanged in a simple way. Anyway, this section is devoted to explain how the operations for qubits are

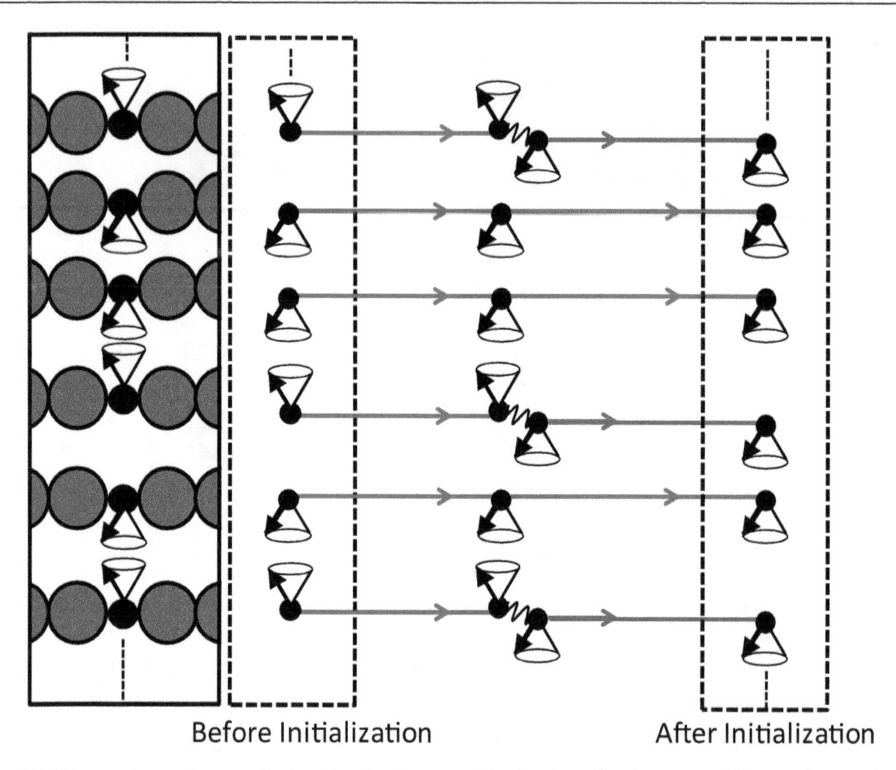

Fig. 12. Dynamic nuclear polarization in the one-dimensional spin array. The nuclear spins are all polarized to the down-spin after the bias applications.

carried out in a selective manner. The nuclear-spin conserved situation is essential for the selective operations.

Figure 13 shows the schematic of the selective operations. The key in the selective operations is that the nuclear-spin conserved situation is realized by applying the *gate* bias voltages even when we apply the source-drain bias voltage to the target nucleus. Although the nucleus does not experience the spin-flip, the nucleus feels an important influence from the bias application. The influence is the redistribution of electron densities around the nucleus, and the redistribution will lead to a modulation of the frequencies for the nuclear magnetic resonance (NMR) with respect only to the target nucleus, that is, the bias induced chemical shift of NMR. In general, the chemical shift of NMR is determined by the chemical environment, but in the nuclear spin array device we can selectively tune the chemical shift of the target nucleus (qubit) by applying the bias voltage to the electrodes connected to the qubit (Fig. 13) . Using a rotating magnetic field of the modulated frequency for the target qubit, we can control the spin direction of the target qubit only (i.e., the unitary operation). The c-NOT operation for a pair of the neighboring qubits will be possible by applying the bais voltage to the pair of electrodes connected to the target qubits. The plausibilities of the selective operations explained here are discussed using the first-principles NEGF calculations in the later section.

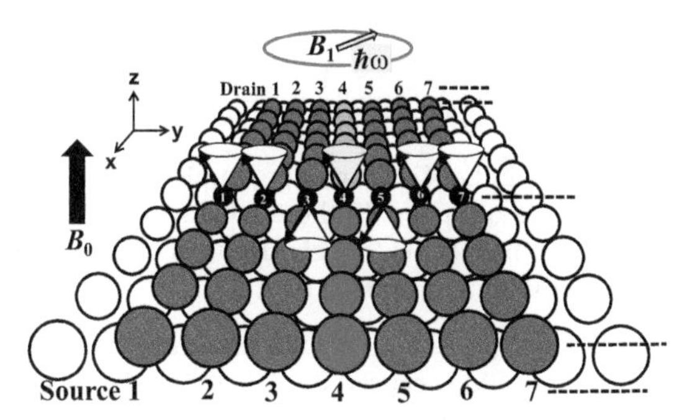

Fig. 13. Schematic of the selective operation. The bias voltage is applied to the electrodes connected to the 4-th qubit to modulate the resonance frequency of the qubit.

5.4 Nonequilibrium Green's function calculations for the robust switching between operation and readout

5.4.1 Computational method

The main issue in the selective readout and operations is the probability of the single nuclear spin-flip process by hyperfine interactions with a spin polarized tunneling electron as described in the previous section. In hyperfine interactions, the relaxation of nuclear spin **I** is mainly caused by the scalar contact and dipolar interactions with electron spin **S**, which are represented as

$$\mathcal{H}^{sc} = -\frac{2\mu_0}{3}\gamma_e\gamma_n\hbar^2\delta(0)(\mathbf{S}\cdot\mathbf{I}) \tag{19}$$

and

$$\mathcal{H}^{di} = -\frac{\mu_0}{4\pi}\gamma_e\gamma_n\hbar^2\left\{\frac{\mathbf{S}\cdot\mathbf{I}}{r^3} - \frac{3(\mathbf{S}\cdot\mathbf{r})(\mathbf{I}\cdot\mathbf{r})}{r^5}\right\}, \tag{20}$$

respectively. γ_e and γ_n are respectively the gyromagnetic ratios of free electron and nucleus, and r is the distance from the target qubit. Note that the target qubit is positioned at the origin in the coordinate. When a bias voltage V is applied to the SNuSF system, the probability w of the nuclear spin-flip by tunneling electron can be written as,

$$w_{(-+)\rightarrow(+-)} = \int_{-\infty}^{\infty} dE\frac{2\pi}{\hbar}|\langle\Psi^{+-}(E)|\mathcal{H}^{sc}+\mathcal{H}^{di}|\Psi^{-+}(E')\rangle|^2\delta(E-E') \times f_L(E)\{1-f_R(E)\}, \tag{21}$$

where Ψ^{-+} is a product state of a scattering wave function with down-spin ψ^- and spin function of nucleus with up-spin (Tada, 2008). The fermi distribution functions of the left (right) electrodes $f_{L(R)}$ in Eq. (21) guarantee that the spin-flip is caused by tunneling electron in the bias window V. The scattering wave function can be expanded with atomic orbitals ϕ_μ as $\psi^{\pm}(E,r) = \sum_\mu C_\mu^{\pm}(E)\phi_\mu(r)$, and thus the matrix elements for the scalar contact and dipolar interactions in Eq. (21) are proportional to

$$\sum_{\mu\nu} C_\mu^{\pm *}(E)C_\nu^{\mp}(E)\phi_\mu^*(0)\phi_\nu(0)(\mathbf{S}\cdot\mathbf{I}), \tag{22}$$

and

$$\sum_{\mu\nu} C_{\mu}^{\pm *}(E) C_{\nu}^{\mp}(E) \left\langle \phi_{\mu}(r) \left| \frac{\mathbf{S}\cdot\mathbf{I}}{r^3} - \frac{3(\mathbf{S}\cdot\mathbf{r})(\mathbf{I}\cdot\mathbf{r})}{r^5} \right| \phi_{\nu}(r) \right\rangle, \tag{23}$$

respectively. The indices μ and ν in Eq. (22) run over atomic orbitals having non-zero values at the qubit position. (e.g., 1s, 2s orbitals). In these equations, we have to take care which terms in $(\mathbf{S}\cdot\mathbf{I})$ and $\{\mathbf{S}\cdot\mathbf{I}/r^3 - 3(\mathbf{S}\cdot\mathbf{r})(\mathbf{I}\cdot\mathbf{r})/r^5\}$ contribute to the nuclear spin-flip processes, depending on the direction of the static magnetic field B_0.

The term $\mathbf{S}\cdot\mathbf{I}$ is equal to $S_xI_x + S_yI_y + S_zI_z$. When the static magnetic field B_0 is directed to the the z-axis, the nuclear spin-flip is caused by $S_xI_x + S_yI_y$ because the term $S_xI_x + S_yI_y$ is equal to $(S_-I_+ + S_+I_-)/2$ using the step-up and -down operators, $S_+ = S_x + iS_y$ and $S_- = S_x - iS_y$. That is, the half of $\sum_{\mu\nu} C_{\mu}^{\pm *}(E) C_{\nu}^{\mp}(E) \phi_{\mu}^{*}(0)\phi_{\nu}(0)$ contributes to the spin-flip process of $|-+\rangle \rightarrow |+-\rangle$ in the scalar contact term. On the other hand, the dipolar term has a more complicated form as

$$\begin{aligned} \left\langle \phi_{\mu}(r) \left| \frac{\mathbf{S}\cdot\mathbf{I}}{r^3} - \frac{3(\mathbf{S}\cdot\mathbf{r})(\mathbf{I}\cdot\mathbf{r})}{r^5} \right| \phi_{\nu}(r) \right\rangle &= \mathbf{S}\cdot\mathbf{A}_{dip}\cdot\mathbf{I} \\ &= \begin{bmatrix} S_x & S_y & S_z \end{bmatrix} \\ &\cdot \begin{bmatrix} \langle\frac{r^2-3x^2}{r^5}\rangle_{\mu\nu} & -\langle\frac{3xy}{r^5}\rangle_{\mu\nu} & -\langle\frac{3xz}{r^5}\rangle_{\mu\nu} \\ -\langle\frac{3xy}{r^5}\rangle_{\mu\nu} & \langle\frac{r^2-3y^2}{r^5}\rangle_{\mu\nu} & -\langle\frac{3yz}{r^5}\rangle_{\mu\nu} \\ -\langle\frac{3xz}{r^5}\rangle_{\mu\nu} & -\langle\frac{3yz}{r^5}\rangle_{\mu\nu} & \langle\frac{r^2-3z^2}{r^5}\rangle_{\mu\nu} \end{bmatrix} \cdot \begin{bmatrix} I_x \\ I_y \\ I_z \end{bmatrix}, \end{aligned} \tag{24}$$

where the notation $\langle\ \rangle_{\mu\nu}$ means the integral including the atomic orbitals ϕ_{μ} and ϕ_{ν}. For example, $\langle\frac{r^2-3x^2}{r^5}\rangle_{\mu\nu}$ corresponds to $\langle\phi_{\mu}(r)|\frac{r^2-3x^2}{r^5}|\phi_{\nu}(r)\rangle$. When the static magnetic field B_0 is directed to the the z-axis, the terms we have to calculate are those related to S_xI_x and S_yI_y as describe in the contact term, and the tensor $\mathbf{A}_{dip}$ can be written as

$$\mathbf{A}_{dip}^{B_0:z} = \begin{bmatrix} \langle\frac{r^2-3x^2}{r^5}\rangle_{\mu\nu} & 0 & 0 \\ 0 & \langle\frac{r^2-3y^2}{r^5}\rangle_{\mu\nu} & 0 \\ 0 & 0 & 0 \end{bmatrix}. \tag{25}$$

In the expression, we also use the fact that the term $S_xI_y + S_yI_x$ does not include both of the spin-flip components S_-I_+ and S_+I_-.

Using the integrals in Eq. (24) and assuming $C^- = C^+$ in Eqs. (22) and (23), which is a reasonable assumption in a spin polarized current by a small bias application, the nuclear spin-flip probability in Eq. (21) is calculated with the lesser Green's function using the relation: $C_{\mu}^{\pm *}(E) C_{\nu}^{\pm}(E) = \frac{-i}{2\pi}[\mathbf{G}^{<,\pm}(E)]_{\mu\nu}$. In the SCF calculations for applied bias cases, a NEGF code (Tada & Watanabe, 2006; Tada, 2008) incorporated in GAUSSIAN03 (Frisch et al., 2003) was employed.

5.4.2 Computational model

The SNuSF system considered in the present work is a simple junction composed of a hydrogen molecule sandwiched between ^{106}Pd one-dimensional metallic electrodes (^{106}Pd(1D)-^{1}H$_2$-^{106}Pd(1D)). Table 1 shows the gyromagnetic ratios of typical nuclei. Since the probability of the nuclear spin-flip depends also on γ_n (see Eqs. (19) and (20)), we adopted the hydrogen molecular junction showing the highest gyromagnetic ratio. The electrodes composed of ^{106}Pd was selected because ^{106}Pd has no nuclear spin: this is one of the conditions required for SNuSF describe in Section 5.1.

nucleus	$\gamma_n(\mathrm{sT})^{-1}$
^{1}H	2.673×10^8
^{7}Li	1.040×10^8
^{13}C	0.673×10^8
^{19}F	2.517×10^8
^{31}P	1.083×10^8

Table 1. The gyromagnetic ratios of typical nuclei

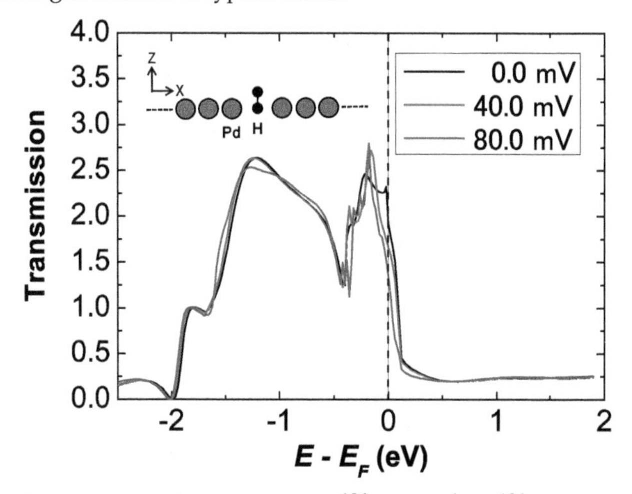

Fig. 14. Calculated transmission functions of the ^{106}Pd(1D)-^{1}H$_2$-^{106}Pd(1D) junction.

We adopted a contact structure in which a hydrogen atom is directly sandwiched between the Pd electrodes as shown in the inset of Fig. 14. Although the adsorption structure is somewhat artificial, we adopted this structure as the first step in the investigation of the nuclear spin-flip probability. The more plausible structures for the hydrogen contact will be investigated in the near future.

5.4.3 Computational results for the robust switching

Figure 14 shows the calculated down-spin transmission function of the Pd(1D)-H$_2$-Pd(1D) junction in the local density approximation (SVWN functional (Hohenberg & Kohn, 1964; Vosko et al., 1980) in GAUSSIAN 03) of the density functional theory with LANL2MB (Pd) (Hay & Wadt, 1985) and cc-pvdz (H) (Dunning, 1989) basis sets. A sharp peak of transmission function appears below 0.2 eV from the Fermi level E_F in 40 and 80 mV applied bias cases. The peak is also confirmed in three-dimensional Pd electrode systems (Khoo et al., 2008) , and thus the shape of the transmission function around the Fermi level is the characteristic property of Pd systems. This is quite important property for the robust switching between the readout (Section 5.1) and operation (Section 5.3) through the control of the hyperfine interaction.

It is well-known that we can modulate the Fermi level of the system by applying the gate bias voltage to the system. When the Fermi level is shifted to the peak level (i.e., $\Delta E_F = -0.2$ eV), the current density of down-spin on the Pd atoms connected to the target hydrogen is significantly enhanced. Since the atomic orbitals of the Pd atoms have non-trivial amplitudes around the target hydrogen, the large electron density in the down-spin current will lead to the enhancement of hyperfine interactions between the nuclear spin and the tunneling current, leading to a fast nuclear spin-flip. The transmission function with up-spin has the same spectra with down-spin, and a conduction channel of up-spin within the bias window opens immediately accompanied by the single nuclear spin-flip process. This is the qubit condition convenient for the selective readout. We use the term *ON-resonance* to express the condition of the Fermi level shifted to the peak position. On the other hand, when the Fermi level is shifted out from the peak level to another level (e.g., $\Delta E_F = 0.1$ eV), we will have a weak hyperfine interaction, leading to an extremely decreased probability of the nuclear spin-flip. This is the qubit condition convenient for the selective operations, because we can rotate the nuclear spins selectively using a rotating magnetic field in this situation as described in Section 5.3. We use the term *OFF-resonance* to express the condition of the Fermi level shifted out from the peak position. Therefore the switching of the hyperfine interactions by shifting the Fermi level will be a robust switching between the readout and operations for qubits in nano-contact systems.

Figure 15 shows the computed spin-flip times (Fig. 15(a)) and significant matrix elements for the scalar contact (Fig. 15(b)) and dipolar interactions (Fig. 15(c)) as a function of the Fermi level shift in 40 mV applied bias case. The spin-flip times clearly depend on the position of the Fermi level: (a) at ON-resonance ($-0.5 < \Delta E_F < -0.1$ eV), the nuclear spin-flip event occurs within 100 - 1000 s because of the enhanced hyperfine interactions, and (b) at OFF-resonance ($0.1 < \Delta E_F < 0.5$ eV), the nuclear spin-flip time is $10^5 - 10^7$ s. The computed spin-flip times indicate that a single nuclear spin-flip in the present molecular junction might be a measurable event at the ON-resonance condition, and that the nuclear spin state is preserved for a long time at the OFF-resonance even when a bias voltage is applied to the single nucleus. The operation-readout robust switching (we call it *hyperfine switching* in the previous study (Tada, 2008)) is strongly related to the electron tunneling through the d-orbitals of Pd, as shown in Fig. 15(b,c). The matrix element for the dipolar interaction of $4d$ orbital shows the drastic variation around $\Delta E_F = 0.0$ eV, whereas the element for the contact interaction in $2s$ orbital of the sandwiched hydrogen shows a moderate variation around $\Delta E_F = 0.0$ eV.

5.4.4 Computational results for the bias induced NMR chemical shifts

The extremely slow relaxation at OFF-resonance is quite useful for the selective operations on NMR qubits as described in Section 5.3. To confirm the modulation of resonance frequency by bias voltage applications at OFF-resonance, the magnetic shielding constant of the target hydrogen in Pd(1D)-H_2-Pd(1D) is calculated using the gauge invariant atomic orbital (GIAO) method (Lee et al., 1995) implemented in the GAUSSIAN03 code. The calculations of shielding constants for Pd(1D)-H_2-Pd(1D) with 40 mV bias application were performed as follows: (i) Hamiltonian matrix of the scattering region (Pd_6-H_2-Pd_6) is calculated from the converged lesser Green's function (density matrix), (ii) molecular orbitals (MOs) for the scattering region are calculated through diagonalization of the Hamiltonian matrix, leading to a set of discrete MO levels, (iii) confirm that the Fermi level used in NEGF calculations is positioned between

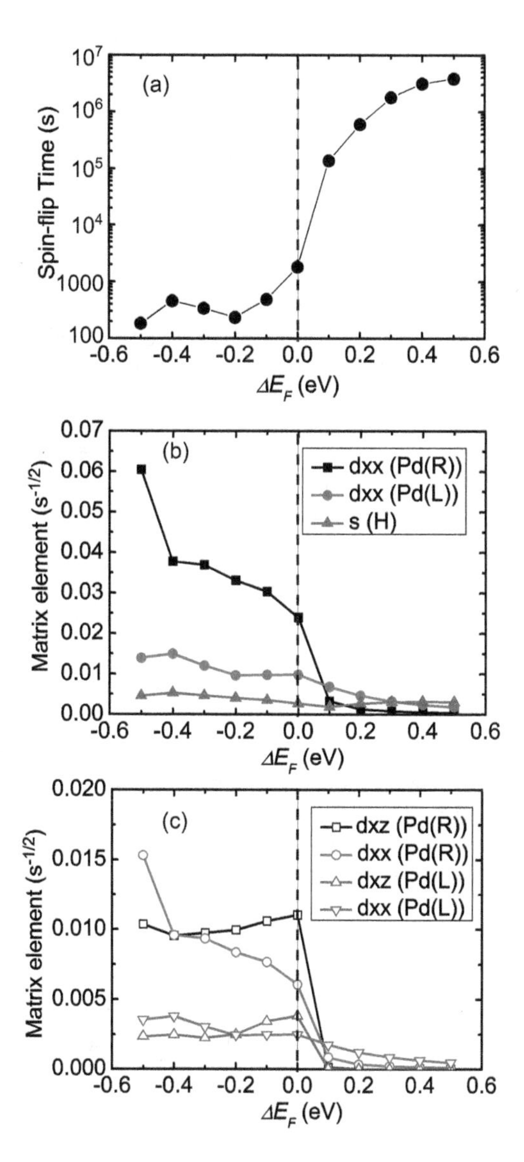

Fig. 15. (a) Computed spin-flip times and (b) significant elements for scalar contact and (c) dipolar interactions as a function of the Fermi level shift ΔE_F.

the highest occupied MO and the lowest unoccupied MO, and (iv) shielding constants are calculated using the discrete MOs in GIAO method. When a magnetic field of 10 T is applied to the SNuSF system, the Larmor frequency $\omega_0/2\pi$ (reference frequency) of proton is 425.7 MHz. The calculated shielding constants of the proton for 0.0 and 40.0 mV bias applications are 157.67 and 35.65 ppm, respectively. The frequency shifts from the reference frequency is thus 67.12 and 15.18 kHz for 0.0 and 40.0 mV bias applications, respectively. The difference of the resonance frequency between 0.0 and 40.0 mV bias applications is large enough for selective operations on qubits by making use of radio-frequency pulses (Vandersypen et al., 2001).

5.4.5 The system temperature

In the previous study of *hyperfine switching* for single molecular junctions, the author discussed the cooling down method to prevent unexpected nuclear spin-flips, and the required temperature was estimated to be $\sim$10 mK even when the static magnetic field of 10T is applied to the nuclear spin array (Tada, 2008). This requirement is extremely terrible. However, we described the dynamic nuclear polarization technique available for the nuclear spin array in Section 5.2, and we can use this technique in stead of the simple cooling down method. The key property for the system condition using the dynamic nuclear polarization is again the nuclear spin-flip probability depending on the system temperature.

When the system temperature is set to be T(K), the electrons in the energy range from $E_F - k_B T/2$ to $E_F + k_B T/2$ can contribute to the nuclear spin-flip processes (Abragam, 1961) , where k_B is the Boltzmann constant. The spin-flip time at $T(K)$ is thus represented as

$$w_{(-+)\rightarrow(+-)} = \int_{E_F - k_B T/2}^{E_F + k_B T/2} dE \frac{2\pi}{\hbar} |\langle \Psi^{+-}(E)|\mathcal{H}^{sc} + \mathcal{H}^{di}|\Psi^{-+}(E')\rangle|^2 \delta(E - E'). \quad (26)$$

Here we assumed that the nuclear spins in the array structure are already polarized to the direction of + by the dynamic nuclear polarization. For instance, when the system temperature is equal to the room temperature (300K), the energy window is about 26 meV. Using Eq. 26 and density matrix at the zero bias condition, we obtained the spin-flip time of 1800 s. In addition, using the OFF-resonance conditions induced by the gate bias, the spin-flip time can be extremely long (e.g., 10^5 - 10^7 s). These results indicate that the nuclear spin states can be preserved in long periods, which are enough for operations and readout.

6. Conclusions

A novel detection mechanism of single nuclear spin-flip by hyperfine interactions between nuclear spin and tunneling electron spin is proposed, and the probability of the nuclear spin-flip is calculated using ab initio non-equilibrium Green's function method. The calculated relaxation times for nano-contact system, Pd(1D)-H_2-Pd(1D), reveal that ON/OFF switching of hyperfine interactions is effectively triggered by resonant tunneling mediated through the d-orbitals of Pd; when the bias voltage of 40 mV is applied to the system, (a) the nuclear spin-flip event occurs within 100 - 1000 s at ON-resonance and (b) the relaxation time of the single nuclear spin-flip is $10^5 - 10^7$ s at OFF-resonance. The effectiveness of bias voltage applications at OFF-resonance for selective operations on qubits is also demonstrated in the calculations of resonant frequencies of proton using the gauge invariant atomic orbital method.

7. Acknowledgments

This work was partially supported by the Grant-in-Aid for Young Scientists (B), MEXT of Japan.

8. References

Abragam, A. (1961). *Principles of Nuclear Magnetism*, Oxford University Press, 019852014X (ISBN), New York.

Aviram, A. & Ratner, M. A. (1974). Molecular Rectifiers. *Chemical Physics Letters*, 29, 277-283, 00092614(ISSN).

Baugh, J.; Kitamura, Y.; Ono, K. & Tarucha, S. (2007). Large Nuclear Overhauser Fields Detected in Vertically Coupled Double Quantum Dots. *Physical Review Letters*, 99, 096804(1)-096804(4), 10797114 (ISSN).

Benenti, G.; Casati, G. & Strini, G. (2004). *Principles of Quantum Computation and Information Volume I: Basic Concepts*, World Scientific, 9812388303 (ISBN), Singapore.

Bruus, H. & Flensberg, K. (2004). *Many-Body Quantum Theory in Condensed Matter Physics*, Oxford University Press, 0198566336 (ISBN), New York.

Brandbyge, M.; Mozos, J. L.; Ordejón, P.; Taylor, J. & Stokbro, K. (2002). Density-functional method for nonequilibrium electron transport. *Physical Review B*, 65, 165401(1)-165401(17), 01631829 (ISSN).

Cappellaro, P.; Jiang, L.; Hodges, J. S. & Lukin, M. D. (2009). Coherence and Control of Quantum Registers Based on Electronic Spin in a Nuclear Spin Bath. *Physical Review Letters*, 102, 21, 210502(1)-210502(4), 10797114 (ISSN).

Carroll, R. L. & Gorman, C. B. (2002). The Genesis of Molecular Electronics. *Angewandte Chemie International Edition*, 41, 4378-4400, 15213773 (ISSN).

Childress, L.; Dutt, M. V. G.; Taylor, J. M.; Zibrov, A. S.; Jelezko, F.; Wrachtrup, J.; Hemmer, P. R. & Lukin, M. D. (2006). Coherent dynamics of coupled electron and nuclear spin qubits in diamond. *Science*, 314, 281-285, 00368075 (ISSN).

Choi, H. J. & Ihm, J. (1999). Ab initio pseudopotential method for the calculation of conductance in quantum wires. *Physical Review B*, 59, 2267-2275, 01631829 (ISSN).

Cui, X. D.; Primak, A.; Zarate, X.; Tomfohr, J.; Sankey, O. F. Moore, A. L.; Moore, T. A.; Gust, D. Harris, G. & Lindsay, S. M. (2001). Reproducible measurement of single-molecule conductivity. *Science*, 294, 571-574, 00368075 (ISSN).

Datta, S. (1997). *Electron Transport in Mesoscopic Systems*, Cambridge University Press, 0521599431 (ISBN-10), Cambridge.

Datta, S. (2005). *Quantum Transport: Atom to Transistor*, Cambridge University Press, 0521631459 (ISBN-10), New York.

Dunning, T. H. (1989). Gaussian-Basis Sets for Use in Correlated Molecular Calculations. 1. The Atoms Boron through Neon and Hydrogen. *Journal of Chemical Physics*, 90, 1007-1023, 00219606 (ISSN).

Emberly, E. G. & Kirczenow, G. (1999). Antiresonances in molecular wires. *Journal of Physics: Condensed Matter*, 11, 6911-6926, 09538984 (ISSN).

Frisch, A.; Frisch, M. J. & Trucks, G. W. (2003). *Gaussian 03 User's Reference*, Gaussian Inc., 0972718702 (ISBN), Carnegie.

Gohda, Y.; Nakamura, Y.; Watanabe, K. & Watanabe, S. (2000). Self-Consistent Density Functional Calculation of Field Emission Currents from Metals. *Physical Review Letters*, 85, 1750-1753, 10797114 (ISSN).

Haug, H. & Jauho, A. -P. (1998). *Quantum Kinetics in Transport and Optics of Semi-conductors*, Springer, 3540616020 (ISBN), Heidelberg.

Hay, P. J. & Wadt, W. R. (1985). Ab initio Effective Core Potentials for Molecular Calculations - Potentials for the Transition-Metal Atoms Sc to Hg. *Journal of Chemical Physics*, 82, 270-283, 00219606 (ISSN).

Hirose, K. & Tsukada, M. (1994). First-Principles Theory of Atom Extraction by Scanning Tunneling Microscopy. *Physical Review Letters*, 73, 150-153, 10797114 (ISSN).

Hirose, K. & Tsukada, M. (1995). First-Principles Calculation of the electronic structure for a bielectrode junction system under strong field and current. *Physical Review B*, 51, 5278-5290, 01631829 (ISSN).

Hohenberg, P. & Kohn, W. (1964). Inhomogeneous Electron Gas. *Physical Review*, 136, B864-B871, 0031899X (ISSN).

Jacques, V.; Neumann, P.; Beck, J.; Markham, M.; Twitchen, D.; Meijer, J.; Kaiser, F.; Balasubramanian, G.; Jelezko, F. & Wrachtrup, J. (2009). Dynamic Polarization of Single Nuclear Spins by Optical Pumping of Nitrogen-Vacancy Color Centers in Diamond at Room Temperature. *Physical Review Letters*, 102, 5, 057403(1)-057403(4), 10797114 (ISSN).

Jelezko, F.; Gaebel, T.; Popa, I.; Gruber, A. & Wrachtrup, J. (2004). Observation of Coherent Oscillations in a Single Electron Spin. *Physical Review Letters*, 92, 7, 076401(1)-076401(4), 10797114 (ISSN).

Jelezko, F.; Gaebel, T.; Popa, I.; Domhan, M.; Gruber, A. & Wrachtrup, J. (2004). Observation of Coherent Oscillation of a Single Nuclear Spin and Realization of a Two-Qubit Conditional Quantum Gate. *Physical Review Letters*, 93, 13, 130501(1)-130501(4), 10797114 (ISSN).

Joachim, C.; Gimzewski, J. K. & Aviram, A. (2000). Electronics using hybrid-molecular and mono-molecular devices. *Nature*, 408, 541-548, 00280836 (ISSN).

Kane B. (1998). Silicon-Based Nuclear Spin Quantum Computer. *Nature*, 393, 133-137, 00280836 (ISSN).

Khoo, K. H.; Neaton, J. B.; Choi, H. J. & Louie, S. G. (2008). Contact dependence of the conductance of H_2 molecular junctions from first principles. *Physical Review B*, 77, 115326(1)-115326(6), 01631829 (ISSN).

Lang, N. D. (1995). Resistance of Atomic Wires. *Physical Review B*, 52, 5335-5342, 01631829 (ISSN).

Ladd, T. D.; Jelezko, F.; Laflamme, R.; Nakamura, Y.; Monroe, C. & O'Brien, J. L. (2010). Quantum computers. *Nature*, 464, 45-53, 00280836 (ISSN).

Lee, A. M.; Handy, N. C. & Colwell, S. M. (1995). The Density-Functional Calculation of Nuclear Shielding Constants using London Atomic Orbitals. *Journal of Chemical Physics*, 103, 10095-10109, 00219606 (ISSN).

Lo, C. C.; Bokor, J.; Schenkel, T.; He, J.; Tyryshkin, A. M. & Lyon, S. A. (2007). Spin-dependent scattering off neutral antimony donors in field-effect transistors. *Applied Physics Letters*, 91, 242106(1)-242106(3), 00036951 (ISSN).

McCamey, D. R.; Huebl, H.; Brandt, M. S.; Hutchison, W. D.; McCallum, J. C.; Clark, R. G. & Hamilton, A. R. (2006). Electrically detected magnetic resonance in ion-implanted Si : P nanostructures. *Applied Physics Letters*, 89, 182115(1)-182115(3), 00036951 (ISSN).

McCamey, D.; van Tol, J.; Morley, G. & Boehme, C. (2009). Fast Nuclear Spin Hyperpolarization of Phosphorus in Silicon. *Physical Review Letters*, 102, 2, 027601(1)-027601(4), 10797114 (ISSN).

Metzger, R. M. (2005). Six Unimolecular Rectifiers and What Lies Ahead. In: *Introducing Molecular Electronics*, Cuniberti, G.; Fagas, G. & Richter, K. (Ed.), 313-349, Springer, 3540279946 (ISBN-10), Heidelberg.

Miyano, K. & Furusawa, A. (2008). *An Introduction to Quantum Computation*, Nippon-Hyoron-sha, 9784535784796 (ISBN), Tokyo.

Morton, J. J. L.; Tyryshkin, A. M.; Brown, R. M.; Shankar, S.; Lovett, B. W.; Ardavan, A.; Schenkel, T.; Haller, E. E.; Ager, J. W. & Lyon, S. A. (2008). Solid-state quantum memory using the ^{31}P nuclear spin. *Nature*, 455, 1085-1088, 00280836 (ISSN).

Nazin, G. V.; Qiu, X. H. & Ho, W. (2003). Visualization and spectroscopy of a metal-molecule-metal bridge. *Science*, 302, 77-81, 00368075 (ISSN).

Neumann, P.; Mizuochi, N.; Rempp, F.; Hemmer, P.; Watanabe, H.; Yamasaki, S.; Jacques, V.; Gaebel, T.; Jelezko, F. & Wrachtrup, J. (2008). Multipartite entanglement among single spins in diamond. *Science*, 320, 1326-1329, 00368075 (ISSN).

Nielsen, M. A. & Chuang, I. L. (2000). *Quantum Computation and Quantum Information*, Cambridge University Press, 0521632358 (ISBN), Cambridge.

Petta, J. R.; Taylor, J. M.; Johnson, A. C.; Yacoby, A.; Lukin, M. D.; Marcus, C. M.; Hanson, M. P. & Gossard, A. C. (2008). Dynamic Nuclear Polarization with Single Electron Spins. *Physical Review Letters*, 100, 067601(1)-067601(4), 10797114 (ISSN).

Press, W. H.; Teukolsky, S. A.; Vetterling, W. T. & Flannery, B. P. (1992). *Numerical Recipes in Fortran 77 Second Edition*, Cambridge University Press, 052143064X (ISBN), Cambridge.

Porath, D.; Lapidot, N. & Gomez-Herrero, J. (2005). Charge Transport in DNA-based Devices. In: *Introducing Molecular Electronics*, Cuniberti, G.; Fagas, G. & Richter, K. (Ed.), 411-444, Springer, 3540279946 (ISBN-10), Heidelberg.

Reed, M. A.; Zhou, C.; Muller, C. J.; Burgin, T. P. & Tour, J. M. (1997). Conductance of a Molecular Junction. *Science*, 278, 252-254, 00368075 (ISSN).

Ruitenbeek, J. v.; Scheer, E. & Weber, H. B. (2005). Conducting Individual Molecules using Mechanically Controllable Break Junction. In: *Introducing Molecular Electronics*, Cuniberti, G.; Fagas, G. & Richter, K. (Ed.), 253-274, Springer, 3540279946 (ISBN-10), Heidelberg.

Ruitenbeek, J. M. v. (2010). Quasi-ballistic electron transport in atomic wires. In: *Oxford Handbook of Nanoscience and Technology: Volume 1: Basic Aspects*, Narlikar, A. V. & Fu, Y. Y. (Ed.), 117-143, Oxford University Press, 0199533040 (ISBN-10), Place of publication.

Sagawa, H. & Yoshida, N. (2003). *Quantum Information Theory*, Springer Japan, 9784431100560 (ISBN), Tokyo.

Smeltzer, B.; Mcintyre, J. & Childress, L. (2009). Robust control of individual nuclear spins in diamond. *Physical Review A*, 80, 5, 050302(1)-050302(4), 10502947 (ISSN).

Smogunov, A.; Corso, A. D.;& Tosatti, E. (2004). Ballistic conductance of magnetic Co and Ni nanowires with ultrasoft pseudopotentials. *Physical Review B*, 70, 045417(1)-045417(9), 01631829 (ISSN).

Stegner, A. R.; Boehme, C.; Huebl, H.; Stutzmann, M.; Lips, K. & Brandt, M. S. (2006). Electrical detection of coherent ^{31}P spin quantum states. *Nature Physics*, 2, 835-838, 17452473 (ISSN).

Stokbro, K.; Taylor, J.; Brandbyge, M. & Guo, H. (2005). Ab-initio Non-Equilibrium Green's Function Formalism for Calculating Electron Transport in Molecular Devices. In: *Introducing Molecular Electronics*, Cuniberti, G.; Fagas, G. & Richter, K. (Ed.), 117-151, Springer, 3540279946 (ISBN-10), Heidelberg.

Tada, T. & Watanabe, S. (2006). Submatrix Inversion Approach to the ab initio Green's function method for electrical transport. *e-Journal of Surface Science and Nanotechnology*, 4, 484-489, 13480391 (ISSN).

Tada, T. (2008). Hyperfine switching triggered by resonant tunneling for the detection of a single nuclear spin qubit. *Physics Letters A*, 372, 6690-6693, 03759601 (ISSN).

Taylor, J.; Guo, H. & Wang, J. (2001). Ab initio modeling of quantum transport properties of molecular electronic devices. *Physical Review B*, 63, 245407(1)-245407(13), 01631829 (ISSN).

Vandersypen, L. M. K.; Steffen, M.; Breyta, G.; Yannoni, C. S.; Sherwood, M. H. & Chuang, I. L. (2001). Experimental realization of Shor's quantum factoring algorithm using nuclear magnetic resonance. *Nature*, 414, 883-887, 00280836 (ISSN).

Ventra, M. D. (2008). *Electrical Transport in Nanoscale Systems*, Cambridge University Press, 9780521896344 (ISBN), New York.

Vosko, S. H.; Wilk, L. & Nusair, M. (1980). Accurate Spin-dependent Electron Liquid Correlation Energies for Local Spin-Density Calculations - A Critical Analysis. *Canadian Journal of Physics*, 58, 1200-1211, 00084204 (ISSN).

Wang, W.; Lee, T. & Reed, M. A. (2005). Intrinsic Electronic Conduction Mechanisms in Self-Assembled Monolayers. In: *Introducing Molecular Electronics*, Cuniberti, G.; Fagas, G. & Richter, K. (Ed.), 275-300, Springer, 3540279946 (ISBN-10), Heidelberg.

Xu, B. & Tao, N. J. (2003). Measurement of Single-Molecule Resistance by Repeated Formation of Molecular Junctions. *Science*, 301, 1221-1223, 00368075 (ISSN).

7

Theoretical Study for High Energy Density Compounds from Cyclophosphazene

Kun Wang, Jian-Guo Zhang*, Hui-Hui Zheng,
Hui-Sheng Huang and Tong-Lai Zhang
State Key Laboratory of Explosion Science and Technology,
Beijing Institute of Technology
China

1. Introduction

The phosphazenes have distinguished ancestry. The reaction between phosphorus pentachloride and ammonia was described by Rose in 1834[1], and in an editorial comment, Liebig [13] reported work carried out in conjunction with Wöhler. The major reaction product was phospham and a small quantity of a stable crystalline compound containing nitrogen, phosphorus, and chlorine was obtained. Gerhardt and Laurent established that the empirical composition was $NPCl_2$, and Gladstone and Holmes and Wichelhaus measured the vapor density and deduced the molecular formula, $N_3P_3Cl_6$[2].

Phosphorus nitrogen compounds are renowned for their ability to form a variety of ring and cage structures. The most prominent P–N ring systems are phosphazanes, featuring single P–N bond[3], and phosphazenes, having multiple P–N bonds[4-8]. The two kinds of systems tend to occur in different ring sizes. The polyphosphazene has been used in medical community widely because its excellent biocompatibility and biological activity. The chemists have synthesized the medical polyphosphazene in 1977 with the substituent of glycine-ethylester[9]. In addition, There are applications of polyphosphazene in membrane separation, dye and catalysts [10].

Cyclophosphazene as a kind of phosphazene compounds attracts many researchers for a long time due to their unique properties. The energetic cyclophosphazene compounds without heavy metal elements are environmentally friendly and have very high energy density. Cyclophosphazene containing amino, nitro, nitramino and azido groups would be a kind of possible high-energy compound. It is a polymer where alternate regularly with the double and single bond between the nitrogen and the phosphorus [11]. The generally accepted "island model"[12]. supposes the σ-bonds in the phosphanzenes being formed by sp^3 hybrid orbital of phosphorus. The orbital available for out-of-plane π-bonding is dyz orbital being combined into sets of three-center π-molecular orbital. These three-center orbital overlap only weakly with one another and the π-electrons are effectively localized in definite three-center-π-bonds. Unusual chemical bonding in P-N backbone causes many

* Corresponding Author

unique properties of phosphazenes. There are nitrogen atoms in the heterocyclic, also the empty *d* orbital of P atom can accommodate the electrons. That made the modification to the ring possible such as adding new nitrogen heterocyclic, azido or modifying the ring by nitrification to increasing the nitrogen content. The Fig. 1.1 has showed the structure of what we talked.

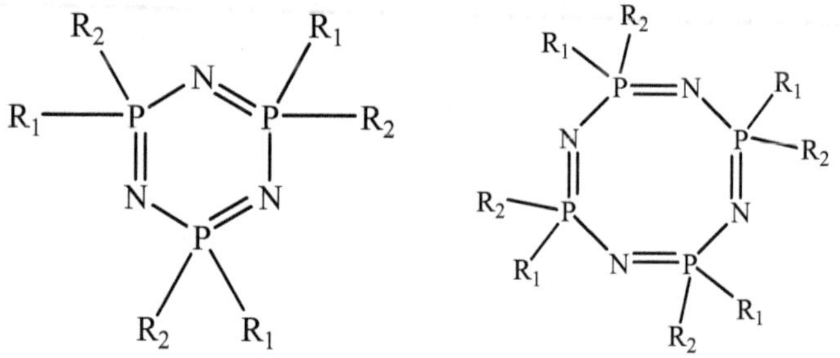

Fig. 1.1. The structure of hexa-cyclophosphazene and octa-cyclophosphazene

This structure can melt the advantages between the ignore materials and organism to form another outstanding compounds which is stable and acid and alkali resistant. Based on the N-P cross structure and the excellent flame retardant, also there will be no poison when it degrades, we can use that for the high temperature resistant. Further, we can introduce the cyclophosphazene into resin to improve this character. For example, replace the chlorine of the chlorinated- cyclophosphazene by the polymeric moiety and melt by the graphite will obtain the composite material which is ought to use in the aerospace industry [10].

Liebig [13] is the first people who had synthesis the phosphazene oligomer through NH_4Cl and PCl_5 in 1834. Then people gradually studied the structure, molecular weight, chemical properties and the synthesis method in the subsequent 100 years. All jobs included the theory of synthesis and theoretical calculation.

In the development of synthesis recent 20 years, In 90s, Tuncer Hökelek of Hacettepe university have synthesized $N_4P_4Cl_4(NEt_2)_4$[14], $N_4P_4Cl_7(OC_6H_2\text{-}2,6\text{-}t\text{-}Bu_2\text{-}4\text{-}Me)$[15], $N_4P_4(NC_4H_8O)_6(NHEt)_2$][16] and $N_4P_4(NC_5H_{10})_6(NHEt)_2$[17] and gave their structure parameters. Christopher W. Allen[4] and Dave[18,19] and many other scientists[20-27] have studied in this field to perfected this system. For the other aspect, it's a very rapid development for the theory development these years due to the computer technology. We got many useful data to predict the experimental result and guide the synthesis from many experts all over the world [28-37].

Our group always paid attention in the energetic materials. So how to increase the energy of this class of compounds is the point of our work. In 2008, we have reported the theoretical study for high nitrogen-contented energetic compound of 1,1,3,3,5,5,7,7-octaazido-cyclotetraphosphazene ($N_4P_4(N_3)_8$) [12]. Molecular structure, vibrational frequencies and infrared intensities of it has been studied in different theoretical method. The structure has been showed in Fig. 1.2. We obtain this is a non-planar structure but there are some special characters in the P-N bonds in the nitrogen-phosphorus ring. Another paper also reported in the same year to compared with the experiment of synthesis of the 1,1,3,3,5,5-hexazaidocyclotetraphosphazene ($N_3P_3(N_3)_6$) by Michael Göbel[27] in 2006. We have anglicized the crystal in different theoretical method. The structure is in Fig. 1.3. We studied this compound about the energy gap by DFT method, the molecular activity by frontier orbital theory. Also the geometric data and the electrostatic potential has been calculated and compared by the experimental data. In 2009, we researched the 1, 1-diaminohexaazido-cyclo-tetraphophazene (DAHA) and its isomers to perfected the theoretical study of azaido-triphosphazene[38]. In this research we point out there is no aromaticity in the ring. And we found the weakest bonds and proved different substituent affect the stability of P-N bonds in the ring. We predicted they will be a kind of right energetic materials since the high heats of formation. Fig. 1.4 has showed the five structures of the isomers. The structure of five isomers for diamino-hexaazido-cyclo-tetraphosphazene have numbered like this:

1,1-Diamino-3,3,5,5,7,7-hexaazidocyclotetra- phosphazene(a);
trans-1,5-diamino-1,3,3,5,7,7-hexaazidocyclotetraphosphazene (b);
cis-1,5-diamino- 1,3,3,5,7,7-hexaazidocyclotetraphosphazene(c);
trans-1,3-diamino-1,3,5,5,7,7-hexaazidocyclotetraphosphazene(d);
cis-1,3-diamino-1,3,5,5,7,7-hexa-azidocyclotetraphosphazene (e).

All the above we have talked was the azido-cyclosphazene. For the other aspect, we have some research of the spiro-cyclotriphosphazene. Our group has synthesized 1,1-spiro(ethylenediamino)-3,3,5,5-tetrachloro-cyclotriphosphazene (ETCCTP)[39] and performed its theoretical study and the nitration product 1,1-Spiro- (N,N'-dinitro-ethylenediamino)-3,3,5,5–tetrachloro-cyclotriphosphazene (DNETCCTP). The molecular structures and crystal structures of ETCCTP have been showed in Fig.1.5. And the Fig. 1.6 showed the structure of DNETCCTP. Their structures were demonstrated by elemental analysis, NMR, MS, and FT-IR methods. We will explain these two compounds in details. Besides, the crystal of these compounds was obtained and characterized by X-ray single-crystal diffraction technique. The obtained results showed that the crystal belongs to Crystal system of Monoclinic with space group of C2/c. Based on the crystal data, the geometries and normal vibrations have been obtained by using the B3LYP method with the 6-31G**, 6-311G** and 6-31++G** basis sets. The calculation results further demonstrate the molecular structure of the compounds.

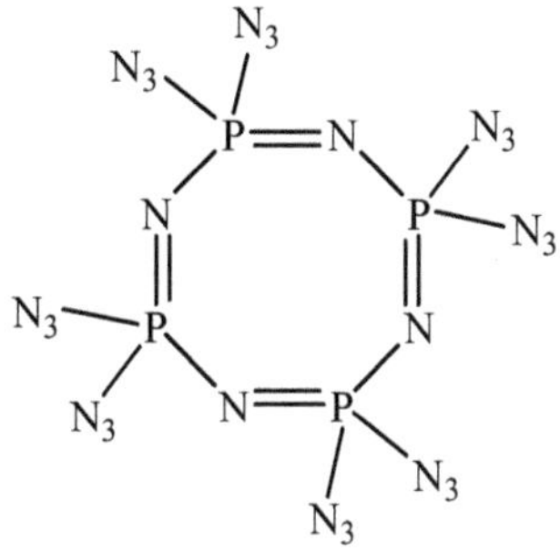

Fig. 1.2. The molecular structure of $N_3P_3(N_3)_6$

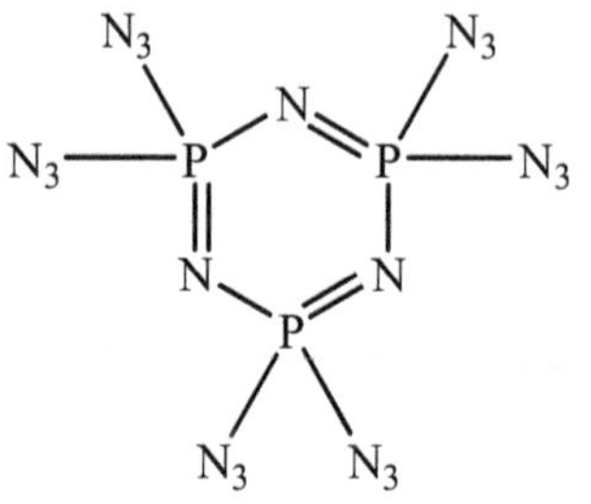

Fig. 1.3. The molecular structure of $N_4P_4(N_3)_8$

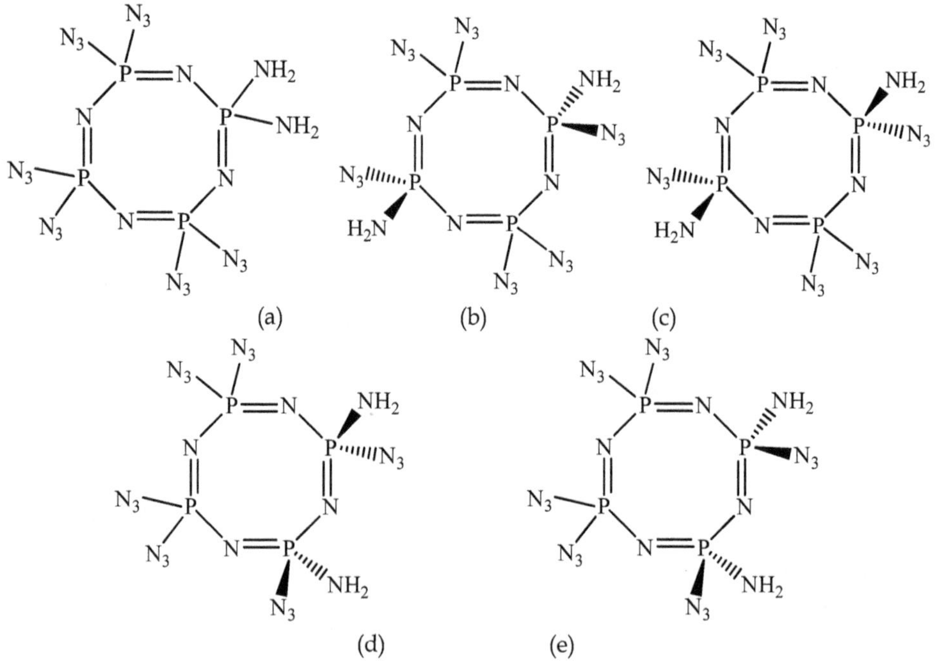

Fig. 1.4. The structure of five isomers for diamino-hexaazido-cyclo-tetraphosphazene

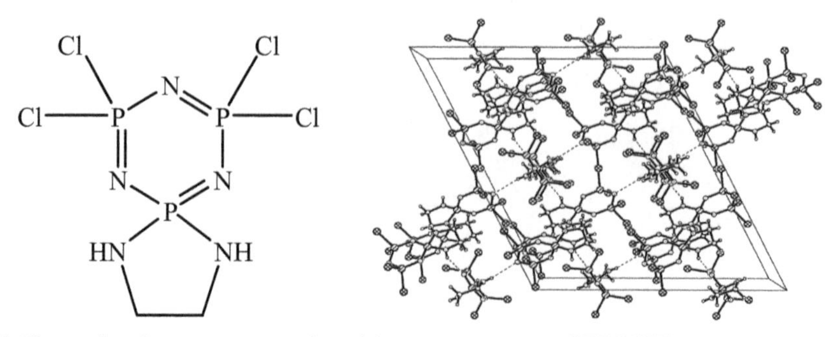

Fig. 1.5. The molecular structure and packing arrangement of ETCCTP

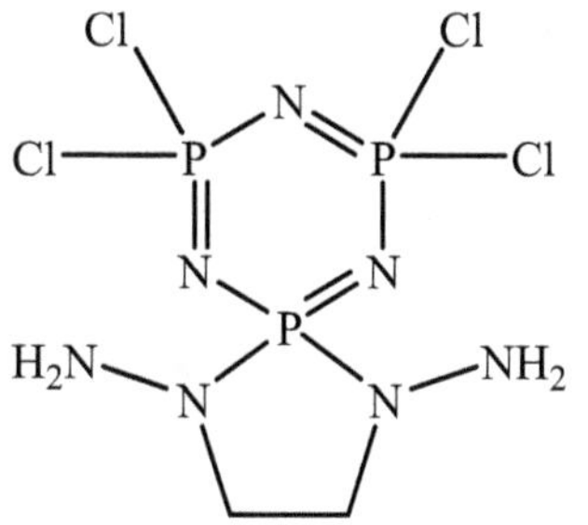

Fig. 1.6. The molecular structure and packing arrangement of DNETCCTP

The two kinds of spiro-(N,N'-dinitro-ethylenediamino)-cyclotriphosphazene compounds: 1,1,3,3,5,5- Tris-spiro-(N,N'-dinitro-ethylenediamino)-cyclotriphosphazene (3-a) and 1,1-spiro-(N,N'-dinitro-ethylene- diamino)-3,3,5,5-tetraazido-cyclotriphosphazene (3-b) have been investigated theoretically using HF, B3LYP and B3PW91 methods with 6-31G* and 6-31G** basis sets. Here are their structures in Fig. 1.7 and Fig. 1.8. The details you can see in section 3.1 and 3.2.

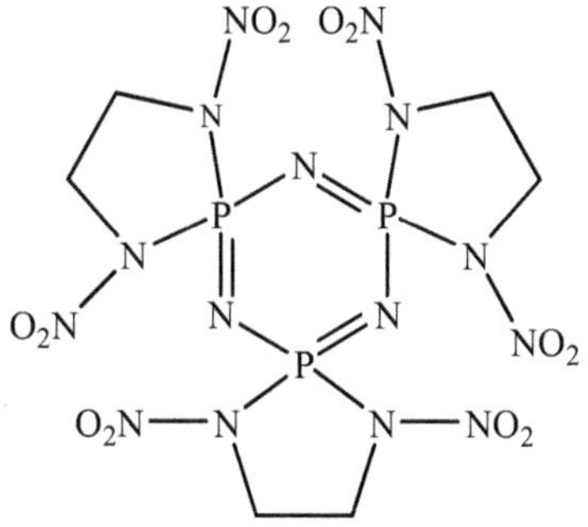

Fig. 1.7. Structure of (3-a)

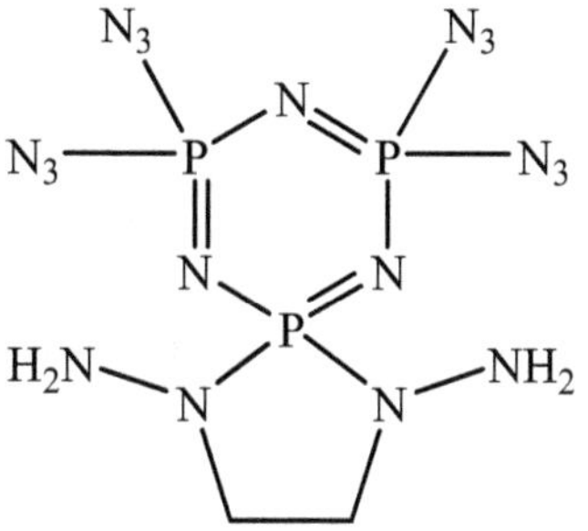

Fig. 1.8. Structure of (3-b)

In 2010, the isomers of 1,1,3,3,5,5-Tris-spiro (1,5-Diamino-tetrazole) Cyclo- triphosphazene (3-c) and (3-d) was pointed out to be a nice application[40]. Fig. 1.9 and Fig. 1.10 showed that. We have explained this at last in this chapter.

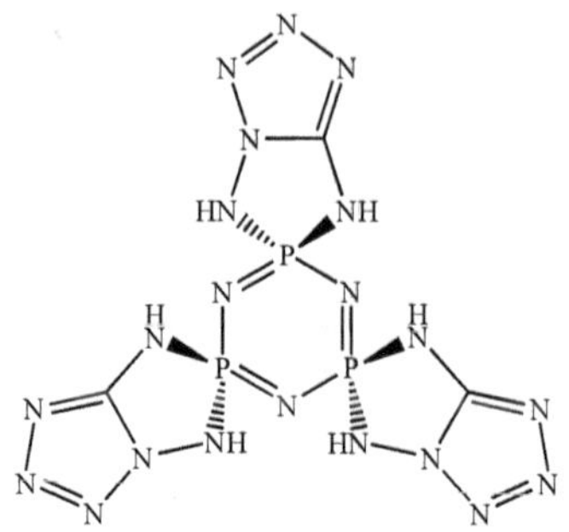

Fig. 1.9. Cis stucture of (3-c)

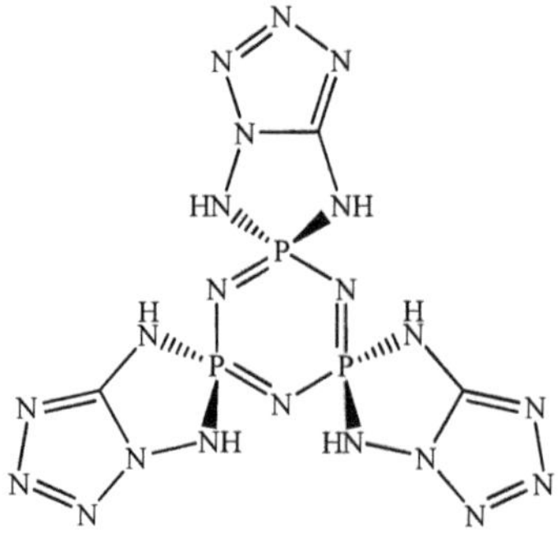

Fig. 1.10. Cis stucture of (3-d)

2. Computational method

2.1 Ab initio methods

This method is an approximate quantum mechanical calculation called Hartree-Fork calculation, in which the primary approximation is the central field approximation[41].

$$\hat{E}\psi_i = \varepsilon_i \psi_i \ (\mathrm{i} = 1,2,\ldots,\mathrm{n}/2) \tag{2.1}$$

This means that the Coulombic electron-electron repulsion is taken into account by integrating the repulsion interaction. This is a variational calculation, meaning that the approximate energies calculated are all equal to or greater than the exact energy. One of the advantages of this method is that it breaks the many-electron Schrödinger equation into many simpler one–electron equations. The other one is the approximation in HF calculations is due to the fact that the wave function must be described by some mathematical function, which is known exactly for only a few one-electron systems. In the HF equation,

$$\hat{E} = \hat{H}^{core} + \sum_{j}(2\hat{J}_j - \hat{K}_j) \tag{2.2}$$

$\hat{H}^{core}$ means single electron Hamilton operator, $\hat{J}_j$ is Coulombic operator, $\hat{K}_j$ is exchange operator. So the solving process is SCF method that is a temptation and iteration. Roothaan[42] combined the atom obital χ_μ to the molecular orbital ϕ_i linearly (LCAO-MO).

$\phi_i = \sum_{\mu=1}^{N} c_{\mu i}\chi_\mu$. After variational calculation of this deduction, we can get a secular equation showed below:

$$\sum_{\mu=1}^{N}(F_{\mu v} - \varepsilon_i S_{\mu v})c_{vi} = 0 \quad (\mu = 1,2,...,N) \tag{2.3}$$

Transfer the equation to the matrix, that is $FC = SC_\varepsilon$. In this equation, C is MO coefficient matrix, εi is the energy of ϕ_i, and F is Fock matrix which is means the average potential field of the electron in each orbital. The solving result is a series of MO coefficient and energy level.

A variation on the HF procedure is the way that orbital is constructed to reflect paired or unpaired electrons. If the molecule has a singlet spin, then the same orbital spatial function can be used for both the α and β spin electrons in each pair. This is called restricted Hartree-Fock method (RHF). This scheme results in forcing electrons to remain paired. This means that the calculation will fail to reflect cases where the electrons should uncouple. We have to say that one Slater matrix wave function as the trial function of a molecular will lead to the HF equation by variation of total energy. In this method, there will be a big error although we use a high level, which is because we haven't consider the electron correlation. So there are many methods have taken account into the electron correlation such as CI, Mφller-Plesset[43] (MP_n). And the methods can give more accuracy results.

The disadvantage of *ab* initio methods is that they are expensive. These methods often take enormous amounts of computer CPU time, memory, and disk space. And presently, the density functional theory (DFT) very popular. We will talk it below.

2.2 Density functional theory

This theory has been developed more recently than other ab initio methods. Because of this, there are classes of problems not yet explored with this theory, making it all the more crucial to test the accuracy of the method before applying it to unknown.

The premise behide DFT is that the energy of a molecular a molecular can be determined from the electron density instead of a wave function. This theory originated with a theorem by Hohenburg and Kohn[44] that stated this was possible. A particle application of this theory was developed by Kohn an Sham[45,46] who formulated a method similar in structure to the HF methods.

A density functional is then used to obtain the energy for the electron density. A functional is a function of a function, in this case, the electron density[44,45].

$$E_T[\rho]=T[\rho]+U[\rho]+E_{xc}[\rho]$$
$$=-\frac{1}{2}\sum_i\int\phi_i(\vec{r_1})\nabla^2\phi_i(\vec{r_1})d\vec{r_1}+\sum_A\int\frac{Z_A}{\left|\vec{R_A}-\vec{r_1}\right|}\rho(\vec{r_1})d\vec{r_1}+\frac{1}{2}\int\frac{\rho(\vec{r_1})\rho(\vec{r_2})}{\left|\vec{r_1}-\vec{r_2}\right|}d\vec{r_1}\,d\vec{r_2}+E_{xc}[\rho] \quad (2.4)$$

ρ means the electron density, $T[\rho]$ means the kinetic energy of the system with no interaction, $U[\rho]$ is the classic Coulomb interaction, $E_{xc}[\rho]$ means the other energy in the total energy such as the exchange correlation energy. The next equation is the detail of the three items. The exchange correlation energy can express as[47-53]

$$E_{xc}[\rho]=\sum_{\gamma}\sum_{\gamma'}-2\pi\frac{\rho_1^{\gamma}\left(\vec{r_1}\right)\rho_x^{\gamma\gamma'}\left(\vec{r_1},s\right)}{s}d\vec{r_1}\,s^2ds \quad (2.5)$$

γ and γ' means two ways of spin. The single electron orbit {$\phi_i(\vec{r_1})$ i=1, 2, …,n} is the solution of the singer electron Kohn- Sham equation.

$$\left[\frac{1}{2}\nabla^2+\sum_A\int\frac{Z_A}{\vec{R_A}-\vec{r_1}}+\int\frac{\rho(\vec{r_2})}{\left|\vec{r_1}-\vec{r_2}\right|}d\vec{r_2}+V_{xc}\right]\phi_i(\vec{r_1})=h_{KS}\phi_i(\vec{r_1})=\varepsilon_i\phi_i(\vec{r_1}) \quad (2.6)$$

E_{xc} on the density of the derivative is V_{xc} that is the exchange correlation potential. This is the same with the molecular orbital theory, the multi-electron wave function equate the linear product of single molecular orbit, which can express like this,

$$\Psi(\vec{r})=\left|\phi_1(1)\overline{\phi_1(1)}\phi_2(1)\overline{\phi_2(1)}\cdots\phi_{n-1}(1)\overline{\phi_{n-1}(1)}\phi_n(1)\overline{\phi_n(1)}\right| \quad (2.7)$$

The information of electron exchange and correlation are contained in the point function $\rho_x^{\gamma\gamma'}\left(\vec{r_1},s\right)$, also it includes the interaction between the electron correlation and kinetic energy. $\rho_x^{\gamma\gamma'}\left(\vec{r_1},s\right)$ means the electron located at $\vec{r_1}$ will repulsed other electrons to close to itself in the range s. The repulsive energy is increased with the $\rho_x^{\gamma\gamma'}\left(\vec{r_1},s\right)$ increasing. $\rho_x^{\gamma\gamma'}\left(\vec{r_1},s\right)$ can be solved by the schrödinger equation of multi-electron system[54]. The approximate processing is start with the singer electron Kohn- Sham equation, we can instead of the $\rho_x^{\gamma\gamma'}\left(\vec{r_1},s\right)$ by model function. It is proved there is some properties of the correlate function [50-52]:

$$4\pi\int \rho_{X}^{\gamma\gamma'}\left(\vec{r},s\right)s^2 ds = 0 \tag{2.8}$$

The related Fermi function $\rho_{X}^{\gamma\gamma'}\left(\vec{r_1},s\right)(r=r')$ is satisfying the normalization condition.

$$4\pi\int \rho_{X}^{\gamma\gamma'}\left(\vec{r_1},s\right)s^2 ds = 0 \tag{2.9}$$

$$\rho_{X}^{\gamma\gamma'}\left(\vec{r_1},s\right) = \rho_{X}^{\gamma'}\left(\vec{r}\right) \tag{2.10}$$

Here 2.8 to 2.10 is the limiting condition of tectonic model function [54].

2.3 Mulliken population and NBO analysis

One of the original and still most widely used population analysis schemes is the Mulliken population analysis. The fundamental assumption used by the Mulliken scheme for partitioning the wave function is that the overlap between two orbitals is shared equally. This does not completely reflect the electro-negativity of the individual elements. However, it does give one a means for partitioning a wave function and has been found to be very effective for a small basis sets. A molecular orbital is a linear combination of basis functions. The integral of a molecular orbital squared is equal to 1 as normalization. The square of a molecular orbital gives many terms, which yield the overlap when integrated. Thus, the orbital integral is actually a sum of integrals over one or two center basis functions. In mulliken analysis, the integral from a given orbital are not added. Instead, the contribution of a basis function in all orbitals is summed to give the net population of that basis function. Likewise, the overlaps for a given pair of basis functions are summed for all orbitals in order to determine the overlap population for that pair of basis functions. The overlap populations can be zero by symmetry or negative, indicating anti-bonding interactions. Large positive overlaps between basis functions on different atoms are one indication of a chemical bond.

Natural bond orbital (NBO) analysis [55-58] has been carried out to complete the picture of the ring bonding system in the nitrogen-phosphorus compounds. In the NBO analysis according to references' method [58,59], in order to complete the span of the valence space, each valence bonding NBO (σ_{AB}) ought to be paired with a corresponding valence anti-bonding NBO (σ^*_{AB}). In the equation 2-11, the coefficient c_A means the contribution of atom A to the bond A-B, and h_A means A's atom orbital.

$$\sigma^*_{AB} = c_A h_A - c_B h_B \tag{2.11}$$

The NBO analysis is carried out by examining all possible interactions between donor Lewis-type NBOs and acceptor non-Lewis NBOs, and estimating their energies. Since these interactions lead to loss of occupancy from the localized NBOs of the idealized Lewis structure into the empty non-Lewis orbital, they are referred to as "delocalization" corrections to the zeroth-order natural Lewis structure. For each donor NBO (i) and acceptor NBO (j), the stabilization energy E associated with delocalization i $\rightarrow$ j is estimated as

$$E = \Delta E_{ij} = q_i \frac{F(i,j)^2}{\varepsilon_j - \varepsilon_i} \tag{2.12}$$

Where q_i is the donor orbital occupancy, ε_j, ε_i are diagonal elements and F(i,j) is the off-diagnole NBO Fock matrix element

2.4 Thermodynamic function calculation

We can calculate the thermodynamic properties such as enthalpy, entropy, Gibbs free energy and chemical equilibrium constant or the composition by using Gaussian 03. Conveniently we can obtain the activation energy, pre-exponential factor and rate constant.

The standard heat of formation of an energetic compound is a very important parameter, which may be used to estimate the explosion pressure and explosion velocity. The calculation of theoretical heats of formation is split into two steps[60]. The first is to calculate the heats of formation of the molecule at 0K. It can be expressed by

$$\Delta_f H^{\ominus}(M,0K) = \sum_{atoms} x\Delta_f H^{\ominus}(X,0K) - \sum D_0(M) \tag{2.13}$$

where M stands for the molecule, and X to represent each element which makes up M, and x will be the number of atoms of X in M. $\sum D_0(M)$ is atomization energy of the molecule, which is readily calculated from the total energies of the molecule ($\varepsilon_0(M)$),the zero point energy of the molecule ($\varepsilon_{ZEP}(M)$) and the constituent atoms:

$$\Delta_f H^{\ominus}(M,0K) = \sum_{atoms} x\Delta_f H^{\ominus}(X,0K) - \sum x\varepsilon_0(X) - \varepsilon_0(M) - \varepsilon_{ZEP}(M) \tag{2.14}$$

The second step is to calculate the heats of formation of the molecule at 298 K.

$$\begin{aligned}\Delta_f H^{\ominus}(M,298K) = \Delta_f H^{\ominus}(M,0K) + \left(H_M^0(298K) - H_M^0(0K)\right) \\ -\sum x\left(H_X^0(298K) - H_X^0(0K)\right)\end{aligned} \tag{2.15}$$

Where $H_M^0(298K) - H_M^0(0K)$ and $H_X^0(298K) - H_X^0(0K)$ are the enthalpy corrections of the molecule and atomic elements, respectively. Here, the enthalpy corrections of atomic elements can be obtained from both the calculated and the experimental data [61,62], and the enthalpy correction for the molecule is $H_{corr} - \varepsilon_{ZEP}(M)$, where H_{corr} is the thermal correction to enthalpy.

The calculation of the Gibbs free energy of a reaction is similar, except that we have to add in the entropy term:

$$\Delta_f G^{\ominus}(M,0K) = \Delta_f H^{\ominus}(298K) - T\left(S^0(M,298K)\right) - \sum S^0(X,298K) \tag{2.16}$$

Where $S^0(M,298K)$ can be given by using the reaction of $S = (H-G)/T$ and $S^0(M,298K)$ can be from the JANAF tables [63].

3. Theoretical study on the spiro derivatives of cyclophosphazene

A large number of spiro compounds formed by the reaction of chloro-cyclosphazene or fluoro-cyclosphazene with difunctional reagents have been reported[64]. Muralidharan[65,66] has studied synthetical ansa-fluorophosphazene and ansa- or spiro- style substituted fluoro-phosphazene. In 1994, compound (3-a) has been synthesized by Dave[19] and its structure was confirmed by X-ray crystallography. The crystal has shown to have moderate impact sensitivity, high melting point and excellent density, and can be applied for the explosive composition. Dave has also studied the synthesis route of compound (3-b).(Fig. 1.7 and 1.8) In 2004, Magdy and others studied the synthesis route of (3-b) and the application in new primary explosive. Compound (b) was analyzed by DSC. In this section, we also design another new spiro-cyclo-phosphazene (c) (Fig. 1.9 and 1.10) maybe a good application prospect. As a part of the series of research works on high-energy-density compounds derived from cyclo-phosphazene, we performed the theoretical calculation about some spiro derivatives of cyclo-phosphazene compared by the experiment data to predict the application in energetic material in the future.

3.1 1,1,3,3,5,5-tris-spiro (N,N'-dinitro-ethylenediamino) cyclotriphosphazene (3-a)

3.1.1 Geometric properties

The structures and the atom serial numbers of (3-a) studied in this work are showed in Fig. 3.1. All the optimized structural characteristics calculated at HF, B3LYP and B3PW91 levels of theory for the compounds with 6-31G* and 6-31G** basis set are also calculated. As can be seen from the result, B3LYP and B3PW91 methods, used in this study, lead to similar values for bond lengths, bond angles and dihedral angles. However, the results are different from those obtained by HF method.

The phosphorus-nitrogen bond length of the ring by HF method is 1.576Å on average, which is consistent with the literature (1.58Å), and it is the significantly shortest. But the averaged length of P=N by the B3LYP and B3PW91 methods is 1.60Å, which is relatively approach to cyclo-phosphazene of azido style studied before. The two P=N bond being separated is equal in the six P=N bond of the ring, and the biggest value of all the P=N bond is 0.013 Å, so we think that the P=N bond is equal in the hexa-phosphazene ring, and we know that the phosphazene ring is a total flat surface according to the bond angles having known. In the six P=N bond out the ring, because in the $-NO_2$ in connect with spiro ring, the position mindset at space is different and the stretch function to the P=N bond is different, so three pentaspiro rings are distortional and not total side, and P=N bond length in connect with the same phosphorus atom is unequal. In the six $-NO_2$ which is connect with penta ring, N-N bond length is nearly equal, with the maximum size being 0.012 Å. N=O bond length in the $-NO_2$ in connect with the same spiro ring is unequal, and N=O bond length in the same ring is unequal, which are 1.221 Å, 1.226 Å, 1.219 Å, 1.224 Å at B3LYP/6-311G** level, respectively. Half of twelve N=O bonds are double bonds in the three spiro ring, and the other are delocalized bonds. Three penta-spiro rings are perpendicular to the cyclo-tri-phosphazene ring. Seen from the dihedral angle, atoms in the heterocyclic made up of N and P are in the same plane, while atoms out the heterocyclic and the heterocyclic are not coplanar.

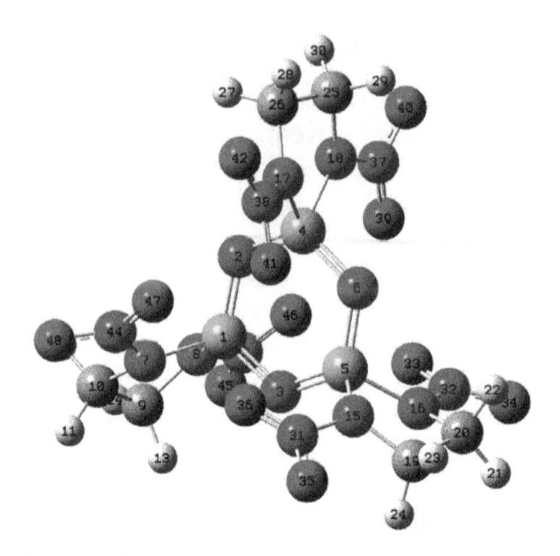

Fig. 3.1. The structure of molecular (3-a)

3.1.2 Vibrational analysis

The vibrational harmonic frequencies of (3-a) have been calculated using the same level of theory and basis set used in the geometry optimization, we only show the vibrational frequencies and their infrared intensities of the stationary point for (3-a) at B3LYP/6-31G* level in Table 3.1. From the calculation, we can see there is no imaginary frequency. The result indicates that all the optimized structures correspond to the minimum point on the potential energy surface. From the result, we can see that the strongest absorption peak located at 1324.52 cm^{-1} ， 1683.43 cm^{-1} and 1202.21 cm^{-1}. What they means -CH_2 in-plane asymmetry wag, -NO_2 in-plane stretching vibrational absorbing and P-N-P ring twist.

3.1.3 Charge distribution and bond order analysis

Table 3.2 summarizes overlap electron population of (3-a) at B3LYP/6-31G* level. From the bond electron population, we can discover that the electron population of P=N bond in the phosphazene ring is the largest. So P=N bond in the ring exists stronger interaction and the whole phosphazene ring is more stable. The interaction of the N=O bond of -NO_2 is more stronger, but the population of the N=N bond in which -NO_2 is connect with three spiro rings are the smallest, so -NO_2 is more lively and splits earliest from the rings. The interaction of the C-N bond in the spiro ring and the P=N bond in connect with the phosphazene ring is weaker, so they also take place to split easily.

Table 3.3 summarized the second-order perturbation estimates of "donor-acceptor" (bond, anti-bond) interactions for the couple [lone pair/N3 (or NO) anti-bond] on the basis of NBO with the limit of 2.09 kJ/mol threshold. In the molecular (3-a), the interaction of the N=O bond in three spiro rings is the strongest. The interaction between the σ N=O anti-bond and the π N=O anti-bond is the strongest stabilization which can up to 30077.86 kJ/mol. The interaction between the different σ N=O anti-bond and π N=O anti-bond or the π anti-bond and the π anti-bond is stronger. The N=O bond between three spiro rings has very strong area function.

v	Frequencies(cm^{-1})	Intensities (km/mol)	Assignment
1	633.53	22.68	P-N-P (ring) in-plane stretching
	634.03	20.92	
2	1043.63	283.93	N-NO_2 symmetry stretching
	1060.87	614.90	-CH_2 symmetrical wag
3	1187.41	194.44	P-N, C-N in-plane twist
	1190.80	123.59	
	1192.48	104.26	
	1194.82	268.45	
4	566.21	236.99	P-N-P ring twist
	1133.55	4.60	
	1200.98	479.36	
	1202.21	760.71	
5	1351.65	80.95	-CH_2 in-plane symmetry wag
	1386.35	65.96	
6	1324.52	1305.60	-CH_2 in-plane asymmetry wag
	1388.22	74.74	
	1389.70	432.92	
7	1677.18	131.88	-NO_2 in-plane stretching vibrational absorbing
	1678.42	195.52	
	1683.43	942.15	
	1689.05	836.33	
	1703.75	22.04	
8	3067.84	17.21	-CH_2 symmetry stretching vibration
	3074.16	13.73	
	3088.27	11.26	
9	3144.39	4.50	-CH_2 asymmetry stretching vibration
	3148.25	3.50	
	3160.99	2.46	
	3167.99	1.75	

Table 3.1. Vibrational harmonic frequencies in cm^{-1} and their IR intensities in km/mol of (3-a) calculated for the optimized structures at B3LYP/6-31G* level

chemical bond	electron population	chemical bond	electron population
N32-O33	0.316	P4-N2	0.467
N32-O34	0.330	P4-N6	0.455
N31-O35	0.331	C19-C20	0.285
N31-O36	0.323	N18-C25	0.216
N17-N38	0.175	N17-C26	0.215
N18-N37	0.182	N24-H25	0.357
P5-N16	0.201	C19-H23	0.381
P5-N15	0.183		

Table 3.2. The overlap electron population of (3-a)

Donor NBO (i)	Acceptor NBO (j)	E(2)/(kJ/mol)
BD*(1)N44-O48	BD*(2)N44-O47	973.856
BD*(1)N44-O48	BD*(1)N44-O47	1174.162
BD*(2)N44-O47	BD*(2)N44-O48	1535.941
BD*(1)N44-O47	BD*(2)N44-O48	30077.859
BD*(1)N43-O46	BD*(2)N43-O45	9292.307
BD*(1)N43-O45	BD*(2)N43-O46	2751.652
BD*(1)N38-O41	BD*(2)N38-O42	12070.168
BD*(2)N37-O40	BD*(1)N37-O39	16672.432
BD*(1)N32-O34	BD*(2)N32-O33	2134.183
BD*(1)N32-O33	BD*(2)N32-O34	8571.383

Table 3.3. NBO analysis results of (3-a)

3.1.4 The total energy and heats of formation from computed atomization energies

The total energies, the heats of formation and the density at 298.15K are computed. The target compounds are definite A, B, C and (3-a) in terms of the number of spiro-dinitro-ethylenediamino contained, respectively. The molecule structure is showed in Fig. 3.2.

According to the data from Table 3.4, it can be found that the number of nitro has the certain influence on heats of formation of the target compounds. Among A, B, C, and 4-a, the heat of formation of A containing three spiro-ethylenediamine is the least, and the full nitration product, (3-a) has the largest value. So, they show that heats of formation of the target compounds increase with the increment of the nitro number. This is because the nitro is a high-energy group, its introduction resulted in the increase in the heat of formation of the target compounds, the level of content energies can also increase, but the target compounds stability will be reduced, become sensitive to hot and impact. We also calculated the density of a series of compounds, calculated the density of (3-a) which is 1.893 g/cm^3. These values indicate that the compound would expect to contain more energy, thus may potentially be used as energetic materials.

	E_0 (kJ/mol)	HOF (kJ/mol)	ρ (g/cm^3)
A	-4606942.85	55.21	1.52
B	-5679675.77	62.09	1.47
C	-6752288.55	94.16	2.23
3-a	-7824988.02	112.05	1.893 (1.887)

Table 3.4. The calculated total energies, heats of formation, density of target compounds at 298.15K

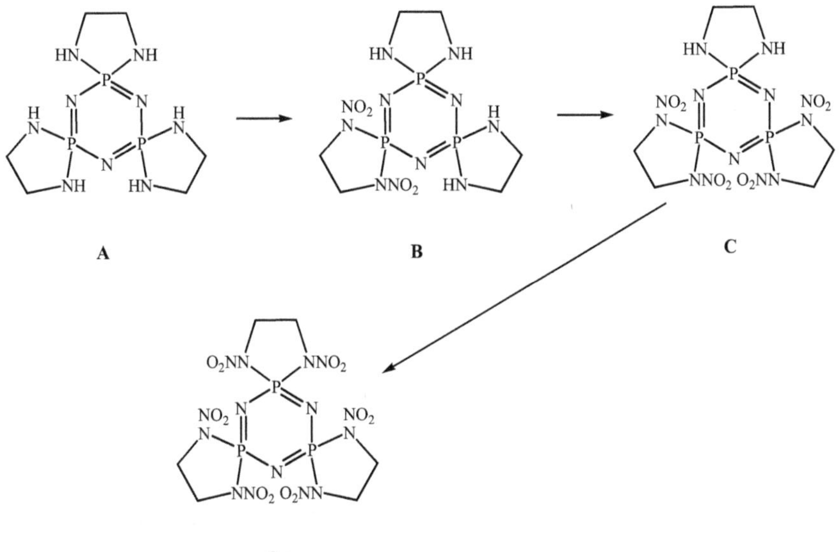

Fig. 3.2. Molecular (3-a) and its related products

3.2 1,1-spiro- (N,N'-dinitro-ethylenediamino)-3,3,5,5- tetraazido- cyclotriphosphazene (3-b)

3.2.1 Geometric properties

The structure of (3-b) have been showed in Fig 3.3. From the result, three P=N bond lengths are equal in the hexa-numbered ring of (3-b), but the P=N bond length (1.594 Å) next to sprio ring is the shortest. Four azido groups have certain regulation because of the equal P=N bond by ones and twos. So they exist the equal $P=N_{\alpha}$, $N_{\alpha}=N_{\beta}$, $N_{\beta}=N_{\gamma}$ bonds by ones and twos. Because the stretch function is different, so that the spiro ring and azido group are the whole cyclophosphazene, the different P=N bonds make existence outside the ring and the P=N bond is obviously longer than the $P= N_{\alpha}$ bond in the spiro ring. The equal N=N bonds make existence because the function that the two $-NO_2$ is to the spiro ring in the pentaspiro ring is same. But two N=O bonds are unequal in a $-NO_2$, and the N=O bonds are equal in the different $-NO_2$. Seen from the dihedral angle, atoms in the heterocyclic made up of N and P are in the same plane, while atoms out the heterocyclic and the heterocyclic are not coplanar.

3.2.2 Vibrational analysis

The vibrational frequencies and their infrared intensities of stationary point have been showed in table 3.5. Compared with (3-a) from the result, we see they are very consistent with the experimental results. Also here is no imaginary frequency, which is proved the structure correspond to the minimum point on the potential energy surface. From the result, we can see that the strongest absorption peak is due to P-N-P ring twist and P-N, C-N in-plane twist.

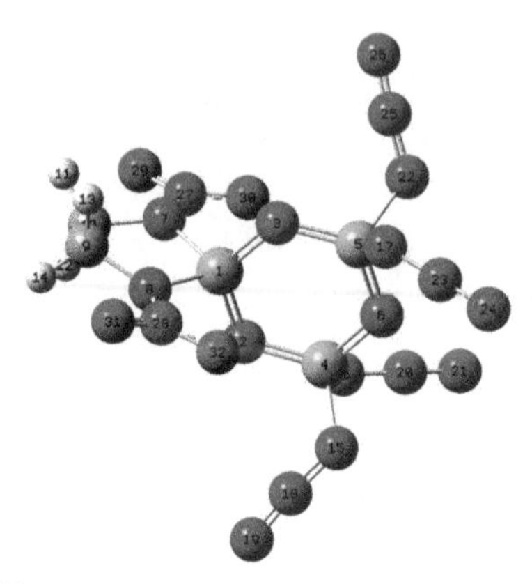

Fig. 3.3. The structure of (3-b)

v	Frequencies(cm^{-1})	Intensities (km/mol)	Assignment
1	605.49	206.54	P-N-P (ring) in-plane stretching
	617.16	23.61	
2	1005.33	2.30	N-NO_2 symmetry stretching
	1068.23	120.55	-CH_2 symmetrical wag
3	1118.38	3.00	P-N, C-N in-plane twist
	1201.91	1048.03	
	1211.08	46.99	
	1213.83	24.89	
4	567.04	68.08	P-N-P ring twist
	1150.12	0.50	
	1224.62	1611.53	
5	1360.21	95.62	-CH_2 in-plane symmetry wag
	1396.03	147.63	
	1322.04	280.35	
	1419.59	320.01	
	1534.97	3.58	
6	1326.64	226.87	N-N-N in-plane stretching
	1327.99	795.58	
	1338.57	89.55	
	1343.62	72.39	
	2296.14	982.23	
	2298.47	630.57	
	2311.56	544.68	
	2318.88	273.00	
7	1669.76	93.44	-NO_2 in-plane stretching vibrational absorbing
	1675.59	64.27	
8	3078.59	14.56	-CH_2 symmetry stretching vibration
	3083.69	11.19	
9	3156.41	1.76	-CH_2 asymmetry stretching vibration
	3163.90	4.29	

Table 3.5. Vibrational harmonic frequencies in cm^{-1} and their IR intensities in km/mol of (3-b)

3.2.3 Charge distribution and bond order analysis

As we have mentioned above, in compound (3-b), the interaction between two N=O bonds in connection with a nitryl is the strongest stabilization, 15168.42 kJ/mol, this indicates that the electronics transferring tendency on of molecule orbits of the N=O bond is bigger, this is mainly because the lone pair electronics of oxygen atom have strong interaction, and two N=O bonds present to leave an area form. The interaction between the N=N bond in the spiro ring and the C-C bond in the ring is weaker stabilization, 7.52 kJ/mol. There exists the stronger interaction between lone pair electrons in the N_α of four azido groups and the πN_β-N_γ anti-bond, but the interaction between the P-N_α bond and the whole phosphazene ring is weaker, this indicates that azido groups split easily. Table 3.6 have showed the overlap electron population of (3-b) at B3LYP/6-31G* level.

In this compound, the interaction at the end of the azido group is the strongest, and the population of they N_β=N_α bond is the largest and the stablest. The P=N bond in the phosphazene ring is the second. The interaction of the P=N bond in the phosphazene ring is the weakest, and split most easily while being stimulated by the external world. The spiro ring opens. The azido group also split easily, but the N=O bond in the spiro ring exists delocalization and more stable. The result of the NBO analysis has been listed in table 3.7.

chemical bond	electron population	chemical bond	electron population
N6-P5	0.483	N8-N28	0.198
N3-P5	0.458	N28-O32	0.328
N22-P5	0.275	N28-O31	0.338
N16-P4	0.278	N23-N24	0.596
N5-P22	0.275	N18-N19	0.595
P1-N8	0.177	N8-C9	0.217
P1-N7	0.177	C9-H13	0.378
N7-N27	0.198		

Table 3.6. The overlap electron population of (3-b)

Donor NBO (i)	Acceptor NBO (j)	E(2)/(kJ/mol)
BD*(2)N28-O32	BD*(2)N28-O31	703.4522
BD*(2)N28-O31	BD*(1)N28-O32	15132.1852
BD*(2)N27-O29	BD*(1)N27-O30	15168.4258
BD*(1)N27-O29	BD*(1)N27-O30	1255.5048
LP(2)N16	BD*(2)N20-N21	429.3278
LP(2)N15	BD*(2)N18-N19	432.2956
BD*(1)N28-O31	BD*(1)N28-O32	1253.9582
BD*(3)N25-N26	BD*(1)P5-N22	27.0446
BD*(3)N23-N24	BD*(1)P5-N17	27.0446
BD*(1)N7-N27	BD*(1)C9-C10	7.524
BD*(1)N7-N27	BD*(1)P1-N2	4.7652

Table 3.7. NBO analysis results of (3-b)

3.2.4 The total energy and Heats of formation from computed atomization energies

Compared with what we have discussed in section 4.1.4, the heat of formation of (3-b) is much larger than the (3-a), containing three spiro-dinitro-ethylenediamino, mainly due to the existence of azido groups. It explains that azido groups content energy is higher and more unstable than nitryl. As talking the density about them, We calculated the density of (3-b) is 1.920 g/cm^3, which is bigger than (3-a) , 1.893 g/cm^3. That is very consistent with the crystal density of the literature. These values indicate that these two compounds would expect to contain more energy, thus may potentially be used as energetic materials. The details have been showed below (Table 3.8).

Parameters	Value
E0 (kJ/mol)	-6382918.40
HOF (kJ/mol)	328.40
ρ (g/cm3) (experimental)	1.920 (1.830)

Table 3.8. The calculated total energies, heats of formation, density of (3-b)

3.3 1,1,3,3,5,5-Tris-spiro (1,5-Diamino-tetrazole) Cyclotriphosphazene and its isomers.

3.3.1 Geometric properties

Two isomers will be produced when 1,5-diamino-tetrazole (DAT) reacted with hexa-chlorin-cyclotri- phosphazene. The reason for this is the different location of C atom of the tetrazole. You can see the molecular structure of them in Fig 3.4. (3-c) and (3-d) are the two isomers of the title compound. We optimized by AM1 in the first time, and at last, we got the optimization by using B3LYP and B3PW91 methods with 6-31G* and 6-311G** basis set. From the calculated data, we proved the two can exist stably.

We see the cyclotriphosphazene ring is nearly coplanar from the dihedral values. Different methods and basis sets would get similar result, the maximum error is 0.01 Å. The basic data of the structure is nearly equal. We will take the cis structure for analysis.

The result told us the length of bond P-N is always equal to each other, the average is 1.607 Å, which is similar to the length of the same bond in $N_3P_3Cl_6$. The hydrogen of the amino of the DAT will lost with two chlorine of $N_3P_3Cl_6$ when react is in the process. So there is no same length to the bond N-P in the quinaryring such as 1.709 Å to P_5-N_{13}, but 1.747 Å to P_5-N_{12}. The bond N-P have the same length when the nitrogen connecting with the phosphorus of the cyclotri- phosphazene ring. The length of P3-N20 is equal to P1-N27, the same to P1-N26 and P3-N19. The length of N-H, N-N and C-N are equal at the corresponding positions. The calculated result of the bond length of the tetrazole is consistent to the experimental data. The length of N18-N14, N14-N15, N15-N16 are 1.362 Å, 1.293 Å, 1.382 Å respectively, the corresponding data of the experiment were 1.363 Å, 1.279 Å, 1.367Å. That's to say our calculation and prediction is correct and credible. Compared the bond of N18-N19 (1.400 Å) and N18-N14 (1.362 Å), the former is greater than the latter that is due to the conjugate function between the quinaryring and the tetrazole. And this effect makes the bond length average and the electron delocalization. Also this is a stable state. The angle of the N20-C17-N18 or N19-N18-N14 are between 93°~117° instead of the 120° caused by sp^2 hybrid of N and C. That is to say there is the tension between the two ring.

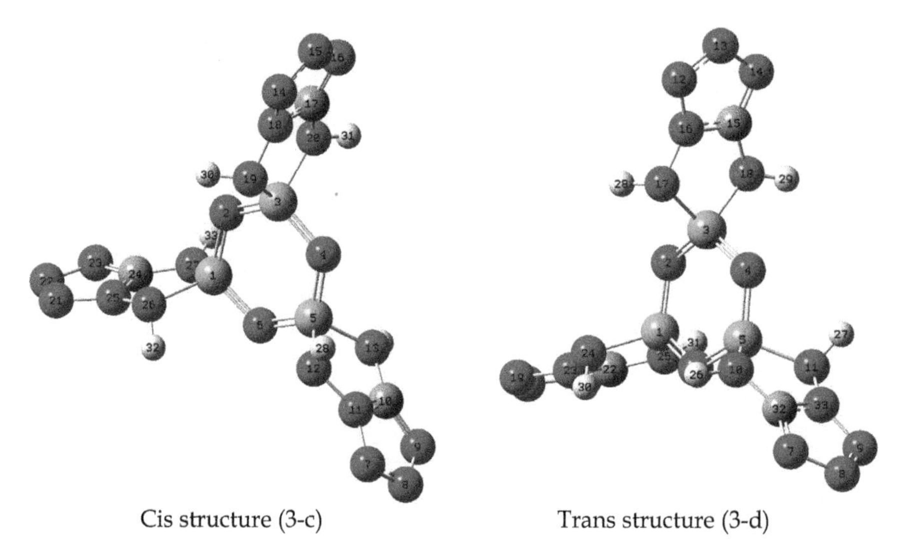

Cis structure (3-c) Trans structure (3-d)

Fig 3.4 The structure of two isomers of the title compound

3.3.2 Vibrational analysis

No imaginary frequency in the vibrational calculation, So they're the minimum point of the potential energy surface. That's to say they are all the stable structure. We have obtained 93 IR frequencies and their intensity, 12 in which has greater intensity. We did the simulation shown in Fig.3.5 and 3.6.

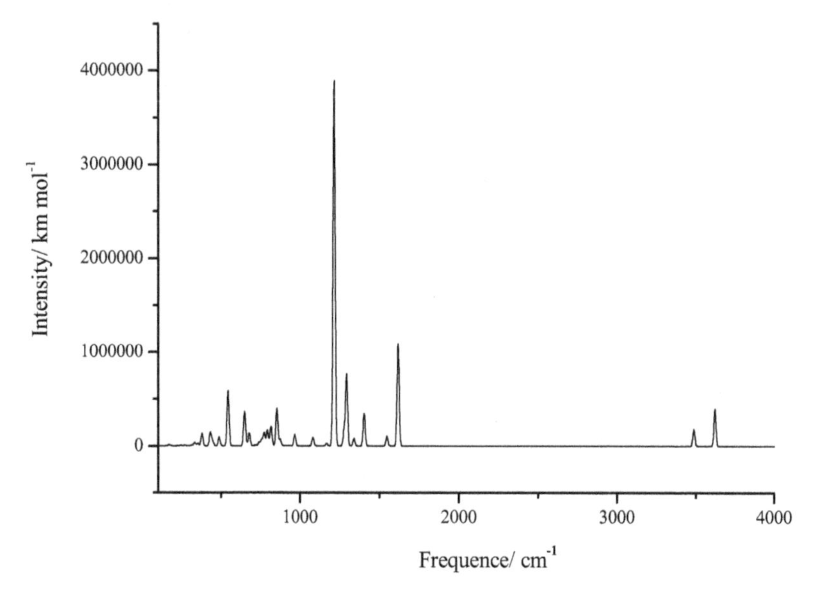

Fig. 3.5. The IR spectrum of (3-c)

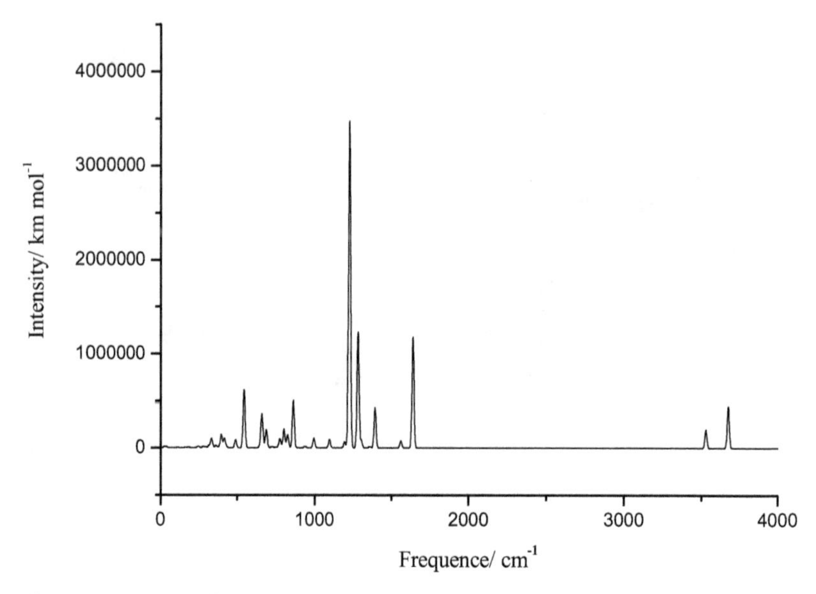

Fig. 3.6. The IR spectrum of (3-d)

We analyzed the cis structure to explain the vibrational frequency and infrared intensities. The N-H stretching vibration is in the high frequency region near by 3400cm^{-1} or 3600 cm^{-1}. So there are 2 lines in the high frequency region. The C-N stretching vibrational intensity between the DAT and the quinaryring is around 1600 cm^{-1}. Its in-plane stretching vibrational region is from 1546 cm^{-1}. In-plane stretching vibrationof bond N-H and the stretching of N-N at the ring is at the region between 1270~ 1400 cm^{-1}. The intensity of the strongest vibration absorption for the bond P-N stretching is up to 1351 km/mol. The twist of the cyclophosphazene ring, bending of the N-H and in-plane rocking located at 900~ 1200 cm^{-1}. In Fig 3.6, we can see a very similar IR spectrum picture with the Fig 3.5.

3.3.3 Charge distribution and bond order analysis

The overlap population has been showed in Table 3.9. The two isomers have the similar performance. The bond N-N between the quinaryring and the DAT has the lower value. The same trend is also appeared on the bond N-N in the DAT. That's means it's easy to destruct when the ring is heat or done by the external force. Also means the bond energy is weak here. The population of bond P-N of the cyclophosphazene is up to 0.465 in average. So there is strong interaction between these bonds, which decided the ring is very stable. But to the contrary, the bond N-P out of the ring is much weaker than the same bond in the ring. For example, The population of the P1-N2 is 0.459, but only 0.259 to P1- N27. We will discuss the this phenomenon from the NBO analysis. As the delocalization, the population of bond C-N is bigger than the it of bond N-N from the result. That's consistent with the structure analysis.

The stabilization energy E(2) for each donor NBO(i) and acceptor NBO (j) are associated with i → j delocalization, which is estimated by the calculation. We can review it in section

3.1.3 what we said before. From the data listed in table 3.10, the interaction ofπ*-N-N andπ*-C-N in the DAT ring can be up to 233.67 kJ/mol. The lone electron of N atom such as N11, N18, N25 used by the tetrazole rings and quinaryring connectting with antibonding orbitals of N-N usually have a big value. This is also prove a existence of the delocalization of the DAT. The electrons between bond N-N or C-N are in domain forms. Conjugate function effects the tetrazole and quinaryring when we see the E(2) value decided by lone electron of N atom of C-N and the π*-C-N is 152.15 kJ/mol. E(2) between the lone electron of N atom of the quinaryring and the σ * P-N is only 2.717 kJ/mol. It's consistent with the conclusion about the stability of the molecular what we talked about before.

bond	B3LYP/6-31G*		B3LYP/6-31G**		B3PW91/6-31G*		B3PW91/6-31G**	
	Cis-	Trans-	Cis-	Trans-	Cis-	Trans-	Cis-	Trans-
N_4-P_3	0.460	0.458	0.457	0.455	0.462	0.46	0.459	0.457
N_4-P_5	0.469	0.470	0.467	0.468	0.471	0.472	0.469	0.470
N_6-P_1	0.469	0.468	0.467	0.467	0.471	0.469	0.469	0.468
N_2-P_1	0.460	0.458	0.458	0.455	0.463	0.460	0.459	0.457
N_{19}-P_3	0.246	0.244	0.236	0.234	0.254	0.250	0.243	0.240
N_{20}-P_3	0.283	0.282	0.277	0.276	0.286	0.286	0.281	0.281
N_{18}-N_{19}	0.168	0.171	0.165	0.168	0.166	0.168	0.162	0.164
C_{17}-N_{20}	0.317	0.317	0.316	0.315	0.316	0.315	0.315	0.315
C_{17}-N_{18}	0.306	0.306	0.302	0.302	0.302	0.301	0.297	0.297
N_{18}-N_{14}	0.167	0.167	0.165	0.165	0.150	0.150	0.149	0.149
N_{14}-N_{15}	0.246	0.247	0.245	0.246	0.239	0.239	0.238	0.238
N_{15}-N_{16}	0.252	0.251	0.252	0.251	0.252	0.251	0.252	0.252

Table 3.9. The selected overlap population of (3-c) and (3-d)

3.3.4 The total energy and heats of formation from computed atomization energies

The HOF has been calculated by B3LYP and B3PW91 method at 298 K. the result is listed in Table 3.11. Since no experimental values can be compared, we choose the DAT and hexa-chloro- cyclophosphazene as contrast. The HOF of hexa-chloro-cyclophosphazene is negative, while it of the title compound and its isomer are positive. That's to say this two structure is metastable in the chemical reaction. The two groups of data is relatively close. The HOF of trans structure is slightly larger than the cis one. That means cis structure is more stable. We also calculate the HOF of DAT lonely to get the conclusion that the two have lower energy. It is mainly due to the N atom which will increase the HOF. So the stability is relatively poor for this reason. We also studied the total energy and the frontier orbital energies of the two isomers, DAT and the hexa-chloro-cyclophosphazene. The data was listed in Table 3.12. The total energy of the cis structure is less than it of the trans one. But the sequence is just the opposite to the energy gap. This also explained the stability of the cis structure is better tan the trans.

Orbit of Donor (i)	Orbit of Acceptor (j)	E(2) (kcal mol-1)
BD*(2)N21-N22	BD*(2)N23-C24	55.81
BD*(2)N14-N15	BD*(2)N16-C17	55.90
BD*(2)N7-N8	BD*(2)N9-C10	55.87
LP(1)N27	BD*(2)N23-C24	36.42
LP(1)N25	BD*(2)N23-C24	52.59
LP(1)N25	BD*(2)N21-N22	34.87
LP(1)N20	BD*(2)N16-C17	36.36
LP(1)N13	BD*(2)N10-C9	36.36
LP(1)N18	BD*(2)N16-C17	52.54
LP(1)N18	BD*(1)N19-P3	0.65
LP(1)N25	BD*(1)N26-P1	0.67
BD(2)N23-C24	BD*(2)N21-N22	23.53
BD(2)N16-C17	BD*(2)N14-N15	23.53
BD(1)N6-P1	BD*(1)N26-P1	2.53
BD(1)N2-P3	BD*(1)P1-N27	1.07
LP(1)N2	BD*(1)P1-N6	11.80
LP(1)N2	BD*(1)P3-N4	12.09

Table 3.10. The selected calculated NBO results of (3-c) at B3LYP/6-31G* level.

Compounds	B3LYP/6-31G*	B3LYP/6-31G**	B3PW91/6-31G*	B3PW91/6-31G**
3-c	1725.82	1670.5	1661.68	1604.31
3-d	1728.49	1673.22	1661.89	1606.99
DAT	442.21	407.9	440.12	405.89
$(NPCl_2)_3$	-812.12 (experimental)			

Table 3.11. The formation heats of (3-c) and (3-d) in different methods and basic sets

Compounds	E_{total}	E_{LUMO}	E_{HOMO}	ΔE_{L-H}
DAT	-967707.99	-9.29	-645.52	636.08
$(NPCl_2)_3$	-10359846.98	-250.10	-809.98	559.75
3-c	-6010672.73	-101.67	-751.22	649.45
3-d	-6010669.84	-105.26	-750.44	644.99

Table 3.12. The total energy and frontier orbital energy of different compounds in B3LYP/6-31G*

4. Conclusion

In this chapter, we have summerized the spiro derivatives of cyclophosphazene. We calculated the geometric, frequency and thermodynamics constant. We analyzed the charge distribution and the national bond orbitals (NBO) and multiple overlap to judge its molecular stability. Some crystal had been synthesized when we do the theoretical study.

We explained the other category in part 3. Three meterials and four stuctures analysized by us. The first two compounds is non-planar. The screw ring distored but the space orientation of the azido is the same with the stucture of that of (3-a). Strong delocalization effect of N-O of the screw ring may lead to the bond breaking between the two ring. The theoretical density of the two is 1.893 g/cm^3 and 1.920g/cm^3 respectively. The thermodynamics analysis showed the molecular (3-b) has higher energy for the proportion of azido compared to (3-a). Two isomers of 1,1,3,3,5,5-tris-spiro (1,5-Diamino-tetrazole) cyclotriphosphazene summarized at last in this chapter. The geometric analysis showed the hexacyclophosphazene is nearly planar. And the angle between the DAT and 5-membered ring is 180°. And they are nearly vertical to the cyclophosphazene ring. And the introduction of DAT increased the nitrogen content of the cyclophosphazene, so to the HOF. The electron analysis showed the cis structure is more stable than the trans one.

5. Acknowledgements

This work has been financially supported by the Program for New Century Excellent Talents in University (NCET-09-0051)

6. References

[1] Rose, H.(1834) *Annalen der Chemie*, Vol.11

[2] R.A. SHAW, B. W. F., B.C.SMITH. (1961) *The phosphazenes.*

[3] Keat, R.(1982) Phosphorus(III)-nitrogen ring compounds.*Topics in Current Chemistry*, Vol.10 pp.289-90.

[4] Allen, C. W.(1991) Regio- and Stereochemical Control in Substitution Reactions of Cyclophosphazenes.*Chem. Rev.*, Vol.91, pp. 119-135.

[5] Allcock, H. R.(1972) *Phosphorus-Nitrogen Compounds*; Academic Press: New York.

[6] Allen, C. W.(1994) Linear, cyclic and polymeric phosphazenes *Coordination Chemistry Reviews*, Vol.130,NO.1-2.pp.137. ISSN 0010-8545

[7] Chandrasekhar V, K. V.(2002) Advances in the chemistry of chlorocyclo- phosphazenes. *Advances in Inorganic Chemistry*, Vol.53, pp.159-211, ISBN 978-0-12-385904-4

[8] Elias A. J, S. J. M.(2001) Perfluorinated cyclic phosphazenes *Advances in Inorganic Chemistry*, Vol.52, pp. 335-358, ISBN 978-0-12-385904-4

[9] Fekete, T. M.; Start, J. F. Continuous process for the production of phosphonitrilic chlorides. In USPaent/4046857, 1977.

[10] ZHENG Hui hui, Z. J. g., ZHANG Tong lai, YANG Li, FENG Lina.(2008) High Energy Density Compounds Cyclophosphazene VI Cyclophosphazene Compounds and Their Application on Energetic Materials.*Chinese Journal Of Energetic materials*, Vol.16, NO.6, pp.758, ISSN 1006-9941

[11] M.J.S. Dewar, E. A. C. L., M.A. Whitehead,.(1960) The structure of the phosphonitrilic halides *Journal of the Chemical Society*, Vol10, NO.0, pp.2423, ISSN 2674-2680

[12] Zhang, J. G.,Zheng, H. H.,Zhang, T. L.(2008) Theoretical study for high energy density compounds from cyclophosphazene. III. A Quantum Chemistry Study: High Nitrogen-contented Energetic Compound of 1,1,3,3,5,5,7,7-octaazido-cyclo- tetra-phosphazene: N4P4(N3)8. *Inorganica Chimica Acta*, Vol.361, pp. 4143-4147.

[13] Liebig J. , W. F.(1834) *Annalen der Chemie*, Vol.11, pp. 139.

[14] T. Hökelek, A. K.(1990) Phosphorus Nitrogen-compounds .1. Structure of 2,Cis-4,trans-6,trans-8-tetrachl-2,4,6,8-tetrakis (diethylamino) Cyclotetra(Phosphazene) *Acta Crystallogr A*, Vol.46, pp. 1519-1521, ISSN 0108-2701

[15] Hokelek, T. K., A; Begec, S; Kilic, Z; Yildiz, M (1996) 2-(2,6-di-tert-butyl-4-methylphenoxy)-2,4,4,6,6,8,8-heptachlorocyclo-2 lambda(5),4 lambda(5),6 lambda(5),8 lambda(5)-tetraphospha- zatetraene *Acta crystallographic A section C*, Vol.52, pp. 3243-3246, ISSN 0108-2701

[16] Hokelek, T. K., E; Kilic, Z (1998) trans-2,6-bis(ethylamino)-2,4,4,6,8,8-hexamorpholinocyclo-2 lambda(5), 4 lambda(5), 6 lambda(5), 8 lambda(5)-tetra-phosphazatetraene *Acta crystallogr C*, Vol.54, pp. 1295-1297, ISSN 0108-2701

[17] Hokelek, T.Kilic, E.&Kilic, Z.(1999) trans-2,6-bis(ethylamino)-2,4,4,6,8,8-hexapiperidinocyclo-2 lambda(5),4 lambda(5),6 lambda(5),8 lambda(5)-tetra-phosphazatetraene.*Acta Crystallographica Section C-Crystal Structure Communications*, Vol.55, pp.983-985, ISSN 0108-2701

[18] Dave, P. R.,Forohar, F.,Axenrod, T.(1994) Novel Spiro Substituted Cyclo-triphosphazenes Incorporating Ethylenedinitramine Units.*Phosphorus, Sulfur & Silicon.*, Vol.90, pp.175-184.

[19] Dave, P. R.,Forohar, F.,Chaykovsky, M. Spiro(N,N'-dinitroe-thylenediamino) cyclotriphosphazenes; C07F 9/547 ed.; The United States of America as represented by the Secretary of the Navy, Washington, DC (US): US Pat., 1994.

[20] Rajendra Prasad Singh, A. V., Robert L. Kirchmeier, and Jean'ne M. Shreeve.(2000) A Novel Synthesis of Hexakis(trifluoromethyl)cyclotriphosphazene. Single-Crystal X-ray Structures of N3P3(CF3)6 and N3P3F6.*Inorganic Chemistry*, Vol.39, pp.375-377.

[21] Michael B. McIntosh, T. J. H., and Harry R. Allcock*.(1999) Synthesis and Reactivity of Alkoxy, Aryloxy, and Dialkylamino Phosphazene Azides.*Journal of the American Chemical Society*, Vol.121,NO.4.pp.884-885.

[22] Hokelek, T.,Ozturk, L.,Isiklan, M.(2002) Crystal structure of trans-2,6-bis(n-propylanino)- 2,4,4,6,8,8-hexapyrrolidinocyclo-tetraphosphazatetraene. *Analytical Sciences*, Vol.18, pp.961-962.

[23] Omotowa, B. A.,Phillips, B. S.,Zabinski, J. S.(2004) Phosphazene-Based Ionic Liquids: Synthesis, Temperature-Dependent Viscosity, and Effect as Additives in Water Lubrication of Silicon Nitride Ceramics. *Inorganic Chemistry*, Vol.43, pp.5466-5471.

[24] Steiner, A.; Richards, P. I.(2004) Cyclophosphazenes as nodal ligands in coordination polymers.*Inorganic Chemistry*, Vol.43,NO.9.pp.2810-2817, ISSN 0020-1669

[25] Richards, P. I.; Sterner, A.(2005) A Spirocyclic System Comprising Both Phosphazane and Phosphazene Rings. *Inorganic Chemistry*, Vol.44, pp.275-281.

[26] Heston, A. J.,Panzner, M. J.,Youngs, W. J.(2005) Lewis Acid Adducts of $[PCl_2N]_3$. *Inorganic Chemistry*, Vol.44, pp.6518-6520.

[27] Michael. G., K. K., et al.(2006) The First Structural Characterization of a Binary P-N Molecule: The Highly Energetic Compound P_3N_{21}. *Angewandte Chemie International Edition*, Vol.45, pp.6037-6040, ISSN 0044-8249

[28] Trinquier, G.(1986) Structure, Stability, and Bonding in Cyclodiphosphazene and Cyclotriphosphazene. *Journal of American Chemical Society*, Vol.108, pp. 568-577.

[29] Paasch, S..Kruger, K.&Thomas, B.(1995) Solid-state nuclear magnetic resonance investigations on chlorocyclophosphazenes. *Solid State Nuclear Magnetic Resonance.*, Vol.4, pp.267-280.

[30] Elass, A.,Vergoten, G.,Dhamelincourt, P.(1997) A scaled quantum mechanical force field for hexachlorocyclophosphazene trimer $(NPCl_2)_3$. Force field transferability to the octachlorocyclophosphazene tetramer $(NPCl_2)_4$ and the decachlorocyclophosphazene pentamer $(NPCl_2)_5$.*Electr. J. Theor. Chem.*, Vol.2, pp.11-23.
[31] Elass, A.,Vergoten, G.,Dhamelincourt, P.(1997) A scaled quantum mechanical force field for hexachlorocyclophosphazene trimer $(NPCl_2)_3$. *Electr. Journal of Theoremotical Chemistry*, Vol.2, pp.1-10.
[32] Breza, M.(2000) The electronic structure of planar phosphazene rings.*Polyhedron*, Vol.19, pp.389-397.
[33] Elias, A. J.,Twamley, B.,Haist, R.(2001) Tetrameric Fluorophosphazene, $(NPF_2)_4$, Planar or Puckered? *Journal of American Chemical Society*, Vol.123, pp.10299-10303.
[34] Sabzyan, H.; Kalantar, Z.(2003) Ab initio RHF and density functional B3LYP and B3PW91 study of $(NPF_2)_n$; n=2; 3; 4 and $(NPX_2)_3$; X = H, Cl, Br cyclic phosphazenes. *Journal of Molecular Structure (THEOHEM)*, Vol.663149-157.
[35] Breza, M.(2004) Comparative study of non-planar cyclotetraphosphazenes and their isostructural hydrocarbon analogues. *Journal of Molecular Structure (THEOHEM)*, Vol.679, pp.131-136.
[36] Vassileva, P.,Krastev, V.,Lakov, L.(2004) XPS determination of the binding energies of phosphorus and nitrogen in phosphazenes. *Journal of Materials Science*, Vol.39, pp.3201-3202.
[37] Gall, M.; Breza, M.(2008) On the structure of hexahydroxocyclotriphosphazene.*Journal of Molecular Structure (THEOHEM)*, Vol.861, pp.33-38.
[38] Jianguo Zhang *, H. Z., Tonglai Zhang and Man Wu.(2009) Theoretical Study for High-Energy-Density Compounds Derived from Cyclophosphazene. IV. DFT Studies on 1,1-Diamino-3,3,5,5,7,7-hexaazidocyclotetraphosphazene and Its Isomers.*Molecular Sciences*, Vol.10, pp. 3502-3516, ISSN 1422-0067
[39] Zhang, J. G.,Zheng, H. H.,Bi, Y. G.(2008) High energy density compounds from cyclophospazene. II. The preparation, structural characterization, and theoretical studies of 1,1-spiro(ethylenediamino)- 3,3,5,5-tetrachlorocyclotriphosphazene and its nitration product. *Structure Chemistry*, Vol.19,NO.2, pp.297-305.
[40] Li, S. J. Z. J.-G. Z. H.-H. Z. T.-L. Y.(2011) High Energy Density Compounds Cyclophosphazene -VII. DFT Study of 1,1,3,3,5,5-Tris-spiro (1,5-Diamino-tetrazole) Cyclotriphosphazene. *Chinese Journal Of Explosives & Propellants*, Vol.34, NO.4, pp.10-16, ISSN 1007-7812
[41] Young, D. C.(2001) *Computational Chemistry*.
[42] J., R. C. C.(1951) New developments in molecular obital theory *Reviews of Modern Physics*, Vol.23, pp.69.
[43] Moller C, P. M. S.(1934) Note on an approximation treatment for many-electron system.*Physical Review*, Vol.4, pp. 6618, ISSN1943-2879
[44] Hohenberg P, K. W.(1964) Inhomogeneous Electron Gas.*Physical Review B*, Vol.1, pp.6864, ISSN 1943-2879
[45] Kohn W, S. L. J.(1965) Self-Consistent Equations Including Exchange and Correlation Effects.*Physical Review A*, Vol.140, pp. 1133, ISSN 1943-2879
[46] Sham L J, K. W.(1966) One-particle properties of an inhomogeneous interacting electron gas.*Physical Review*, Vol.145, pp. 561, ISSN 1943-2879

[47] D, B. A.(1988) Correlation energy of an inhomogeneous electron gas *Journal of Chemical Physics*, Vol.88, pp.1053, ISSN 0021-9606
[48] D, B. A.(1989) Density functional theories in quantum chemistry.*ACS Symposium Series*, Vol.394.
[49] McWeeney R, S. B.(1969) T. *Methods of molecular Quantum Mechanics*; Academic Press.
[50] Gunnarsson O, L. I., Wilkins J W.(1974) Contribution to be the cohesive energy of simple metals: Spin dependent effect.*Physical Review B*, Vol.10, pp.1319, ISSN 1943-2879
[51] Gunnarsson O, L. I.(1976) Exchange and correlation in atoms, moleculars and solids by the spin-density-functional formalism.*Physical Review B*, Vol.13, pp.4274, ISSN 1943-2879
[52] Gunnarsson O, J. M., Lundquist I, .(1979) Descriptions of exchange and correlation effects in inhomogeneous electron systems.*Physical Review B*, Vol.20, pp.3136. , ISSN 1943-2879
[53] Luken W L, B. D. N.(1982) localized orbitals and the Fermi hole.*Theoretica Chimica acta*, Vol.61, pp.265 , ISSN 0040-5744
[54] Buijse M, B. E. J., Snijders J G.(1989) Analysis of correlation in terms of exact local potentials: applications to two-electrons systems.*Physical Review A*, Vol.40, pp. 4190, ISSN 1943-2879
[55] Reed, A. E.; Weinhold, F.(1983) NBO.*J. Chem. Phys.*, Vol.78, pp.4066 , ISSN 1063-1079
[56] Reed, A. E.Weinsrock, R. B.&Weinhold, F.(1985) NBO. *Journal of Chemical Physics*, Vol.83,pp. 735., ISSN 1063-1079
[57] Carpenter, J. E.; Weinhold, F.(1988) NBO.J. Mol. Struct. (Theochem), Vol.1169, pp.41,
[58] Reed, A. E..Curtiss, L. A.&Weinhold, F.(1988) Intermolecular Interactions from a Natural Bond Orbital, Donor-Acceptor Viewpoint. *Chemical Review*, Vol.88, pp.899-926,
[59] Ebrahimia, A.Deyhimib, F.&Roohi, H.(2003) Natural bond orbital (NBO) population analysis of the highly strained central bond in [1.1.1]propellane and some [1.1.1]heteropropellane compounds. *Journal of Molecular Structure* (HEOCHEM), Vol.626, pp.223-229,
[60] Zhang, J. G.Li, Q. S.&Zhang, S. W.(2005) A theoretical study on the structures and heats of hydrogenation of the BN-analogs of barrelene. *Chemical Physics Letters*, Vol.407, NO.4-6, pp.315-321,
[61] Curtiss, L. A.,Raghavachari, K.,Redfern, P. C.(1997) Assessment of Gaussian-2 and density functional theories for the computation of enthalpies of formation.*Journal of Chemical Physics*, Vol.106, NO.3, pp.1063-1079,
[62] Joseph, W. O..George, A. P.&Wiberg, K. B.(1995) A Comparison of Model Chemistries. *Journal of the American Chemical Society*, Vol.117, pp.11299-11308,
[63]Chase Jr, M. W.,Davies, P. G.,Downey Jr, J. R.(1985) J. Phys. Ref. Data 14 (Supp. No. 1). Vol.,
[64] Mark, J. E..Allcock, H. R.&West, R. Inorganic Polymers: An lnlroduction; Prentice-Hall: Hoboken, NJ, 1992.
[65] Muralidharan K, E. A.(2003) Preparation of the first examples of ansa-spiro substituted fluorophosphazenes and their structural studies: Analysis of C-H center dot center dot center dot F-P weak interactions in substituted fluorophosphazenes *Inorganic Chemistry*, Vol.42, NO.23, pp.7535-7543, ISSN 0020-1669
[66]J, M. K. R. N. D. E. A.(2000) Syntheses of novel exo and endo isomers of ansa-substituted fluorophosphazenes and their facile transformations into spiro isomers in the presence of fluoride ions *Inorganic Chemistry*, Vol.39, NO.18, pp.3988-3994, ISSN 0020-1669

8

Charge Carrier Mobility in Phthalocyanines: Experiment and Quantum Chemical Calculations

Irena Kratochvilova
Institute of Physics, Academy of Sciences of the Czech Republic, Prague, Czech Republic

1. Introduction

The main goal of this chapter is to show how and why quantum chemistry modeling can /should be applied on class of organic materials with relatively high carrier mobility - phthalocyanines (H_2Pc, NiPc and $NiPc(SO_3Na)_x$) . It will be shown how Density Functional Theory (DFT) can be used to calculate/model main parameters that influence the group of material properties that are crucial from practical point of view.

With the ongoing miniaturization of microelectronics, functional elements in electronic circuits may soon consist of only a couple of electrons or molecules. We therefore address the question how the physical laws which hold for macroscopic solids become modified when one deals with very small structures. It turns out that down-scaling of electronic properties from the macro world to the atomic or molecular level does not work at all. For example, the famous Ohm's law does not hold anymore, because the resistance does not scale with the length of a "quantum wire" [1-15].

Once having realized this fundamental issue, one immediately conceives this as a chance to develop new concepts. Instead of continuously scaling down, as in industrial chip designs, one considers building up electronic circuits with tailored properties "bottom up". The first questions which have to be answered now are: "Which atoms or molecules have to be combined in which way to achieve the desired properties?" and: "Which physical and chemical properties determine the electrical conductance of atomic-size or molecular-size circuits?".

In most areas of science, there are free major steps that need to be taken for understanding to be gleaned: the first is synthesis and preparation, the second is measurements and characterization and third is theory and modelling [12-20]. The great advance in measurement and characterization was clearly the advent of scanning probe microscopy which permitted measurements both of structure and of transport at the level of one to a few molecules. The critical advances in theory and modelling came with adaption of the coherent tunnelling models originally developed by Landauer to the study of transport in molecular tunnel junctions [20-31].

Current research focuses, among others, on the design and implementation of nanometer scale electronic systems which exhibit new classical and quantum mechanical effects. The motivation for creating such elements has been two-fold: first, to create nanoscale laboratories to explore physics in a new way, and second, to develop novel devices with significant applications. The architecture of molecular-scale electronic devices can be designed starting from molecular segments whose properties have been known from experiment and/or suitable theoretical models. The field of "Molecular Electronics" has been opened by the seminal proposition of A. Aviram and M. Ratner in 1974 to build a diode from a single molecule [1-2]. Since then, it took almost 20 years before the first molecular diode was experimentally realized. By now, molecular electronics is a broad field of research world-wide [3-21].

Some examples of molecular nanostructures that might be used as switching units, memories, logic elements and devices embodying a negative resistance have recently been demonstrated (e.g., [1-5]). A pioneering construction of a molecular switch was based on the electron tunneling principle [1]. An electron travels along a 'molecular wire' (e.g., a conjugated polymer chain) containing a finite series of periodic potential walls. The tunnel switch is 'on' if the transmission coefficient of the electron is close to unity, i.e., if the electron energy matches pseudostationary energy levels of the walls, and can be turned off by either changing a barrier height or the depth of a potential well, which can be controlled by the dipole moment of polymer side groups.

As we have already mentioned the charge transport in molecular electronic materials is a very complex process which can be affected by many physical and chemical parameters; the knowledge of its nature is crucial for the development and optimization of molecular-based devices. Each step forward in understanding and controlling the charge transport is extremely important for the practical applications of molecular systems in electronics [1-16].

One of the reasons for the difficulties in charge-transport phenomena/conditions description is the lack of a well-understood mechanism of charge transport in organic materials. The charge-carrier transport in molecular systems is a very complicated and comprehensive event affected by many parameters. In the early days, band theory was applied to predict the charge-carrier mobility in organic materials. However, it was repeatedly pointed out by several authors [18-21] that this approach is not suitable. Comparing organic and inorganic semiconductors, the materials from the latter class are usually less disordered and have molecular sites closer to one another. For disordered materials, the hopping mechanism seems to be more appropriate for the description of the charge-carrier behaviour. In this approach, the charge carrier is localised on the molecular site and jumps to the other site by overcoming some energy barrier. However, for materials possessing a preferred direction of charge-carrier transport like phtalocyanines we need a combination of different approaches for the description of the charge-carrier motion along the preferred paths and for hopping among them.

Molecular materials are generally not very good conductors. Main limitation follows from low charge carrier "on-chain" mobility which mostly using the microwave photoconductivity was found to be 10^{-5} m^2 V^{-1} s^{-1} and 10^{-4} m^2 V^{-1} s^{-1}, for σ- and π-conjugated molecular materials, respectively. The mobility is limited by polaron formation and by the dispersion of transfer integrals among the monomer units the wire. Electrical current through a single molecule is influenced by charge tunnelling – the Fowler-Nordheim model seems to be a good approximation for the description of charge transport. The presence of dipolar species results in the mobility decrease due to the increase of the transfer integral dispersion. Polar group chemically attached to the molecular wire, can cause the orbital localization. The charge transport in 3D samples can be described by the theory of disordered polarons which postulates that the activation energy of the charge carrier mobility is composed of contribution both from the dynamic disorder, i.e. the polaronic barrier, and from the static disorder, i.e. the variation of the energy of transport states as a result of the environment. The main contribution to the polaron binding energy results from molecular deformation; electron-phonon term makes for 20 % only. Dipolar additives make the distribution of hopping states broadened and new localized states for charge carriers are formed; it results in the reduction of charge mobility [17-19].

Experimentally measured conductivity is a macroscopic phenomenon and reflects on-stack mobility plus intra-stack hopping mobility. Intra-stack transport mainly contributes to the whole resistivity – i.e. the main charge-transport obstruction is between stacks.

From the charge carrier mobility point of view phtalocyanines seem to be promising materials - in devices with vacuum evaporated phthalocyanine thin films, the values reached from 10^{-5} to 10^{-4} cm^2 V^{-1} s^{-1}, when phthalocyanine was evaporated on hot substrates the value increased up to 10^{-2} cm^2 V^{-1} s^{-1} [15-22]. Preliminary results obtained on sulphonated phthalocyanines, i.e., materials containing both electronic and electrolytic segments, seem to be very promising. The charge carrier mobility determined from the dependences of the source-drain current vs. the source-drain voltage of the OFET was surprisingly high. In the case of $NiPc(SO_3Na)_4$, the field-effect mobility was 0.02 cm^2 V^{-1} s^{-1} [12].

Phthalocyanines can be organized in columns at a supramolecular level, giving rise to conducting properties. The cofacial stacking of metallophthalocyanines enables electron delocalization along the main axis of the column through π-π orbital overlapping. Metallophthalocyanines generally crystallize in an inclined stacked insulating arrangements called α or β-modifications that do not allow an appropriate overlap of π-orbitals and hence no formation of a conduction band. Only in few cases stacked arrangements are found, being the most representative the nonplanar cone-shaped phthalocyaninatolead (II) (PbPc) in its monoclinic modification.

In the last few years, phthalocyanines (Fig. 1) are being intensively studied as targets for optical switching and limiting devices, organic field effect transistors, sensors, light-emitting devices, low band gap molecular solar cells, optical information recording media, photosensitizers for photodynamic therapy, and nonlinear optical materials, among others. Phthalocyanines will burst also in a very near future into the nanotechnology field.

Phthalocyanine is an intensely blue-green coloured macrocyclic compound. Phthalocyanines form coordination complexes with most elements of the periodic table. Phthalocyanines are structurally related to other macrocyclic pigments, especially the porphyrins. Four pyrrole-like subunits are linked to form a 16-membered ring. The pyrrole-like rings within H_2Pc are closely related to isoindole. Both porphyrins and phthalocyanines function as planar tetradentate dianionic ligands that bind metals through four inwardly projecting nitrogen centers. Such complexes are formally derivatives of Pc^{2-}, the conjugate base of H_2Pc.

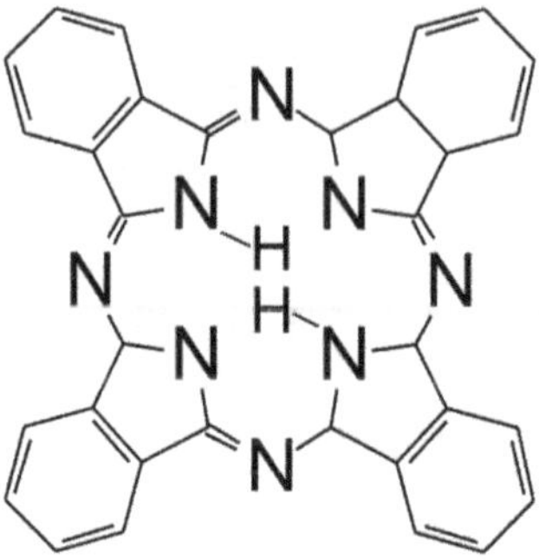

Fig. 1. Chemical structure of H_2 phthalocyanine (H_2Pc).

Many derivatives of the parent phthalocyanine are known, where either carbon atoms of the macrocycle are exchanged for nitrogen atoms or where the hydrogen atoms of the ring are substituted by functional groups like halogens, hydroxy, amino, alkyl, aryl, thiol, alkoxy, nitro, etc.

Due to phthalocyanine stack formation and the presence of ionic groups charge transport consists of an electronic feature through the stack, charge hopping among the stacks and ionic type of the transport.

In order to model the physical properties of H_2Pc layer, quantum chemical calculations on H_2Pc dimers and tetramers were performed at DFT level. The α and β types of polymorph of H_2Pc dimer structures were optimized using several density functionals. The optimizations lead to geometry with almost parallel orientation of H_2Pc planes. The optimised conformer structures are depicted in Fig. 2, the structural parameters calculated for different functional and basis sets are listed in Table 1.

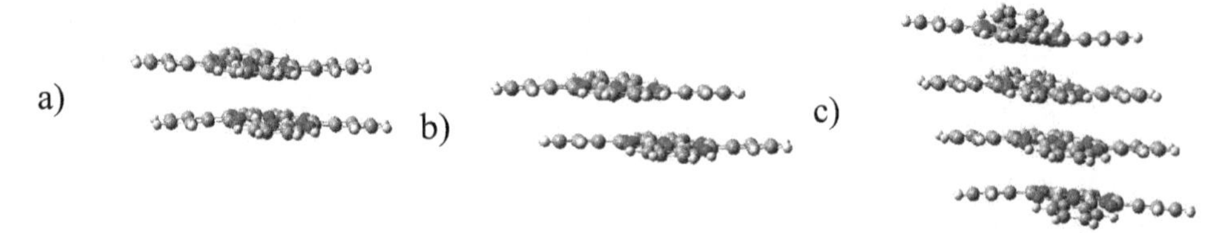

Fig. 2. DFT calculated optimized structures of H_2Pc dimers and tetramer. In the figure, a) α-modification of H_2Pc dimer b) β-modification of H_2Pc dimer c) α-modification of H_2Pc tetramer, respectively.

Calculated geometry of H_2Pc dimer reasonably well represents the local part of the experimental structure of phthalocyanine films [15-19]. The best performance was found for MPW1B95/6-31+G* calculations, where calculated structures fit the phthalocyanine stacking in the crystal. For α-modification, the calculated structural parameter b = 3.77 Å, derived as averaged distances of corresponding atoms at individual monomers well reproduces the experimental value of 3.81 Å . The calculated approximate interplane distance (3.48 Å) is close to experimental value of 3.4 Å [15] as well. For β-modification, the calculations give the interplane distance 3.46 Å (experiment 3.4 Å) and parameter b = 4.87 Å (experiment 4.72). The inclusion of diffusion functions (6-31+G* basis) does not substantially change the structural parameters. As shown in Table 1, for α-modification the functional MPW1B95 slightly better describes the distance parameter b than MPWB1K. For β-modification, the performance of both functional is comparable. Standard B3LYP functional strongly overestimates the separation of molecular planes and fails to predict the real structure.

DFT Functional	MPW1B95		MPWB1K		Experiment
Basis Set	6-31G*	6-31+G* (Å)	6-31G*	6-31+G* (Å)	(Å)
		H_2Pc dimer polymorph α			
neutral					
a[1]	3.463	3.478	3.442	3.466	3.4
b[2]	3.744	3.768	3.721	3.752	3.81
ΔE[3]	5.1	4.6			
cation					
a[1]	3.440				
b[2]	3.665				
anion					
a[1]	3.450				
b[2]	3.804				
		H_2Pc dimer polymorph β			
neutral form					
a[1]	3.428	3.456	3.428	3.432	3.4
b[2]	4.849	4.871	4.840	4.841	4.72
ΔE[3]	3.6	3.0			
cation					
a[1]	3.289				
b[2]	5.255				
anion					
a[1]	3.325				
b[2]	4.678				
		H_2Pc tetramer polymorph α			
a[1]	3.388				3.4
b[2]	3.715				3.81
		H_2Pc tetramer polymorph β			
a[1]	3.399				3.4
b[2]	4.833				4.72

[1] interplanar distance
[2] the average distance of corresponding atoms
[3] stabilization energy of neutral dimers in kcal/mol

Table 1. Calculated structural parameters (in Å) and stabilization energies (kcal/mol) of phthalocyanine dimers and tetramers.

Calculated stabilization energies listed in Table 1 are typical for $\pi - \pi$ stacking interactions and indicate that the polymorph α is more stable than the β-modification; the calculated difference of stabilization energies is 1.5 kcal/mol for the functional MPW1B95. Table 1 shows that the withdrawal of electron from α polymorph leads to geometry with more closely lying interplanar arrangements of monomeric subunits characterized by the distances a = 3.44 Å and b = 3.67 Å. Optimized geometry of anionic form is characterized by similar interplane distance 3.45 Å and larger parameter b = 3.80 Å. Cationic form of β polymorph is characterized by the distances a = 3.29 Å and b = 5.26 Å, anionic form of this polymorph by distances a = 3.33 Å and b = 3.68 Å. All following single point calculations were done at optimized geometries.

On the basis of the dimer calculations, the geometry optimization of H_2Pc tetramer was done by MPW1B95/6-31G* calculations on both polymorphs. The middle part of the optimized tetramer structure interprets well the H_2Pc crystal structure (the calculated interplane distance is 3.39 Å [15]).

The optical spectroscopy was used for the characterization of the phthalocyanine layer. Figure 3 shows the experimental spectrum in the region of the phtalocyanine Q band. The comparison of characteristic features of this spectrum with the previously measured spectra [36] of different polymorphs of H_2Pc layers indicates the presence of the α polymorph.

TD DFT calculated transitions on this polymorph reasonably well reproduce the strong features of characteristic Q band and the effect of aggregation. The experimental transition energies measured at 1.77 and 1.93 eV are slightly overestimated by TD DFT calculations (obtained at 2.15 and 2.21 eV) with oscillator strengths 0.52 and 0.58, respectively. Transitions with low oscillator strength calculated at 1.65 and 1.71 eV describe the appearance of the week features in the long wave region. The TD DFT calculation on the β polymorph describes correctly the shift of the intense transitions to the longer wavelengths.

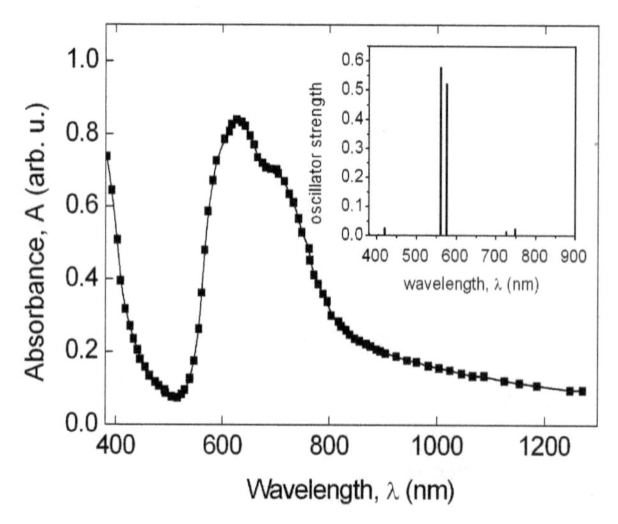

Fig. 3. The experimental spectrum of H_2Pc layer in the region of the Q band. Inset shows calculated transitions for H_2Pc dimer.

At room temperature, the charge carrier transport can be described by a hopping within the Marcus theory. Subsequently, the rate expression of a self-exchange process is provided by the expression [19]

$$k_{et} = \frac{2\pi}{\hbar}\frac{1}{\sqrt{4\pi\lambda^{+}kT}}(t^{+})^{2}\exp\left(-\frac{\lambda^{+}}{4kT}\right), \tag{1}$$

where $\hbar$ is the reduced Planck constant, k is the Boltzmann constant, T is the temperature an λ^+ is the reorganisation energy and t^+ is the electronic coupling matrix element (the measure of charge transport probability - electron transfer integral) for the charge transfer. In order to estimate the electronic coupling between individual monomeric units, electron transfer integrals were calculated for α and β crystallographic modifications of the phthalocyanine dimer in HF approximation.

For the configuration with doubly occupied HOMO on the first monomer and singly occupied HOMO of the second one, calculation gives electron transfer integral (for otimized geometry MPW1B95/6-31G*) values for non-relaxed cation radical 11.7 meV and 5.7 meV (hole transfer) for the α and β modification, respectively. It is supposed that in the course of electron or hole hopping, the atoms can relax in order to reach the minimum of energy at the potential surface. Transfer integral for optimized cation radical α crystal form with relaxed atoms (UMPW1B95/6-31G*) is 18.7 meV. The calculations indicate that electron mobility is smaller than hole mobility because optimized geometry of anionic form is characterized by larger interplane parameter b (see Tab. 1) - for anionic form b=3.804 Å, for cationic form b=3.665 Å.

The transfer integral t^+ was calculated according to

$$t^{+} = \frac{H_{RP} - S_{RP}(H_{RR} + H_{PP})/2}{1 - S_{RP}^{2}}, \tag{2}$$

H_{RP} is the interaction energy between reactant and product states, S_{RP} is the overlap between the reactant and product states and H_{RR} is electronic energy of the reaktant state andH_{PP} is electronic energy of the product state. All these terms were obtained via the direct coupling of localised monomer orbitals. On the molecular level, three factors, the electronic coupling (transfer integral t^+) between the individual parts of the molecule, the reorganisation energy λ^+ during charge transport and the effective length of hole transfer L, are usually considered to be important for charge transport in organic materials. The reorganization energy λ^+ consists of the sum of λ_1^+ and λ_2^+, $\lambda^+ = \lambda_1^+ + \lambda_2^+$, where the deformation energy of the system was calculated as the difference between the vertical and cationic state: $\lambda_1^+ = E_+(Q_N) - E_+(Q_+)$ and $\lambda_2^+ = E_+(Q_+) - E_+(Q_N)$. Here, $E_+(Q_N)$ is the total electronic energy of the cationic state in the neutral geometry, $E_+(Q_+)$ is the total energy of the cationic state in the cationic state geometry, $E_N(Q_+)$ is the total energy of the neutral state in the cationic state geometry and $E_N(Q_N)$ is the total energy of the neutral state in the neutral geometry. λ_1^+ is frequently called as deformation energy of the system.

The diffusion coefficient D of charge carriers can be expressed using the Einstein-Smoluchowski equation

$$D = \frac{L^2 k_{et}}{2}. \tag{3}$$

This makes it possible to evaluate the drift mobility of the charge carriers using the Einstein relation $\mu = eD/kT$. It should also be noted that the calculated charge mobility mentioned here represents the zero electric field approximation value.

To explain the differences between the charge carrier mobility in sulphonated and non-sulphonated Ni phthalocyanines (NiPc), quantum chemical calculations were performed on the DFT (density functional theory) level for H_2Pc, NiPc and $NiPc(SO_3Na)_x$ (x = 1, 2) dimers and their cationic and anionic forms. Quantum chemical modeling was found to be very useful - we were able to see states of various molecular systems from new and comprehensive perspectives.

The optimized structure of the $[Ni(Pc(SO_3Na)_2)]_2$ is depicted in Fig.4. The calculated geometry of the NiPc dimer and tetramer reasonably well represents the local part of the NiPc crystal structure [34]. The best performance for the NiPc dimer was found by the MPW1B95/6-31+G* method. The calculated approximate interplane distance (3.383 Å) is close to the experimental value of 3.4 Å [15,19].

The calculated stabilization energies are typical for π–π stacking interactions. The calculated stabilization energies with the BSSE correction are 4.1, 5.4, 7.4 and 11.3 kcal/mol for the $[H_2Pc]_2$, $[NiPc]_2$, $[NiPc(SO_3Na)]_2$ and $[NiPc(SO_3Na)_2]_2$ dimers, respectively. The MPWB1K/6-31G* calculations yield similar results, whereas the calculated stabilization energies are slightly higher. The increasing stabilization energies in the series of dimers going from $[H_2Pc]_2$ to $[NiPc(SO_3Na)_2]_2$ indicate the better stability and organization of the $NiPc(SO_3Na)_2$ layers in comparison with the others.

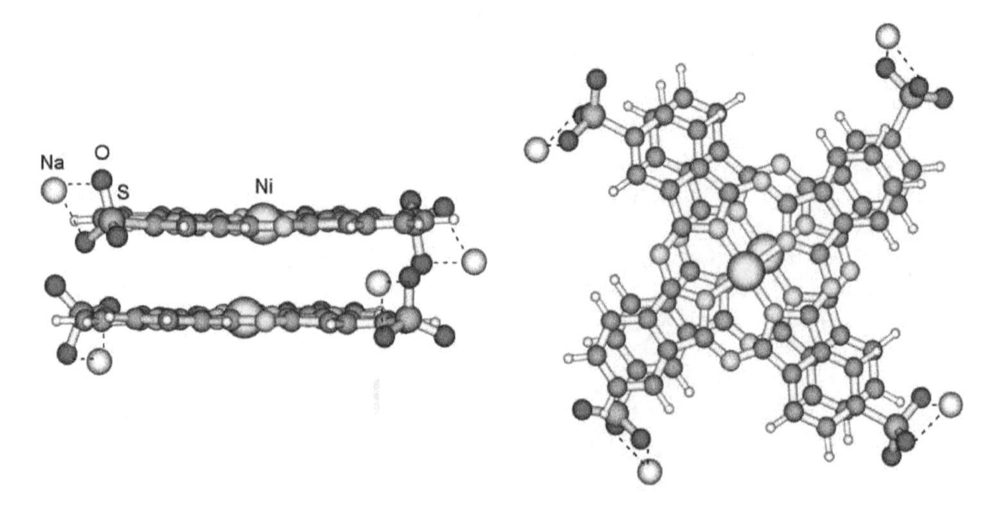

Fig. 4. The DFT-calculated optimized structure of the dimer $[NiPc(SO_3Na)_2]_2$. The dashed line indicates the shortest distance between the O and Na atoms.

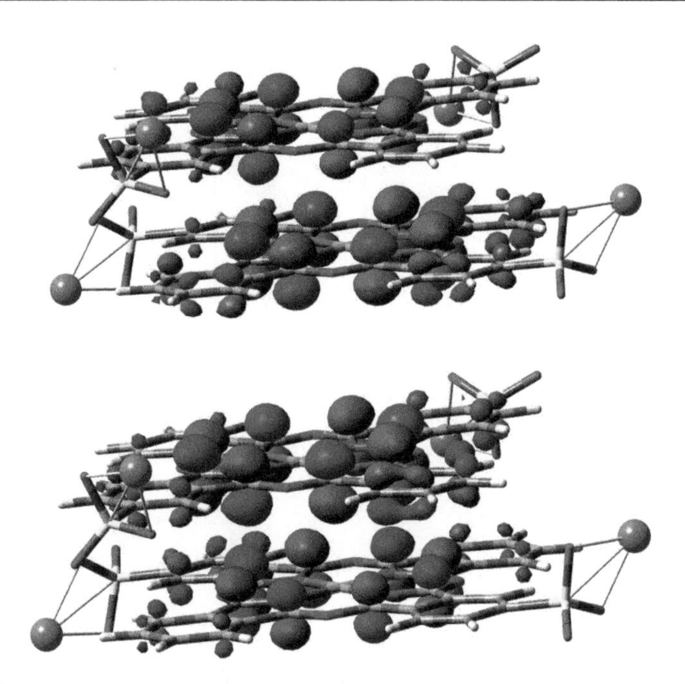

Fig. 5. The schematic representation of the HOMO (top) and HOMO-1 (bottom) of the dimer $[NiPc(SO_3Na)_2]_2$.

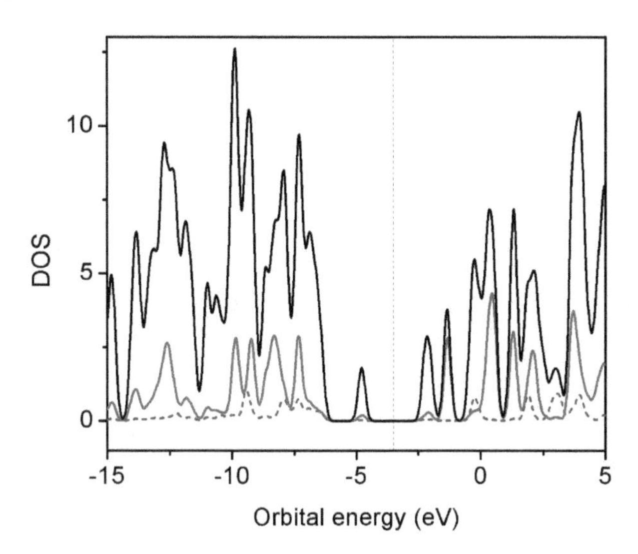

Fig. 6. The density of the states (DOS) for $[NiPc(SO_3Na)_2]_2$. The black line indicates the total density of the states of the whole system, the blue dashed line the contributing Ni orbitals and the red one the contribution from the SO_3Na group. The vertical dashed line indicates an approximate midpoint of the HOMO-LUMO levels.

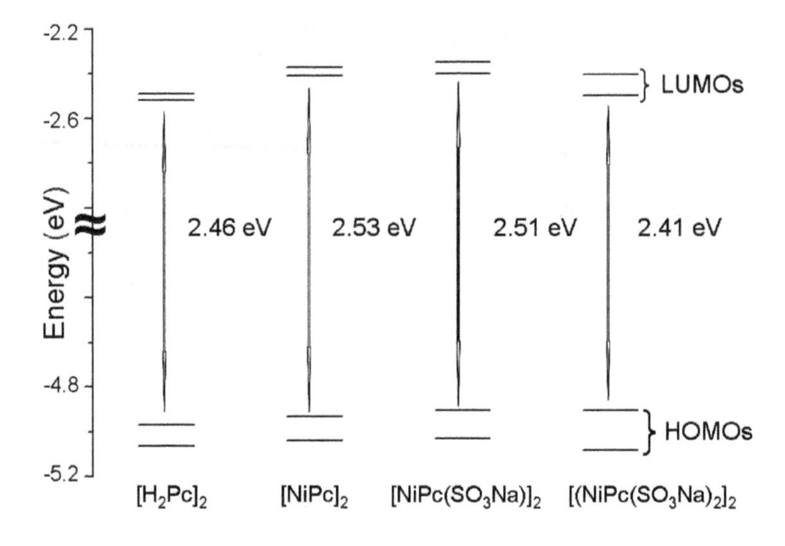

Fig. 7. The energies of the frontier orbitals of the studied phthalocyanine dimers.

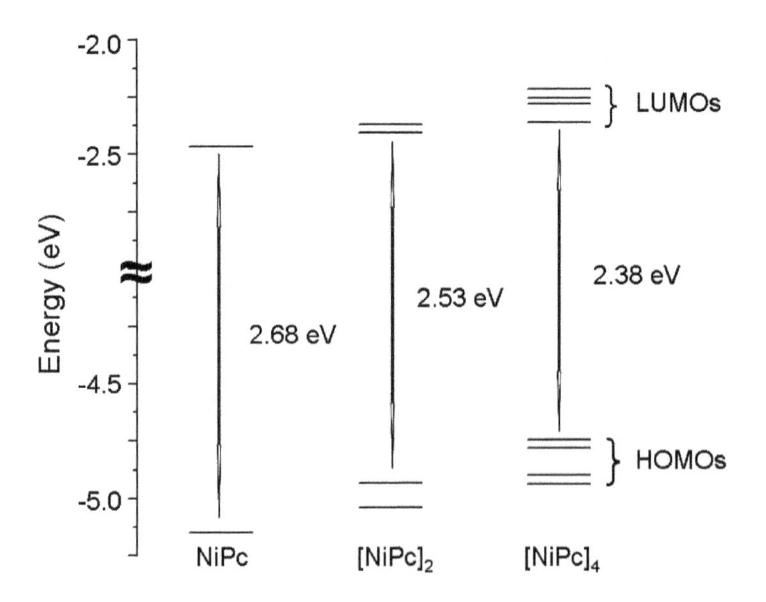

Fig. 8. The energies of the frontier orbitals of the NiPc monomer, dimer and tetramer.

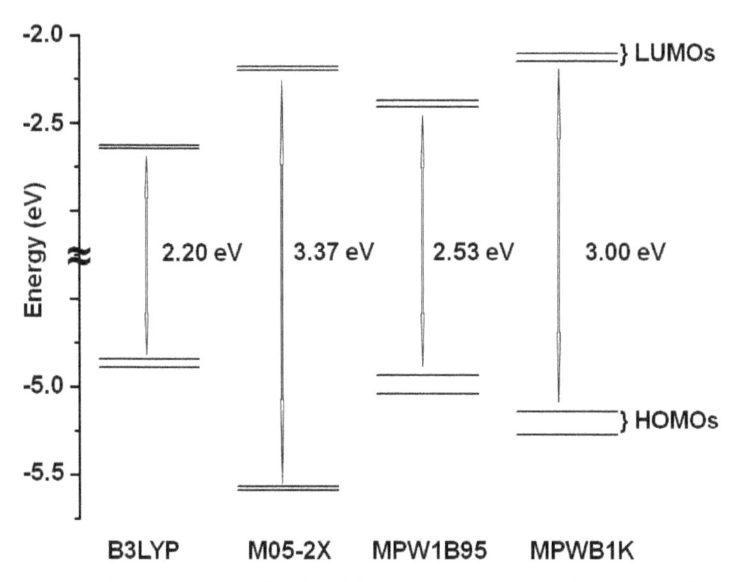

Fig. 9. The energies of the frontier orbitals of the NiPc dimer, calculated with different density functional for geometry optimized with corresponding functional.

Figure 5 depicts two highest occupied molecular orbitals (HOMO, HOMO-1) of $[NiPc(SO_3Na)_2]_2$, the character of which can influence the hole conductivity. The plot of the density of the states in Fig. 6 shows the distribution of one-electron molecular orbitals and indicates that the Ni and SO_3Na orbitals do not substantially contribute to the frontier molecular orbitals of the dimer $[NiPc(SO_3Na)_2]_2$. These orbitals are formed by Pc π orbitals. Owing to the mutual interaction of the π monomer orbitals, the frontier orbitals of the supersystem form groups of two closely lying π molecular orbitals. The contribution of the 3d metal orbitals is negligible as in the case of the monomeric NiPc. The second pair of lower-lying occupied orbitals lies about 1.57 eV lower. Thus, it can be supposed that these orbitals do not strongly influence the charge carrier transport. The calculated HOMO–HOMO-1 separations are visualized in Figs. 7 and 8 while HOMO–LUMO gaps are listed in Tables 4 and 5. Figure 7 depicts the splitting of the frontier orbitals and their mutual position in the series of the systems studied; Fig. 8 compares orbital splitting in the series of the NiPc monomer, dimer and tetramer. The splitting of the frontier orbitals reflects the mutual interaction and thus the probability of charge transfer between individual monomeric subsystems. It should be mentioned that calculated HOMO-LUMO gap (Fig. 8) is diminishing in the series NiPc, $[NiPc]_2$ and $[NiPc]_4$ and approaching to the experimental value of 1.8 eV [13]. Fig. 9 shows how the variation of functional influences the energies of frontier orbitals.

The rate constant k_{ET} of the electron transfer between molecular orbitals was expressed - see (1). The electronic coupling between the individual monomeric units V_{RP}, the electron-transfer integral, can be calculated by several procedures [8-13]. Koopman's approximation, where the electronic coupling is estimated as half of the corresponding orbital energies, was used for the matrix-element estimation [14-18]. This method has already been used in the case of interacting metal-containing dimmers [22].

The values of the electron-transfer integrals for the closed-shell systems were estimated from the separation of HOMO and HOMO-1 for the optimized structures. The values of the electron-transfer integrals based on the MPW1B95/6-31G* functional model are 47.5 meV, 54.3 meV, 63.1 meV and 88.6 meV for $[H_2Pc]_2$, $[NiPc]_2$, $[NiPc(SO_3Na)]_2$ and $[NiPc(SO_3Na)_2]_2$, respectively (see Table 2).

	MPW1B95/6-31G*			
Dimer	ΔE	ΔE(BSSE)	ET coupling	Gap
	kcal mol^{-1}		meV	eV
$[H_2Pc]_2$	14.8	4.1	47.5	2.46
$[NiPc]_2$	19.4	5.4	54.3	2.53
$[NiPc(SO_3Na)]_2$	22.1	7.4	63.1	2.51
$[NiPc(SO_3Na)_2]_2$	27.6	11.3	88.6	2.41
$[(NiPc(SO_3Na)_2)$ $(NiPc(SO_3Na)\ SO_3\ )]^-$	31.2[a]	13.2[a]		2.34[a]
$[NiPc(SO_3Na)(SO_3)]_2^{2-}$	6.3[a]	-13.1[a]		2.41[a]
$[NiPc(SO_3)_2]_2^{4-}$	-79.4	-100.5		2.44

[a] an average of the calculated values for the two different Na ions locations

Table 2. The calculated stabilization energies – ΔE, (kcal/mol), electron-transfer integrals – ET coupling (meV) and magnitude of the gap between the HOMO a LUMO orbitals of the various types of phthalocyanine dimers.

From these values, it follows that the electronic contribution to the rate constant of the electron transfer (proportional to the power of the electronic coupling – V^2_{RP}) for $[NiPc(SO_3Na)_2]_2$ should be 2.7 times larger than for $[NiPc]_2$ and 3.5 times larger than for $[H_2Pc]_2$, whereas the electronic contribution for $[NiPc(SO_3Na)]_2$ is only about 1.4 times larger than for $[NiPc]_2$. The dependence of the electron-transfer integrals on the Ni-Ni separation for $[NiPc]_2$ and $[NiPc(SO_3Na)_2]_2$ dimers is depicted in Fig. 10. The electron-transfer integral value varies strongly with the intermolecular separation, nevertheless the ratio between V_{RP} calculated for SO_3Na substituted and unsubstituted systems is close to this ratio obtained from equilibrium intermolecular separation.

Both water molecules surrounding or the application of an electric field can cause the dissociation of the ionic Na-SO_3 bond. In order to explain the effect of the Na^+ dissociation, calculations were performed on the model dimeric systems, where Na^+ ions were stepwise removed from the $NiPc(SO_3Na)_2$ dimer . The stabilization energy of the system with one Na^+ removed (13.2 kcal mol^{-1}) is even larger than that of $[NiPc(SO_3Na)_2]_2$. From Table 2, it can be seen that $[Ni_2Pc_2(SO_3Na)_3(SO_3)]^-$ is the most stable and thus the best organized of all the structures investigated. The density of the states plot depicted in Fig. 11 shows how the withdrawal of one Na^+ influences the electronic structure. Projected Densities of States reported in this chapter were broadened using discrete molecular orbital levels – the broadening parameter used in our PDOS calculations (the full width at half maximum) was 0.3 eV.

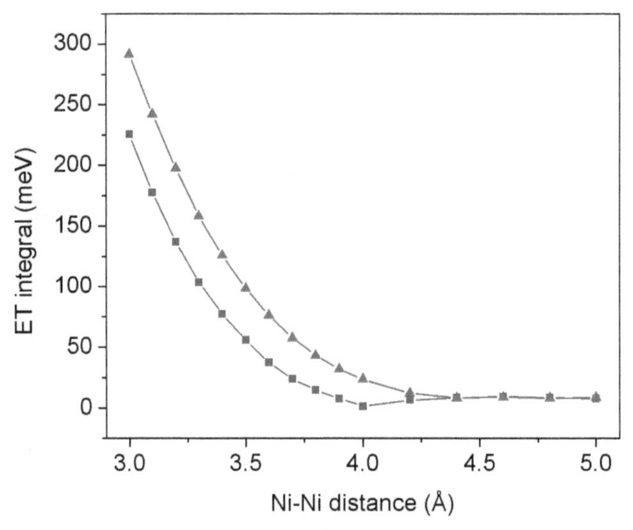

Fig. 10. The dependence of electron-transfer integrals on the Ni-Ni separation. $[NiPc]_2$ - blue squares, $[NiPc(SO_3Na)_2]_2$ – red triangles.

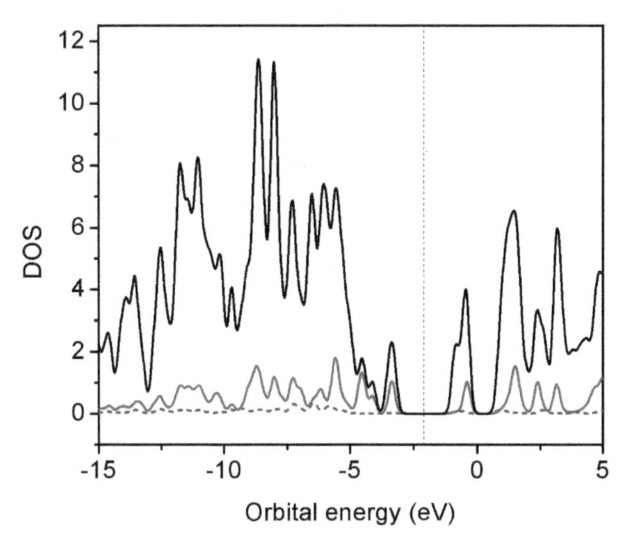

Fig. 11. The density of the states (DOS) for the system $[Ni_2Pc_2(SO_3Na)_3(SO_3)]^-$. the black line indicates the total density of the states of the whole system, the blue dashed line the contributing Ni orbitals and the red one the total contribution from the SO_3Na and SO_3 groups; the vertical dashed line indicates an approximate midpoint of the HOMO–LUMO levels.

In comparison with the unperturbed system, the set of HOMOs is no longer separated from the lower-lying occupied orbitals and contains contributions from the SO_3^- group. It can be supposed that after Na^+ ion dissociation the whole layer contains mobile Na^+ and fixed $[NiPc(SO_3Na)(SO_3)]^-$ ions surrounded by water molecules. In this case, the highest molecular orbitals, which contain contributions from the SO_3^- group and are close to the lower-lying occupied orbitals, allow the whole system to contain electronic states that are more extended and therefore more conductive in character.

Due to the withdrawal of the remaining Na^+ ions, the stabilization energies decrease. The least stable dimer in the series $[NiPc(SO_3Na)_2]_2$, $[NiPc(SO_3Na)(SO_3)]_2^{2-}$ and $[NiPc(SO_3)_2]_2^{4-}$ is the last, completely dissociated dimer, i.e. $[NiPc(SO_3)_2]_2^{4-}$. This fact can induce reduced layer organization and thus lower the conductivity of the systems with more than one Na^+ ion removed.

How can the selected chemical modifications of Ni phthalocyanine affect on the molecular level the charge carrier mobility?

At zero-gate bias, the HOMO band is filled. As the gate voltage is swept more negative, the positions of the LUMO and HOMO bands rise in energy, allowing mobile holes to sustain a source-drain current via the HOMO band. Materials based on NiPc and $[NiPc(SO_3Na)_x]$ are hole semiconductors, which means that the source of the charge carriers in the structure electrode-semiconductor – electrode is not the injecting metal electrode but the semiconductor. In the case of $[NiPc(SO_3Na)(SO_3)]^-$, the contribution from the SO_3 anions to the whole HOMO orbital system enables the mobile holes to be more delocalized; thus the change charge transport (hopping) mechanism becomes more ballistic (less scattered and more effective).

The stability of dimers (the calculated stabilization energy), which induces better layer organization and consequently better charge mobility[37] grows in the series $[H_2Pc]_2$, $[NiPc]_2$, $[NiPc(SO_3Na)]_2$, $[NiPc(SO_3Na)_2]_2$ and $[(NiPc(SO_3Na)_2)(NiPc(SO_3Na)SO_3)]^-$. The withdrawal of more than one Na cation leads to destabilization and consequently to a decrease in layer organization. The layer of $[Ni_2Pc_2(SO_3Na)_3(SO_3)]^-$ should be the most regular (is supposed to be the most conductive) of all the structures examined.

In the case of $NiPc(SO_3Na)_{3.3}$ the charge carrier mobility (0.02 $cm^2\ V^{-1}\ s^{-1}$) is lower than for $NiPc(SO_3Na)_{1.5}$ (1.08 $cm^2\ V^{-1}\ s^{-1}$) - see Tab. 3.

Phthalocyanine	Film preparation	Mobility ($cm^2\ V^{-1}\ s^{-1}$)
NiPc	evaporation	10^{-5}
$NiPc(SO_3Na)_{3.3}$	spin-coating	0.02
$iPc(SO_3Na)_{1.5}$	spin-coating	1.08

Table 3. The mobilities for the various types of nickel phthalocyanines.

From calculating spatial models it was figured out that due to the sterical communication hindrance caused by third and fourth SO_3Na groups the $NiPc(SO_3Na)_{3.3}$ dimers are less stable. This fact results in reduced organized layer structure with reduced charge mobility.

Using DFT calculations, several aspects that influence the charge-carrier mobility in the case of the materials were compared:

1. Charge transfer probability grows in the series $[H_2Pc]_2$, $[NiPc]_2$, $[NiPc(SO_3Na)]_2$ and $[NiPc(SO_3Na)_2]_2$ because of the increasing electronic coupling. The presence of (SO_3Na) groups increases the probability of electron transfer between the composing units.
2. At the same time, the stabilization energies grow in the series $[H_2Pc]_2$, $[NiPc]_2$ and $[NiPc(SO_3Na)_2]_2$. Due to larger stabilization energies of the systems with (SO_3Na)

groups, it can be assumed that layers with (SO_3Na) groups are organized more regularly - resulting in better π-π interaction and consequently better charge carrier mobility than in the case of nonsubstituted ones. The withdrawal of one Na^+ ion leads to a further stabilization of the structure and thus a higher probability of charge-carrier transport. The dissociations of more than one Na^+ ion strongly destabilize the layer structure. The probability of charge transfer through sulphonated Ni-phthalocyanine dimers with totally removed Na^+ ions should, therefore, be lower than in the case of standard sulphonated Ni phthalocyanines dimers.

3. The system is strongly affected by an external electric field and water molecules in the vicinity, which can cause the dissociation of the ionic Na-SO_3 bond. After the Na-SO_3 bond's dissociation, the whole layer contains mobile Na^+ and fixed $[Ni_2Pc_2(SO_3Na)_3(SO_3)]^-$ anions which allows the mobile holes to delocalize and increase the probability of hole transfer through the layer structure. We found that the highest molecular orbitals in $[Ni_2Pc_2(SO_3Na)_3(SO_3)]^-$ contain new electronic contributions to the density of the states from the SO_3^- group, which should lead to an increase in the hole mobility and thus conductivity.

Other aspects which can increase charge carrier mobility in of the OFET structures based on sulphonated phthalocyanine derivatives should be mentioned: 1. Better resistivity to oxygen and thus the stability of the system in atmospheric air. 2. The interface $SiO_2/[NiPc(SO_3Na)_2]_x$ is between two hydrophilic groups and thus adheres materials enabling the formation of a molecular layer on the SiO_2 surface with proper and regular molecular orientation which increases the conductivity .

The objective of this chapter was to present results of DFT studies of class of organic materials with relatively high carrier mobility - phthalocyanines. It was shown how DFT can be used to calculate/model some parameters that influence charge carrier mobility on essentially level [18]. The calculations indicate that electron mobility in H_2Pc is smaller than hole mobility because optimized geometry of anionic form is characterized by larger interplane parameter.

In the case of $[NiPc(SO_3Na)(SO_3)]^-$, the contribution from the SO_3 anions to the whole HOMO orbital system enables the mobile holes to be more delocalized; thus the change charge transport (hopping) mechanism becomes more ballistic (less scattered and more effective). The withdrawal of one Na^+ ion leads to a further stabilization of the structure and thus a higher probability of charge-carrier transport.

- Larger stabilization energies of the systems with (SO_3Na) groups - layers with (SO_3Na) groups are organized more regularly - resulting in better п-п interaction and consequently better charge carrier mobility than in the case of nonsubstituted ones.
- The withdrawal of one Na+ ion leads to a further stabilization of the structure and thus a higher probability of charge-carrier transport.
- The dissociations of more than one Na+ ion strongly destabilize the layer structure. The probability of charge transfer through sulphonated Ni-phthalocyanine dimers with totally removed Na+ ions is lower than in the case of standard sulphonated Ni phthalocyanines dimers.

2. References

[1] M. A. Reed, *Proc. IEEE* , 1999 , 87, 652.
[2] D. Braga, G. Horowitz, *Adv. Mater.*, 2009, 21, 1.
[3] J. Health, M. Read, Molecular Electronics, Physics Today, 2003, 43

[4] C.P. Collier et al., Science, 2000, 289, 1172
[5] A.J. Heinrich et al. Science, 2002, 298, 1381
[6] J. Park et al. Nature, 2002, 417, 722
[7] M. Taniguchi and T. Kawai, Physica E 2006, 33 1-12.
[8] S. Datta, W. Tian, S. Hong, R. Reifenberger, I. Henderson, and C. Kubiak, Phys.Rev.Lett. 1997, 79 2530-2533.
[9] F. Zahid, M. Paulsson and S. Datta, Chapter published in "Advanced Semiconductors and Organic nano-Techniques", edited by H. Morkoc, Academic Press 2003. See also arXiv: Cond-Mat/0208183.
[10] K. Müllen, G. Wegner (Eds.), *Electronic Materials: The Oligomeric Approach* (Wiley-VCH, Weinheim, 1998).
[11] J. Simon, P. Bassoul, *Design of Molecular Materials. Supramolecular Engineering* (Wiley, Chichester, 2000).
[12] R.G. Enders, D. L. Cox, and R.R.P. Singh, Colloquium: Rev. Modern Phys.2004, 76 195-217.
[13] L. Cai, H. Tabata, and T. Kawai, Self-assembled DNA networks and their electrical conductivity, Appl. Phys. Let. 2000, 77 3105-3106.
[14] I. Kratochvílová, S. Nešpůrek, J. Šebera, S. Záliš, M. Pavelka, G. Wang, J. Sworakowski, *Eur. Phys. J. E*, 2008, 25, 299.
[15] I. Kratochvílová, K. Král, M. Bunček, A. Víšková, S. Nešpůrek, A. Kochalska, T. Todorciuc, M. Weiter, B. Schneider, *Biophys. Chem.*, 2008, 138, 3.
[16] I. Kratochvílová , K. Král, M. Bunček, S. Nešpůrek, T. Todorciuc, M. Weiter, J. Navrátil, B. Schneider and J. Pavluch, *Cent. Eur. J. Phys.*, 2008, 6, 422.
[17] S. Záliš, I. Kratochvílová, A. Zambova, J. Mbindyo, T. E. Mallouk, T. S. Mayer, *Eur. Phys. J. E*, 2005, 18, 201.
[18] M. F Craciun, S Rogge, M - J. L. Den Boer, S Margadonna, K Prassides, Y Iwasa, A. F Morpurgo, Advanced Materials 18, (2006) 320.
[19] I.I. Fishchuk, A. Kadashchuk, V.N. Poroshin, N. Volodymyr, H. Bassler, Philosophical Magezine, 90 (2010) 129.
[20] J. Šebera, S. Nešpůrek, I. Kratochvílová, S. Záliš, G. Chaidogiannos, N. Glezos, European Physical Journal B 72 (2009) 385.
[21] M. C. Reese, M. Roberts, M. Ling and Z. Bao, *Mater. Today*, 2004, 7, 20.
[22] C. D. Dimitrakopoulos, P. R. L. Malenfant, *Adv. Mater.*, 2002, 14, 99.
[23] R. Zeis, T. Siegrist and C. Kloc, *Appl. Phys. Lett.*, 2005, 86, 022103.
[24] S. Nešpůrek, G. Chaidogiannos, N. Glezos, G. Wang, S. Böhm, J. Rakušan, M. Karásková, *Mol. Cryst. Liq. Cryst.*, 2007, 468, 355.
[25] F. Yang, M. Shtein and S. R. Forrest, *Nat. Mater.*, 2005, 4, 37.
[26] Jakub Šebera, Stanislav Nešpůrek, Irena Kratochvílová, Stanislav Záliš, George Chaidogiannos, Nikos Glezos, *European Physical Journal B - Condensed Matter and Complex Systems* 72 (2009) 385-395.
[27] S. M. Bayliss, S. Heutz, G. Rumbles, T. S. Jones, Phys. Chem. Chem. Phys. 1, 3673 (1999)
[28] R.D. Gould, Coord.Chem. Rev. 156, 237 (1996)
[29] M. Ashida, N. Uyeda, E. Suito, Bull. Chem. Soc. Jpn. 39, 2616 (1966)
[30] A. Farazdel, M. Dupuis, E. Clementi, A. Aviram, J. Am. Chem. Soc. 112, 4206 (1990)
[31] V. Coropceanu, J. Cornil, D.A. da Silva Filho, Y. Olivier, R. Silbey, J. L. Brédas, Chem. Rev. 107, 926 (2007)
[32] Hynek Nemec, Irena Kratochvilova , Petr Kuzel, Jakub Sebera, Anna Kochalska, Juraj Nozar and Stanislav Nespurek , Phys. Chem. Chem. Phys., 2011, 13, 2850–2856

Permissions

The contributors of this book come from diverse backgrounds, making this book a truly international effort. This book will bring forth new frontiers with its revolutionizing research information and detailed analysis of the nascent developments around the world.

We would like to thank Tomofumi Tada, for lending his expertise to make the book truly unique. He has played a crucial role in the development of this book. Without his invaluable contribution this book wouldn't have been possible. He has made vital efforts to compile up to date information on the varied aspects of this subject to make this book a valuable addition to the collection of many professionals and students.

This book was conceptualized with the vision of imparting up-to-date information and advanced data in this field. To ensure the same, a matchless editorial board was set up. Every individual on the board went through rigorous rounds of assessment to prove their worth. After which they invested a large part of their time researching and compiling the most relevant data for our readers. Conferences and sessions were held from time to time between the editorial board and the contributing authors to present the data in the most comprehensible form. The editorial team has worked tirelessly to provide valuable and valid information to help people across the globe.

Every chapter published in this book has been scrutinized by our experts. Their significance has been extensively debated. The topics covered herein carry significant findings which will fuel the growth of the discipline. They may even be implemented as practical applications or may be referred to as a beginning point for another development. Chapters in this book were first published by InTech; hereby published with permission under the Creative Commons Attribution License or equivalent.

The editorial board has been involved in producing this book since its inception. They have spent rigorous hours researching and exploring the diverse topics which have resulted in the successful publishing of this book. They have passed on their knowledge of decades through this book. To expedite this challenging task, the publisher supported the team at every step. A small team of assistant editors was also appointed to further simplify the editing procedure and attain best results for the readers.

Our editorial team has been hand-picked from every corner of the world. Their multi-ethnicity adds dynamic inputs to the discussions which result in innovative outcomes. These outcomes are then further discussed with the researchers and contributors who give their valuable feedback and opinion regarding the same. The feedback is then collaborated with the researches and they are edited in a comprehensive manner to aid the understanding of the subject.

Apart from the editorial board, the designing team has also invested a significant amount of their time in understanding the subject and creating the most relevant covers. They scrutinized every image to scout for the most suitable representation of the subject and create an appropriate cover for the book.

The publishing team has been involved in this book since its early stages. They were actively engaged in every process, be it collecting the data, connecting with the contributors or procuring relevant information. The team has been an ardent support to the editorial, designing and production team. Their endless efforts to recruit the best for this project, has resulted in the accomplishment of this book. They are a veteran in the field of academics and their pool of knowledge is as vast as their experience in printing. Their expertise and guidance has proved useful at every step. Their uncompromising quality standards have made this book an exceptional effort. Their encouragement from time to time has been an inspiration for everyone.

The publisher and the editorial board hope that this book will prove to be a valuable piece of knowledge for researchers, students, practitioners and scholars across the globe.

List of Contributors

Masoud Saravi
Islamic Azad University, Nour Branch, Nour, Iran

Seyedeh-Razieh Mirrajei
Education Office of Amol, Amol, Iran

Nelson Henrique Morgon
Universidade Estadual de Campinas, Brazil

Aline Thaís Bruni
Departamento de Química, Faculdade de Filosofia, Ciências e Letras de Ribeirão Preto, Universidade de São Paulo, Brazil

Vitor Barbanti Pereira Leite
Departamento de Física, Instituto de Biociências, Letras e Ciências Exatas, Universidade Estadual Paulista, São José do Rio Preto, Brazil

Fatma Kandemirli
Niğde University, Turkey

M. Iqbal Choudhary and Sadia Siddiq
University of Karachi, Pakistan

Murat Saracoglu
Erciyes University, Turkey

Hakan Sayiner
Kahta State Hospital, Turkey

Taner Arslan
Osmangazi University, Turkey

Ayşe Erbay and Baybars Köksoy
Kocaeli University, Turkey

Ol'ha O. Brovarets and Dmytro M. Hovorun
Department of Molecular and Quantum Biophysics, Institute of Molecular Biology and Genetics, National Academy of Sciences of Ukraine, Kyiv, Ukraine
Research and Educational Center, "State Key Laboratory of Molecular and Cell Biology", Kyiv, Ukraine
Department of Molecular Biology, Biotechnology and Biophysics, Institute of High Technologies, Taras Shevchenko National University of Kyiv, Kyiv, Ukraine

Iryna M. Kolomiets
Department of Molecular and Quantum Biophysics, Institute of Molecular Biology and Genetics,
National Academy of Sciences of Ukraine, Kyiv, Ukraine

Tomofumi Tada
Department of Materials Engineering, Global COE for Mechanical System Innovation, The University of Tokyo, Japan

Kun Wang, Jian-Guo Zhang and Hui-Hui Zheng
Hui-Sheng Huang and Tong-Lai Zhang, State Key Laboratory of Explosion Science and Technology,
Beijing Institute of Technology, China

Irena Kratochvilova
Institute of Physics, Academy of Sciences of the Czech Republic, Prague, Czech Republic